European Mineralogical Union Notes in Mineralogy

Commissioning Editor: R. Oberti

Volume 19

MINERALOGICAL CRYSTALLOGRAPHY

UNIVERSITY TEXTBOOK

Edited by

JAKUB PLÁŠIL, JURAJ MAJZLAN
and SERGEY KRIVOVICHEV

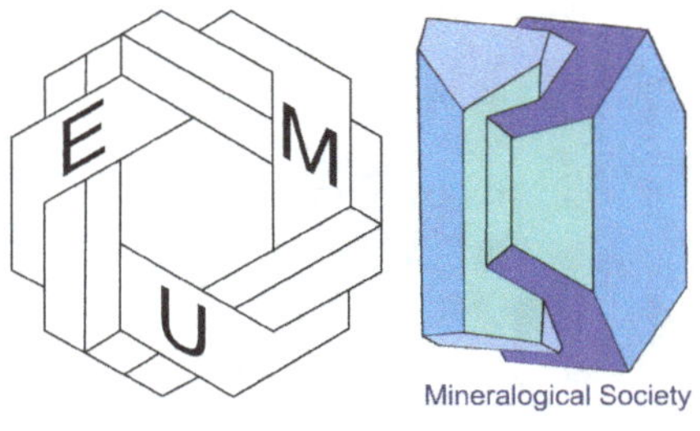

Published by the European Mineralogical Union and the
Mineralogical Society of Great Britain & Ireland, London, 2017

The publication of this textbook is supported by the European Mineralogical Union

EMU Notes in Mineralogy
A series published under the auspices of the European Mineralogical Union (EMU).

Initiator of the EMU Schools and the EMU Notes in Mineralogy
Giovanni Ferraris, Torino, President of the EMU 1992–1996

Commissioning Editor: R. Oberti, Pavia (previous editors: Giovanni Ferraris, Torino, Tamás G. Weiszburg and Gábor Papp, Budapest)

Editors of this Volume
Jakub Plášil, Institute of Physics of the AS CR, Prague, Czech Republic
Juraj Majzlan, Institut für Geowissenschaften, Friedrich-Schiller-Universität Jena, Germany,
Sergey Krivovichev, Department of Crystallography, Institute of the Earth Sciences, St. Petersburg State University, Russia

Managing Editor and Indexer: Kevin Murphy, London

Front cover design: Michel H. Guay

On the front cover: The beauty of electron diffraction: A schematic diagram of precession electron diffraction with an electron diffraction (SAED) image of icosahedrite.

ISSN: 1417 2917
ISBN: 978-0903059-5

Published by the European Mineralogical Union and the Mineralogical Society of Great Britain & Ireland (12, Baylis Mews, Amyand Park Road, Twickenham TW1 3HQ, UK)
Printed by Lightning Source, USA

The EMU Notes in Mineralogy Series

Published volumes

Volume	*Year*	*Editors*	*Title*
1	1997	S. Merlino	*Modular aspects of minerals*
2	2000	D.J. Vaughan R. Wogelius	*Environmental mineralogy*
3	2001	C.A. Geiger	*Solid solutions in silicate and oxide systems*
4	2002	C. Gramaccioli	*Energy modelling in minerals*
5	2003	D.A. Carswell R. Compagnoni	*Ultrahigh pressure metamorphism*
6	2004	A. Beran E. Libowitzky	*Spectroscopic methods in mineralogy*
7	2005	R. Miletich	*Mineral behaviour at extreme conditions*
8	2010	F.E. Brenker G. Jordan	*Nanoscopic approaches in earth and planetary sciences*
9	2010	G. Christidis	*Advances in the characterization of industrial minerals*
10	2010	M. Prieto	*Ion-partitioning in ambient-temperature aqueous systems*
11	2011	M.F. Brigatti A. Mottana	*Layered mineral structures and their application in advanced technologies*
12	2012	J. Dubessy M.-C. Caumon F. Rull	*Applications of Raman Spectroscopy to earth sciences and cultural heritage*
13	2013	D.J. Vaughan R.A. Wogelius	*Environmental mineralogy II*
14	2013	F. Nieto K.J.T. Livi	*Minerals at the nanoscale*
15	2015	M.R. Lee H. Leroux	*Planetary mineralogy*
16	2017	W. Heinrich R. Abart	*Mineral reaction kinetics: Microstructures, textures, chemical and isotopic signatures*
17	2017	I.A.M. Ahmed K.A. Hudson-Edwards	*Redox reactive minerals: properties, reactions and applications in clean technologies*
18	2017	A.F. Gualtieri	*Mineral fibres: crystal chemistry, chemical-physical properties, biological interaction and toxicity*
19	2017	J. Plášil J. Majzlan S. Krivovichev	*Mineralogical crystallography*

Copies of the EMU Notes (volumes 1–7) are distributed in Europe by the larger member societies of the European Mineralogical Union:

Società Italiana di Mineralogia e Petrologia:
www.socminpet.it

Mineralogical Society of Great Britain & Ireland:
www.minersoc.org

Société Française de Minéralogie et de Cristallographie:
www.sfmc-fr.org

and by the Secretary of the EMU, Prof. Juraj Majzlan:
juraj.majzlan@uni-jena.de

in America by the Mineralogical Society of America:
www.minsocam.org

Institutional orders as well as individual requests from outside Europe and America should be sent to the Secretary of EMU, Prof. Juraj Majzlan:
juraj.majzlan@uni-jena.de

For volumes 8 onwards, the Mineralogical Society of Great Britain acts as co-publisher and copies may be ordered from www.minersoc.org

or from:
Mineralogical Society
12 Baylis Mews,
Amyand Park Road,
Twickenham TW1 3HQ
UK

E-mail: admin@minersoc.org
Tel. + 44 (0)20 8891 6600
Fax: + 44 (0)20 8891 6599

Contents

Preface

Crystallography and mineralogy grew up next to each other, mineralogy as the older and crystallography as the younger sister science. Georgius Agricola (1494–1555), in his *De Re Metallica*, gave instructions for growth of crystals of a number of salts. Crystals and crystal growth were considered and discussed by many natural scientists, with remarkable advances and mistakes on the way. Massive efforts in terms of crystal measurements with optical goniometers, combined with the recognition that the number of point groups for crystals is limited to 32, were intended to discover the reason for the perfect geometric shapes of the crystals. These efforts failed in that respect but crystallography was alive and well even before the discovery of X-rays and the advent of diffraction.

At the dawn of structural crystallography, Walther Friedrich, Paul Knipping and Max von Laue carried out the first experiments and developed the theory of X-ray diffraction. From the early days, when even the simpler inorganic structures filled an entire PhD study, structural crystallography evolved at its own pace and found new partners in chemistry, physics, materials science, biology and other fields of physical sciences. Both morphological and structural crystallography, however, have remained as important instruments in the mineralogist's toolbox until today. Efforts to enhance the existing instrumentation, to improve our understanding of the theory of diffraction, to study nanoparticulate or poorly ordered materials, and to master large, complex structures continue in all fields of physical sciences. Mineralogy can thus use the fruits of this labour and include them in its toolbox.

In this volume, we did not intend to create a textbook of structural crystallography. There are many excellent examples of such textbooks available. Therefore, there is no chapter on the principles of diffraction (including X-ray diffraction), nor is there a chapter on the practical aspects of structure solution. Instead, we present selected topics on powerful tools that may help us to understand the structures of minerals and synthetic compounds. We did not wish to include a chapter on neutron diffraction, not because this method would not be of interest to mineralogists, but because there is a relatively recent entire volume on neutron diffraction (Reviews in Mineralogy and Geochemistry, volume **63**, *Neutron Scattering in Earth Sciences*).

Crystal structures, owing to their periodicity, can be described by fairly simple mathematical models. These models condense the information on positions of moles of atoms in a crystal to bytes or kilobytes of data. The structures and their mathematical representation are therefore excellent subjects for classification, generalization and prediction. In chapter 1 of this volume, Sergey Krivovichev discusses how crystal structures are described, interpreted and classified. He focuses on the structural hierarchies and modularity of the crystal structures. The chapter presents general ideas and concepts on the crystal structures, applicable to all groups of minerals and synthetic compounds. There are many structural classifications of specific groups of minerals available; for some of them, the reader may also consult the first volume of the EMU Notes in Mineralogy (*Modular Aspects of Minerals*, edited by S. Merlino, 1994) and many other reviews.

Powder X-ray diffraction, even though not a new technique, offers the opportunity to investigate samples for which good quality crystals are not available. This is the case for many mineralogical and geological samples and this 'routine' method should not be omitted from the repertoire of a mineralogist. As the quantitative treatment of powder X-ray diffraction data becomes automated in some applications, one should not forget the need to understand this technique (instead of using it as a black box) and to grasp its potential.

In chapter 2, Angela Altomare provides a thorough overview of the advances made in powder X-ray crystallography over recent decades. She shows clearly that powder X-ray diffraction is not only a simple finger-printing method but can be used to solve crystal structures, quantify phase compositions of mixtures, extract structural data at extreme conditions and much more.

Precession electron diffraction (PED) allows us to determine crystal structures from very small particles, down to a few nanometers. In chapter 3, Lukáš Palatinus, Mauro Gemmi and Mariana Klementová explore this method, explain the principles of electron crystallography and related techniques, and give a number of examples, most of them mineralogical in nature. The very small size of the crystals probed, together with the strong interaction between electrons and matter, opens up an entire new field for structural studies on nanocrystalline materials. The door to the paradise where the structures of many fine-grained minerals from the surface of the Earth, but also from rocks (*e.g.* charoite), is open, but only a few have entered so far. We hope that this chapter will help to overcome this hesitation.

Pair distribution function (PDF) analysis is a method that can be used to track structures of poorly ordered materials. In chapter 4, Marc Michel reviews the basics and the application of this method the potential of which has not yet been used to its full extent in mineralogy. Natural materials may be mixtures of several closely or distantly related structures and, therefore, the data may be exceedingly difficult to interpret. On the other hand, many poorly crystalline substances can be synthesized easily in the laboratory under controlled conditions and subjected to PDF analysis. Knowing the structures of poorly crystalline iron, aluminium and manganese oxides can certainly further our understanding of many processes at or near the surface of our planet.

Extraction of the structural data from tiny crystals or poorly ordered substances is an end-member the counter-pole of which is the understanding of large, complex or aperiodic structures. In chapter 5, Luca Bindi and Gervais Chapuis explore the world of modulated, composite and quasicrystal structures. The concept of superspace is described and considered. The chapter gives a number of examples of natural substances with such complicated structures, hinting that there may be many more of those and the concept of ideal, three-dimensional crystals may be a restrictive one, even if defects (point, line, plane) are not considered as elements that violate the perfect periodicity.

Jakub Plášil, Juraj Majzlan and Sergey Krivovichev
Prague, July 2017

EMU Notes in Mineralogy, Vol. 19 (2017), Chapter 1, 1–77

Structure description, interpretation and classification in structural mineralogy

SERGEY V. KRIVOVICHEV

Department of Crystallography, Institute of the Earth Sciences, St. Petersburg State University, University Emb. 7/9, 199034 St. Petersburg Russia
Nanomaterials Research Centre, Kola Science Centre of Russian Academy of Sciences, 14 Fersman Str., Apatity 184200, Murmansk Region, Russia,
e-mail: s.krivovichev@spbu.ru

This review provides a summary of the state-of-the-art in terms of description, interpretation and classification of crystal structures in mineralogy. Among the various methods, the focus is on atomic packing (including both anion and cation arrays), coordination polyhedra (both cation- and anion-centred), the concept of fundamental building blocks and related ideas and the use of networks, graphs and tilings (space partitions). The basic concepts under discussion in modern structural mineralogy include structure hierarchy (specification and compositional hierarchies are considered separately), modularity (representation of crystal structures as constructed from modules extracted from simple archetype structures) and complexity (with emphasis on static or informational and algorithmic complexities). Short historical notes are given for all of the topics considered.

1. Introduction

Since the discovery of X-ray diffraction by crystals one hundred years ago, the accummulation of data on crystal structures of minerals has been accompanied by the development of the methods of their description and interpretation in order to provide a coherent framework for their classification and understanding of their occurrences in different geochemical environments. Crystal structures are three-dimensional discrete arrangements of atoms linked by chemical bonds and, compared to biological objects, their description is based upon the finite portion of space called a unit cell. Neverthless, the task of structure description and interpretation is not as simple as it may appear from outside the field, and various methods and concepts have been devised in order to understand the structural architecture of minerals at the atomic level. In this review, I shall try to overview the field of structure description and interpretation in structural mineralogy as broadly as possible. Inevitably, some areas receive more attention than others, based on the author's beliefs and understanding of their importance.

In general, crystal structures of minerals are complex, and various methods of describing them have been developed over the last hundred years: atomic packings (anion and cation), coordination polyhedra, fundamental building blocks (and related concepts), graphs, and tilings (Fig. 1). Obviously, these approaches are closely related to each other. In many cases, a crystal structure can be presented with the same success

DOI: 10.1180/EMU-notes.19.2

from different points of view and it is the choice of the authors to decide which description is better. There is no universal recommendation as to which representation is the best, but there is a certain feeling that a particular method provides the most transparent and efficient interpretation of the underlying principles of structure organization. In this chapter, I shall provide a brief review of the application of the methods mentioned above to the crystal structures of minerals, together with some short historical introductions.

Crystal structures of minerals are complex systems as they consist of a (sometimes very large) number of interacting parts (atoms) and the interactions between these parts are not in any sense simple. Therefore I will pay special attention to such properties of mineral structures as complex systems as: hierarchy, modularity and complexity (Fig. 1). These concepts are highly relevant to the understanding of minerals and their formation and evolution in nature.

2. Methods

2.1. Atomic packings

2.1.1. Anion packings

The first attempts to explain the structure of crystals using sphere packings are due to Kepler (1611), who devised the structural model of snowflakes on the basis of a hexagonal packing of equal spherical particles. Kepler has many followers and the

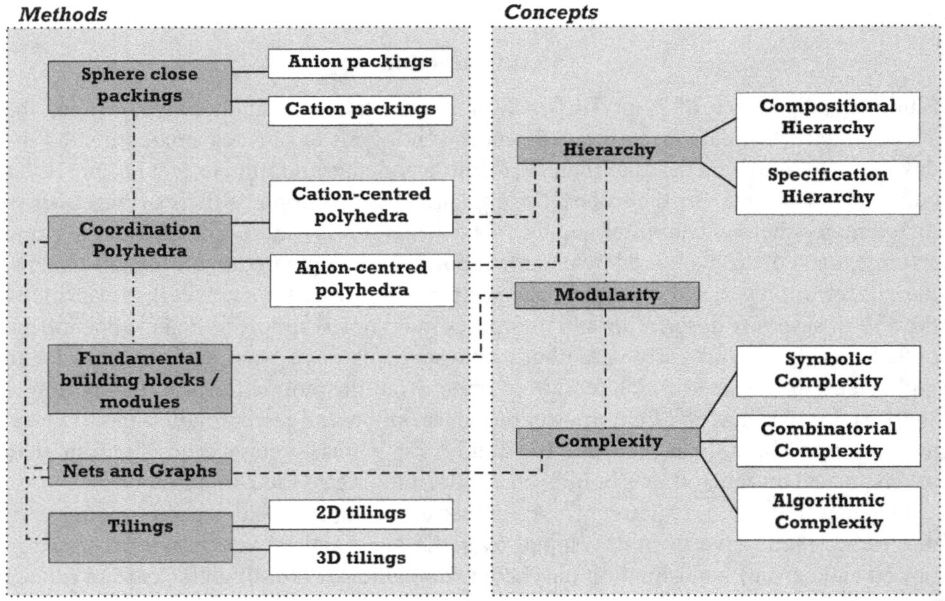

Figure 1. Basic methods of structure description in mineralogy and related concepts used in consideration of crystal structures of minerals as complex systems.

development of this idea was reviewed in detail by Shafranovsky (1978), Lima-de-Faria (2012) and Baur (2014), among others. Nowadays sphere close-packings are taught in introductory courses on crystallography and mineralogy and there is no need to recall basic definitions and concepts here. The two basic closest-packings are 3-layer face-centred cubic (*ccp*) and 2-layer hexagonal (*hcp*).

Within the standard approach, structures of ionic crystals based upon closest packings of spheres are considered as packings of anions with cations filling tetrahedral and/or octahedral interstices. If n is the number of spheres in a packing, then the numbers of tetrahedral and octahedral interstices are $2n$ and n, respectively. Description of crystal structures in terms of anion packings has been very popular since the first days of X-ray diffraction structure analysis (see, *e.g.* Belov, 1947; Bragg and Claringbull, 1965). Indeed, many crystal structures of minerals that we today consider as relatively simple can be adequately described in terms of anion packings.

The approach was exploited systematically by Lima-de-Faria (1994, 2001, 2003, 2004, 2012), who worked out the enormous task of presenting crystal structures of almost all minerals in terms of close packings. As an epigraph to his 1994 book, he had chosen the quotation from Belov (1947): "In spite of the variety of the mineral crystalline world, the whole 'mineralogical game' just reduces to various modes of filling gaps in uniform close packing with various corresponding patterns." It is absolutely true that most of the basic, relatively simple structure types of minerals are based upon close packings of anions. However, modern structural mineralogy is dominated by minerals with complex hierarchical arrangements of coordination polyhedra, building blocks and modules, that cannot be understood easily as close-packed arrangements. The closest packings are not very popular in our days, though it might be that, as noted by Moore (1994), "...most structures which are in fact based on principles of that fundamental law of nature, closest-packing, are misrepresented". Recent developments of packing principles, together with the analysis of distortions induced by various additional parameters, are due to Thompson and Downs (2001, 2010) and Thompson *et al.* (2012).

2.1.2. Cation packings and cation arrays

An alternative approach to the crystal structures of complex ionic minerals is to consider them in terms of cation closest-packings (or eutaxies) with interstices filled by anions. This idea has been expressed by many authors, was elaborated systematically by O'Keeffe and Hyde (1985), Borisov (1982, 1986, 1992, 1996), Krivovichev and Filatov (1999), and is currently being developed by Vegas (2000, 2011) as an Anionic in Metallic Matrices (AMM) model (see, *e.g.* Vegas *et al.*, 2006).

Lebedev (1972) pointed out that the arrays of Ca atoms in Ca metal and fluorite, CaF_2, are identical and correspond to cubic closest-packing (*ccp*). Moreover, both Ca metal and fluorite crystallize in the same space group, $Fm\bar{3}m$, and their unit-cell parameters differ only slightly (5.588 and 5.463 Å, respectively). Thus, fluorite may be considered as Ca metal with F atoms inserted into tetrahedral interstices, which is associated with the decrease in the unit-cell parameters. Another illustrative example of the same kind are the structures of larnite, β-Ca_2SiO_4, and dicalcium silicide, Ca_2Si. It turns out that the Ca_2Si in these two compounds are identical and very similar from the geometrical point of view. The unit-cell volumes are almost the same and are 343.9 and 331.2 Å^3 for larnite and Ca_2Si, respectively (see O'Keeffe and Hyde (1985) for more details).

The description of oxides and oxysalts as anion-stuffed metals and alloys has found many applications in the literature. For instance, Ercit and Hawthorne (1995) described the octahedral framework in the crystal structure of murataite, $(Y,Na)_6(Zn,Fe)_5Ti_{12}O_{29}(O,F)_{10}F_4$, in terms of a graph with vertices corresponding to the centres of Ti octahedra. The topology of the graph is identical to the topology of the boron network in the crystal structure of UB_{12} (Fig. 2). Santamaría-Pérez and Liebau (2011) reported interesting direct topological analogies between the structures of tetrahedral frameworks in zeolites and clathrates and cation arrays in intermetallic compounds. For instance, the array of Al and P atoms in the crystal structure of metavariscite, $Al(PO_4)(H_2O)_2$, is identical to the array of B atoms in CrB_4. Table 1 provides a number of selected examples of the correspondence between the arrangements of cations in some complex oxysalt minerals with the identical arrangements of atoms in metal, alloys, borides, silicides, *etc*. The very occurrence of such a correspondence can be assigned to "Nature's parsimony in selecting appropriate geometrical patterns" (O'Keeffe and Hyde, 1985) or the complexity-limited number of geometrical configurations that can be realized in a three-dimensional space under particular requirements of local coordinations.

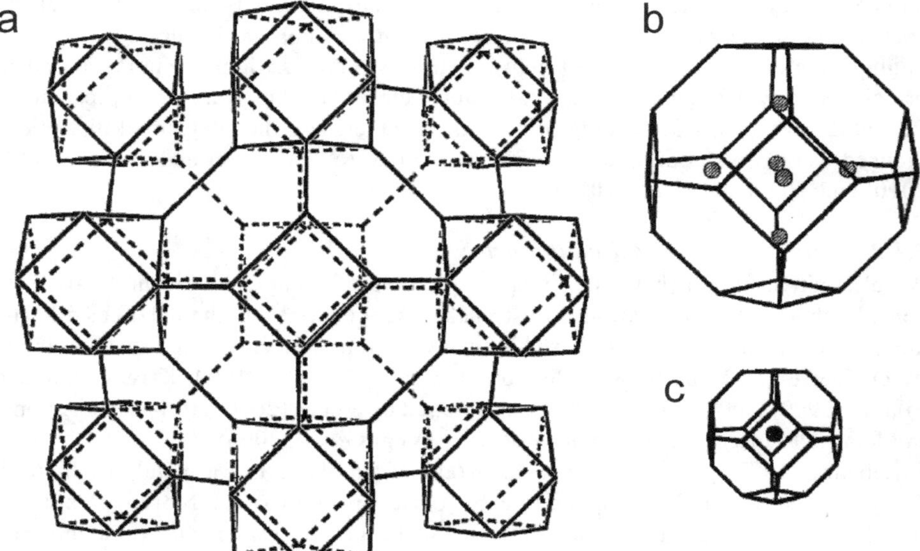

Figure 2. The array of octahedrally coordinated Ti atoms in the crystal structure of murataite-(Y) (a) is identical to the array of B atoms in UB_{12}; the octahedral cage in murataite-(Y) contains six Y atoms (b), whereas the analogous cage in UB_{12} contains one U atom only (c).

Table 1. Selected examples of oxide and oxysalt minerals with cation arrays and the arrangements of atoms in metals, alloys or other simple compounds (after O'Keeffe and Hyde, 1985 with some modifications).

	— Anion-stuffed derivatives —	
Parent compound	Mineral	Chemical formula
Cu (*fcp* array)	Periclase	MgO
	Fluorite	CaF_2
	Thorianite	ThO_2
	Litharge	PbO
	Tenorite	CuO
	Cuprite	Cu_2O
	Paramelaconite	Cu_3O_4
	Bixbyite	Mn_2O_3
NaCl	Sheelite	$CaWO_4$
Mg (*hcp* array)	Zincite	ZnO
$BaPb_3$	Hatrurite	$Ca_3(SiO_4)O$
β-Hg	Rutile	TiO_2
	Zircon	$ZrSiO_4$
CuZn	Tausonite	$SrTiO_3$
$MoSi_2$	Koechlinite	$Bi_2O_2(MoO_4)$
Cu_2MnAl	Elpasolite	K_2NaAlF_6
$MgCu_2$	Spinel	Mg_2AlO_4
FeB	Baryte	$BaSO_4$
Mn_5Si_3	Apatite-(caoh)	$Ca_5(PO_4)_3(OH)$*
NiAs	Aragonite	$CaCO_3$
Th_3P_4	Langbeinite	$K_2Mg_2(SO_4)_3$
W_2CoB_2	Kotoite	$Mg_3(BO_3)_2$
U_3Si_2	Åkermanite	$Ca_2Mg(Si_2O_7)$
Ca_2Si	Larnite	Ca_2SiO_4
Ni_2In	Forsterite	Mg_2SiO_4
$CuAl_2$	Brenkite	$Ca_2F_2(CO_3)$
Cr_3Si	Grossular	$Ca_3Al_2(SiO_4)_3$
Re_3B	Norbergite	$Mg_3(SiO_4)F_2$
~ Mo_3CoSi	Vesuvianite	$Ca_{19}Al_{10}Mg_3(SiO_4)_{10}(Si_2O_7)_4(OH)_{10}$
~ UB_{12}	Murataite-(Y)	$(Y,Na)_6(Zn,Fe)_5Ti_{12}O_{29}(O,F)_{10}F_4$

* the analogy first reported by Wondratchek *et al.* (1964).

2.2. Coordination polyhedra

2.2.1. General notes

The importance of nearest-neighbour interactions in the crystal-structure organization was recognized as early as the 1920s, when the first systematic papers appeared on structural systematics of silicates (Machatschki, 1928; Bragg, 1930) and the "principles determining the structure of complex ionic crystals" were formulated by Pauling (1929). It was natural at that time to propose the idea of inorganic crystal structures on the basis of local geometrical environments of particular atoms, *i.e.* on their coordination numbers and coordination polyhedra. Hazen (1985, 1988) emphasized that the polyhedral approach is a "useful fiction" to estimate the physical properties of minerals by summation of properties of separate polyhedra. He noted that, in many cases, individual polyhedra have the same properties invariant from structure to structure (such as shape, size, heat capacity, thermal expansivity, compressibility, *etc.*). This idea was employed in a number of papers, where authors (with different degrees of success) attempted to derive thermodynamic and other mineral properties from the contributions of particular polyhedra (Chermak and Rimstidt, 1989; Chen *et al.*, 1999).

The majority of crystal structures of minerals can probably be considered better in terms of coordination polyhedra of cations, *i.e.* cation-centred polyhedra. There are two major reasons for the preference of cation coordination over anion coordination: (1) the average strength of bonds within cation-centred polyhedra is usually higher than that within anion-centred polyhedra; (2) the chemistry and coordination geometry of cations are generally much more diverse than those of anions (in fact, the majority of minerals are oxysalts, *i.e.* contain O^{2-} anions only; as pointed out by Hazen (1988): "...our focus on cation polyhedra is largely an artifact of the mineralogical data base"). There are many cases, however, in which description of crystal structures in terms of anion-centred polyhedra is more reasonable and transparent than that in terms of cation-centred polyhedra. Below, both approaches are considered in more detail. First, structure descriptions and systematics in terms of cation-centred polyhedra will be considered with their subdivision into homo- and heteropolyhedral structures following Hawthorne (1985), *i.e.* structures with one and several types of polyhedra, respectively.

2.2.2. Cation-centred polyhedra

Homopolyhedral structures. The idea of classifying crystal structures according to the topology and dimensionality of structural units formed by the condensation of coordination polyhedra of cations was first formulated by Machatschki (1928), Bragg (1930) and Náray-Szabó (1930) on the basis of analysis of crystal structures of silicate minerals. The main classification principle was dimensionality of the condensation of (SiO_4) tetrahedra. Later Liebau (1956, 1962) suggested using, as an additional parameter, periodicity of extended silicate anions. Zoltai (1960) developed a classification of tetrahedral structural units with no limits imposed upon the topology of linkage of tetrahedra: in contrast to silicates, edge- and face-sharing of tetrahedra was allowed with the possibility of one O atom being shared by more than two tetrahedra. The tetrahedral structures within Zoltai's classification must conform with two rules: (a) the difference between the smallest and the largest number of tetrahedra participating in the sharing of a tetrahedral corner cannot be more than one and (b) no edges of tetrahedra can be shared unless the corners are shared between more than four tetrahedra, and no faces can be shared unless the corners are shared between more than eight tetrahedra. As a characteristic of such tetrahedral units, Zoltai suggested using the average number of tetrahedra sharing a corner or sharing coefficient (SC) of a tetrahedral unit $[T_mO_n]$:

$$SC = 2s + 1 - n(s^2 + s)/4m \quad (1)$$

where s is the integral part of the ratio $4m/n$. Obviously, SC is equal to 1 for an isolated tetrahedron, 1.25 for the (Si_2O_7) group of two corner-sharing tetrahedra, 1.50 for the pyroxene-type $[Si_2O_6]$ chains, 1.75 for the mica-like $[Si_2O_5]$ sheets, 2.00 for the $[SiO_2]$ tetrahedral framework in quartz, *etc.*

The most elaborate scheme of classification of tetrahedral structural units is probably that developed by Liebau (1980, 1985). The list of parameters used in his classification is given in Table 2. Liebau (1985) was able to demonstrate that the values of these parameters correlate with crystal-chemical properties of cations present in a structure along with silicate anions.

Krivovichev *et al.* (1997) proposed a set of general parameters to classify any polyion based upon linked polyhedra. Their scheme can be seen as an extension of Liebau's system with additional tools and parameters that take into account the possibility of edge- and face-sharing of polyhedra. In fact, the scheme developed by Krivovichev *et al.* (1997) was elaborated for the classification of structural units based on anion-centred tetrahedra that display much higher topological variations than those in 'traditional' chemistry of cation-centred tetrahedral anions. The most important feature of this system is in distinguishing between crystallographic, topological and configurational equivalence of polyhedra, with the last being based upon the use of Schlegel diagrams (more on this in Section 2.2.4). Even though the system devised by Krivovichev *et al.* (1997) is able to account for heteropolyhedral as well as for monopolyhedral units, to date it has been applied only to monopolyhedral structures and thus is considered in this subsection.

As far as we know, no other detailed classifications were proposed for monopolyhedral units in minerals except those proposed by Pabst (1950) and later Hawthorne (1984) for the systematics of octahedral structures in aluminofluorides. In these works, structural units of linked (AlX_6) octahedra were classified according to their degree of polymerization into separate octahedra, finite clusters, chains, sheets and frameworks. No parameters other than dimensionality have been used in this classification scheme, probably because of the relatively low structural diversity of aluminofluorides.

Heteropolyhedral structures – General notes. The very term heteropolyhedral structural unit should be assigned to Hawthorne (1985), though the concept was

Table 2. Classification parameters and terms in Liebau's systematics of silicates (after Liebau, 1985).

Notation	Parameter/term	Definition
CN	Coordination number of silicon	Number of oxygen atoms bonded directly to a silicon atom.
L	Linkedness	Number of oxygen atoms shared between two adjacent $[SiO_n]$ polyhedra. ***L*** = 0 (isolated), ***L*** = 1 (corner-sharing), ***L*** = 2 (edge-sharing), ***L*** = 3 (face-sharing).
s	Connectedness	Number of $[SiO_n]$ polyhedra that share oxygen atoms with the $[SiO_n]$ polyhedron considered.
B	Branchedness	***uB*** = unbranched anions containing $[SiO_n]$ polyhedra with ***s*** < 3; ***oB*** = open-branched; ***lB*** = loop-branched; ***olB*** = mixed-branched; ***hB*** = hybrid.
D	Dimensionality	Number of dimensions of infinite extension of an anion. ***D*** = 0 (groups), ***D*** = 1 (chains), ***D*** = 2 (layers), ***D*** = 3 (frameworks).
M	Multiplicity	Number of single polyhedra, rings, chains or layers which are linked to a complex anion of the same dimensionality.
P	Chain periodicity	Number of $[SiO_n]$ polyhedra, excluding branches, in the repeat unit of a fundamental chain.
P^r	Ring periodicity	Number of $[SiO_n]$ polyhedra, excluding branches, of a fundamental ring.
N_{an}	Number of different silicate anions in a silicate	N_{an} = 1 for uniform-anion silicates, N_{an} > 1 for mixed-anion silicates.

implicitly employed much earlier, when it was recognized that structures can be described as arrays of linked coordination polyhedra with various degrees of bond-strengths inside them. Povarennykh (1966) made systematic use of the idea of heteropolyhedral complexes in his crystal-chemical systematics of minerals and Wells (1970) provided many examples of mineral structures that can be described as based on linked coordination polyhedra of different kinds. Mixed-polyhedra complexes have been used in classifications of sulfates (Smirnova *et al.*, 1967, 1968; Bokii and Gorogotskaya, 1969; Sabelli and Trosti-Ferroni, 1985) and borates (Christ and Clark, 1977), and in numerous papers dealing with crystal structures of particular minerals and mineral groups. Below is an overview of systematic works in this area that were focused on relatively large classes of structures with similar crystal chemical features.

The concept of mixed anionic radicals (MARs) and its development. This concept originated in the N.V. Belov school of crystallography and crystal chemistry, mostly concentrated in Moscow State University. According to Sandomirskiy and Belov (1984), the first idea of a complex anionic function of mixed octahedral-tetrahedral structural units was formulated by Guan *et al.* (1963) after solution of the structure of bafertisite, $BaFe_2Ti[Si_2O_7]O_2$ (Fig. 3a). The structure was considered as based on the complex $[TiO_2(Si_2O_7)]^{6-}$ anion formed by condensation of TiO_6 octahedra and Si_2O_7 groups (Fig. 3b). Bakakin and Belov (1964) noticed similarity between the structures of bafertisite and mica-group minerals and pointed out that TiO_6 octahedra in bafertisite, along with silicate tetrahedra, have an acidic function as in the structures of titanates such as $K_2Ti_6O_{13}$, where the octahedral anion plays a role analogous to that of tetrahedral anions in silicates. They pointed out that minerals such as narsarsukite, $Na_2TiSi_4O_{11}$, baotite, $Ba_4(Ti,Nb)_8(Si_4O_{12})O_{16}Cl$, benitoite, $BaTiSi_3O_9$, catapleiite, $Na_2ZrSi_3O_9(H_2O)_2$, *etc.*, should be considered not as complex oxides, but as titano- or zirconosilicates based upon complex mixed octahedral-tetrahedral anions.

The concept of mixed (heteropolyhedral) frameworks (and structures of lower dimensionality) was formalized by Voronkov *et al.* (1973, 1974, 1975). They considered frameworks that satisfy the following requirements: (1) framework-forming cations are functionally equivalent from the chemical point of view; (2) anions at the vertices of cation-centred polyhedra can be shared by no more than two polyhedra; and (3) bond-valence sums incident upon the anion sites should correspond to their formal oxidation states.

As an example of their approach, Voronkov *et al.* (1974, 1975) considered frameworks built from two types of polyhedra, (MX_p) and (TX_q), formed by the M and T

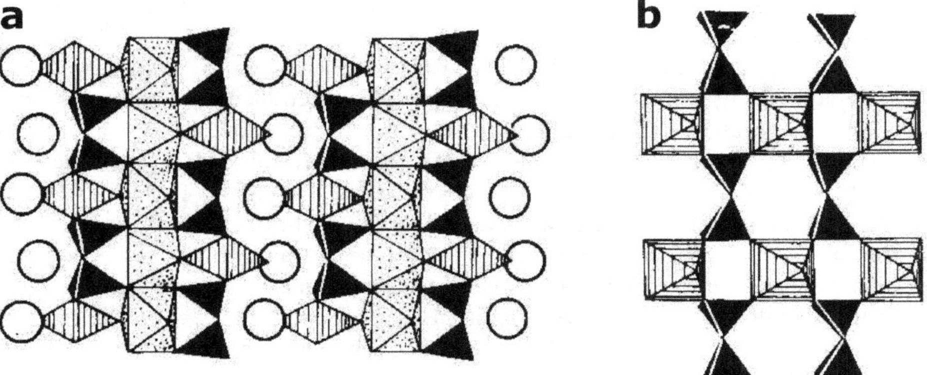

Figure 3. The crystal structure of bafertisite, $BaFe_2Ti(Si_2O_7)O(OH)$ (a) and its $[TiO(OH)(Si_2O_7)]^{8-}$ octahedral-tetrahedral layer (b) (after Guan *et al.*, 1960). Legend: Si tetrahedra = black; Ti octahedra = striated; Fe octahedra = pointed; Ba atoms are shown as circles.

cations, respectively (here X is an unspecified anion, and p and q are coordination numbers of M and T, respectively). The condition (2) requires that the (MX_p) and (TX_q) polyhedra have, as their own, p/2 and q/2 X anions, respectively. Therefore, the formula of the framework with the ratio M:T = m:n can be written as

$$m[MX_{p/2}]\cdot n[TX_{q/2}] = [M_mT_nX_{(mp+nq)/2}]^Q \quad (2)$$

where Q is the charge of the framework calculated as

$$Q = [m(v_M - pv_X/2) + n(v_T - qv_X/2)] \quad (3)$$

where v_M, v_T and v_X are the valences of M, T and X, respectively.

According to formulae 2 and 3, octahedral-tetrahedral frameworks in the crystal structures of oxysalts (p = 6, q = 4, $v_X = -2$) should have the following formula:

$$[M_mT_nO_{3m+2n}]^Q \quad (4)$$

where

$$Q = [m(v_M - 6) + n(v_T - 4)] \quad (5)$$

Equations 2–5 derived by Voronkov *et al.* (1975) allow us to predict the possibility of formation and particular stoichiometries of mixed frameworks in various classes of minerals and inorganic compounds. For instance, in phosphates ($v_T = 5$) of tetravalent elements ($v_M = 4$), the formation of electroneutral frameworks is possible only for m:n = 1:2, e.g. in ZrP_2O_7 and SiP_2O_7.

For octahedral-tetrahedral frameworks with all anions shared between precisely two coordination polyhedra, the m:n ratio has its critical value at 2:3. If all (TO_4) tetrahedra share all their vertices with the adjacent (MO_6) octahedra and *vice versa*, the framework has the formula [$M_2T_3O_{12}$] or [$M_2(TO_4)_3$] (langbeinite, $K_2[Mg_2(SO_4)_3]$ can be given as an example). In this framework, tetrahedra are isolated from each other, and the same is true for octahedra. For m:n >2:3, the framework must have a polymerized octahedral substructure, whereas, for m:n <2:3, the framework must contain a polymerized tetrahedral substructure. In the structure of titanite, $Ca[TiO(SiO_4)]$, with m:n = 1:1 (>2:3), (TiO_6) octahedra share corners to form infinite chains linked by isolated (SiO_4) tetrahedra, whereas, in the structure of benitoite, $Ba[Ti(Si_3O_9)]$, with m:n = 1:3 (<2:3), isolated (TiO_6) octahedra are linked into a three-dimensional framework by three-membered (Si_3O_9) rings of corner-sharing (SiO_4) tetrahedra. Along this line, Krivovichev (2005) suggested classifying octahedral-tetrahedral frameworks according to dimensionalities of octahedral ($\boldsymbol{D}_{oct}$) and tetrahedral ($\boldsymbol{D}_{tetr}$) substructures, respectively. The classification can be represented as a 4 × 4 table (Table 3), where to each octahedral-tetrahedral framework the ($\boldsymbol{D}_{oct}$; $\boldsymbol{D}_{tetr}$) pair of values is assigned.

Voronkov *et al.* (1975) noted that the idea of mixed frameworks can be extended easily to include complexes of lower dimensionality (layers, chains and finite clusters). In order to do this, one has to reject condition (2) mentioned above, which leads to the appearance of pendant (non-shared) vertices of cation-centred coordination polyhedra. Thus, the (MX_p) polyhedra may have p_1 shared and p_2 pendant vertices, and the same holds for the (TX_q) polyhedra that may have q_1 shared and q_2 pendant vertices. The framework (or structural unit of lower dimensionality) has the formula

Table 3. Classification of minerals based on octahedral-tetrahedral frameworks

D_{tetr} / D_{nt}	0	1	2	3
0	wadeite, benitoite, kostylevite, petarasite, keldyshite	hilairite, umbite, terskite, vlasovite, elpidite, gaidonnayite, zektzerite	lemoynite, armstrongite, penkvilksite	–
1	belkovite, labuntsovite-group minerals	batisite-shcherbakovite series, zorite	–	–
2	komarovite	–	–	–
3	–	–	–	–

$$[M_mT_nX_{[(mp_1\ +\ nq_1)/2\ +\ mp_2\ +\ nq_2]}]^Q \tag{6}$$

where

$$Q = m[v_M - v_X(p_2 + p_1/2)] + n[v_T - v_X(q_2 + q_1/2)] \tag{7}$$

For octahedral-tetrahedral units in oxysalts ($v_X = -2$) and $\{p_1 = 4; p_2 = 2; q_1 = 3; q_2 = 1\}$, their formula can be written as $[M_mT_nO_{4m+2.5n}]$, which is realized, *e.g.* in titanosilicates such as bafertisite, which contain the titanosilicate layer $[TiO_2(Si_2O_7)]^{6-}$ (Fig. 3).

The approach proposed by Voronkov *et al.* (1974, 1975) was further developed in many papers and two monographs (Pyatenko *et al.*, 1976; Voronkov *et al.*, 1978) and generalized by Sandomirskiy and Belov in the 205-page monograph published in Russian in 1984 and almost completely unknown in the West. According to Sandomirskii and Belov, MAR is the part of a crystal structure of an inorganic oxysalt that is anionic and composed of crystal chemically different complexes polymerized according to the principles of condensed oxyacid anions (*e.g.* silicates). The most important points of this definition are the following:

(1) Radicals should be anionic, *i.e.* negatively charged. In other words, the structure should contain other cations with basicity greater than that of the radical-forming cations. For instance, an octahedral-tetrahedral framework in the structure of TiP_2O_7 is not considered as a MAR, as the framework is not anionic. Instead, it is classified as a titanium diphosphate with Ti^{4+} playing the role of a charge-compensating cation.

(2) MARs should be mixed, which means that they should consist of crystal-chemically different polyhedra, *i.e.* polyhedra different either in terms of chemistry or geometry or both. For example, borate anions built from BO_3 triangles and BO_4 tetrahedra are MARs, whereas purely tetrahedral borate anions are not.

(3) The coordination polyhedra should be polymerized, *i.e.* should have at least one common ligand.

Sandomirskiy and Belov (1984) proposed a global structural classification of MARs (heteropolyhedral complexes) on the basis of geometrical features of coordination polyhedra. First, the latter are classified into ten types: triangles (*Tr*: trigon), tetrahedra (*T*: tetrahedron), 5-vertex-polyhedra (*P*: pentagon), octahedra (*O*: octahedron), 7-vertex-polyhedra (*H*: heptagon), 8-vertex-polyhedra (*Oc*: octagon), 9-vertex-polyhedra (*En*: ennegon), 10-vertex-polyhedra (*D*: decagon), 11-vertex-polyhedra (*U*: undecagon), 12-vertex-polyhedra (*Do*: dodecagon). Then the MARs (polyhedral structural units) are classified into classes according to the number of types of polyhedra that form their basis: monopolyhedral, bipolyhedral and tripolyhedral (no tetrapolyhedral MARs were known in 1984). Further, each class is subdivided into subclasses in accord with the particular types of radical-forming polyhedra (Fig. 4). Within each subclass, the MARs are subdivided, according to their dimensionality, into finite clusters, chains, sheets and frameworks. Finite clusters are subdivided additionally into those with and without rings, and the same is applied to chains. Sandomirskiy and Belov (1984) provided detailed systematics of 101 subgroups of *TT* radicals, *i.e.* mixed tetrahedral units, by means of the modified Zoltai sharing coefficient (SC). They also discussed in detail polymorphic relations and temperature-induced transformations of *TT* radicals.

Voronkov *et al.* (1983) and Rastsvetaeva *et al.* (1983) applied the theory of MARs to sulfates and a systematic treatment of sulfates along this line was published by Rastsvetaeva and Pushcharovskii (1989). The concept of MARs is still used widely by the crystallographers of the N.V. Belov school. In particular, G.D. Ilyushin (a former student of V.V. Ilyukhin) used this idea in his reconstructions of self-assembly pathways in complex titano- and zirconosilicate systems (Ilyushin, 2003, 2012, and references therein).

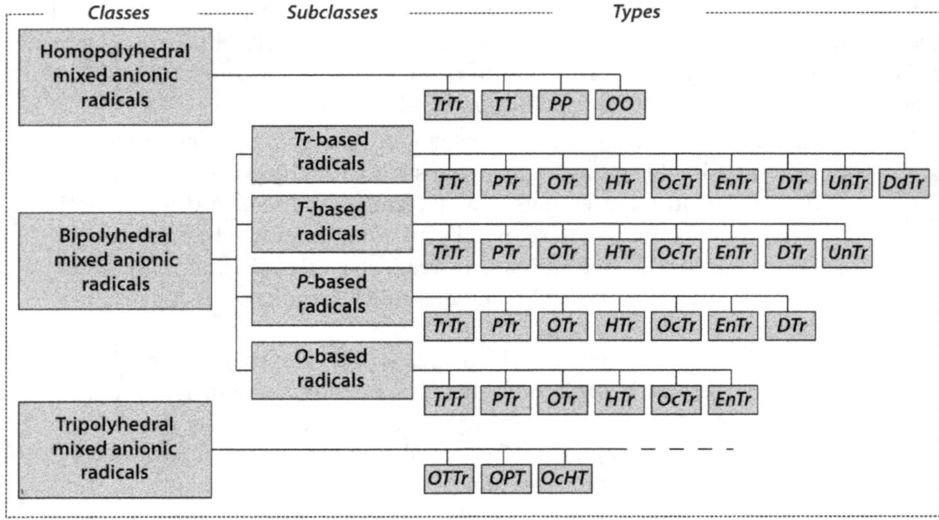

Figure 4. Classification of MARs (modified after Sandomirkii and Belov (1984)).

Crystal chemistry of minerals with heteropolyhedral units in works by F.C. Hawthorne. Hawthorne (1983) established a rigorous quantitative basis for the description of crystal structures in terms of heteropolyhedral units, when he proposed that "crystal structures may be ordered or classified according to the polymerization of those coordination polyhedra (not necessarily of the same type) with the higher bond valences". The bond valences can be calculated directly from the respective bond lengths using empirical relations and the set of parameters derived for the given cation-anion pair (Brown and Shannon, 1973; Brown, 1981, 2002, 2009; Urusov and Orlov, 1999, and references therein). The principle was used by Hawthorne and co-authors to construct structural classifications (specification structural hierarchies, see Section 3.1) for different classes of minerals, including sulfates, phosphates, borates, copper minerals, and beryllium minerals (Hawthorne, 1985, 1986, 1990, 1994, 1998; Eby and Hawthorne, 1993; Burns *et al.* 1995; Hawthorne *et al.*, 1996, 2000; Grice *et al.*, 1999; Hawthorne and Huminicki, 2002). Along the same line, P.C. Burns constructed a structural hierarchy of hexavalent uranium minerals and inorganic compounds (Burns *et al.*, 1996; Burns, 1999, 2005), using the anion-topology approach to be considered in detail in Section 2.5.1.

2.2.3. Anion-centred polyhedra

Structure description of crystal structures in terms of coordination of cations is one of the most efficient and natural approaches for the majority of minerals. However, there are cases when this kind of structure interpretation reflects neither basic principles of structural architecture nor relationships between different structures and between crystal structure and physical properties.

In 1968, Bergerhoff and Paeslack considered a series of compounds with 'additional' oxygen atoms, *i.e.* atoms that do not participate in strongly bonded 'acid residue' complexes or ions such as sulfate, silicate, germanate, chloride, fluoride, *etc.* For instance, they noted that, in the structure of dolerophanite, $Cu_2O(SO_4)$, there are two types of O atoms: those that belong to sulfate groups (O_S) and those that do not (O_a). The Cu:O_a ratio is equal to 2:1, and the O_a atoms are coordinated solely by Cu atoms. The coordination number of the O_a atom is four, which means that it can be considered as a central atom for the (OCu_4) tetrahedron. The adjacent (OCu_4) tetrahedra share edges and corners to produce 2-dimensional (2D) layers with the composition $[OCu_2]^{2+}$ (Fig. 5). These layers can be considered as an extended polycationic structural unit; its positive charge in dolerophanite is compensated by the presence of tetrahedral $(SO_4)^{2-}$ anions. Thus the structure of $Cu_2O(SO_4)$ was considered to consist of two basic units: the $[OCu_2]^{2+}$ layers of oxocentred (OCu_4) tetrahedra and $(SO_4)^{2-}$ tetrahedral oxyanions. Bergerhoff and Paeslack (1968) included in their scheme some other compounds such as $Zn_4O(BO_2)_6$ and $Be_4O(CH_3COO)_6$ (both containing isolated M_4O tetrahedra), kyanite, $Al_2O(SiO_4)$ (containing double $[OAl_2]^{4+}$ chains), $Bi_2O_2(GeO_3)$, La_2O_2S, and Pb_2OF_2. The structure descriptions in terms of anion-centred tetrahedra have been developed mainly by inorganic chemists (see historical remarks and references by Krivovichev *et al.*, 2013). Effenberger (1985) reported structure

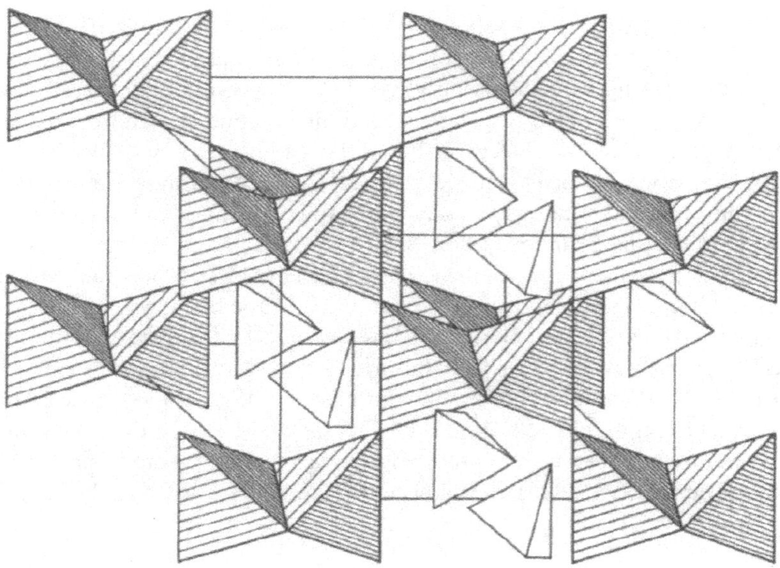

Figure 5. The structure of dolerophanite, $Cu_2O(SO_4)$, consisting of the $[OCu_2]^{2+}$ layers of edge- and corner-sharing (OCu_4) tetrahedra (lined) and (SO_4) tetrahedra (after Bergerhoff and Paeslack, 1968).

refinement of dolerophanite, $Cu_2O(SO_4)$, and provided a brief review of $(OCu(II)_4)$ tetrahedra in inorganic compounds. She noticed that the $Cu-O_a$ bonds are usually shorter and therefore stronger than other Cu–O bonds in the crystal structures, which suggested that consideration in terms of oxocentred tetrahedra is not "simply another way to divide space" (Caro, 1968), but may also have important influences upon the distribution of chemical bond strength and, as a consequence, upon certain physical properties.

In 1975–76, the Kamchatka peninsula (Far East, Russia) witnessed one of the largest basalt volcanic eruptions in modern history, which was named the Great fissure Tolbachik eruption (GFTE). The eruption was followed by extensive post-eruption activity, which was expressed as considerable fumarolic activity. Volcanic gases of the GFTE were enriched in copper, and mineral-formation processes in fumaroles resulted in a unique suite of Cu minerals containing 'additional' O atoms with dolerophanite among the others. Similar processes, but at much smaller scale, were detected in other volcanoes such as Vesuvius (Italy) and Isalco (Salvador). Table 4 provides a list of mineral species contaning both Cu and 'additional' O atoms in the same structure. Structural studies demonstrated that, by analogy with dolerophanite, the O_a atoms in the structures of these minerals are tetrahedrally coordinated by Cu (and sometimes by other metals) with a variety of oxocentred tetrahedral complexes. This led Filatov *et al.* (1992) to propose that oxocentred (OCu_4) tetrahedra may have played the role of carriers of Cu by volcanic gases from a magmatic chamber to the surface of the Earth. This proposal was later confirmed by a series of modelling experiments, where mineral

Table 4. Cu(II) oxysalt minerals of fumarolic origin containing 'additional' O atoms.

Mineral name	Chemical formula	Locality
Allochalcoselite	$PbCu^{+}Cu_5^{2+}O_2(SeO_3)_2Cl_5$	Tolbachik, Russia
Alumoklyuchevskite	$K_3Cu_3FeO_2(SO_4)_4$	Tolbachik, Russia
Atlasovite	$KFe^{3+}Cu_6BiO_4(SO_4)_5Cl$	Tolbachik, Russia
Averievite	$Cu_5O_2(VO_4)_2(CuCl)$	Tolbachik, Russia
Burnsite	$KCdCu_7O_2(SeO_3)_2Cl_9$	Tolbachik, Russia
Chloromenite	$Cu_9O_2(SeO_3)_4Cl_6$	Tolbachik, Russia
Coparsite	$Cu_4O_2[(As,V)O_4]Cl$	Tolbachik, Russia
Cupromolybdite	$Cu_3O(MoO_4)_2$	Tolbachik, Russia
Dolerophanite	$Cu_2O(SO_4)$	Vesuvius, Italy; Tolbachik, Russia
Ericlaxmanite	$Cu_4O(AsO_4)$	Tolbachik, Russia
Euchlorine	$NaKCu_3O(SO_4)_3$	Tolbachik, Russia; Vesuvius, Italy
Fedotovite	$K_2Cu_3O(SO_4)_3$	Tolbachik, Russia
Fingerite	$Cu_{11}O_2(VO_4)_6$	Isalco, Salvador
Georgbokiite	α-$Cu_5O_2(SeO_3)_2Cl_2$	Tolbachik, Russia
Ilinskite	$NaCu_5O_2(SeO_3)_2Cl_3$	Tolbachik, Russia
Kamchatkite	$KCu_3O(SO_4)_2Cl$	Tolbachik, Russia
Klyuchevskite	$K_3Cu_3FeO_2(SO_4)_4$	Tolbachik, Russia
Kozyrevskite	$Cu_4O(AsO_4)$	Tolbachik, Russia
Melanothallite	Cu_2OCl_2	Vesuvius, Italy; Tolbachik, Russia
Nabokoite	$Cu_7TeO_4(SO_4)_5Cl$	Tolbachik, Russia
Nicksobolevite	$Cu_7O_2(SeO_3)_2Cl_6$	Tolbachik, Russia
Parageorgbokiite	β-$Cu_5O_2(SeO_3)_2Cl_2$	Tolbachik, Russia
Piypite	$K_4Cu_4O_2(SO_4)_4(Cu_{0.5}Cl)$	Tolbachik, Russia; Vesuvius, Italy
Ponomarevite	$K_4Cu_4OCl_{10}$	Tolbachik, Russia
Prewittite	$KPb_{1.5}ZnCu_6O_2(SeO_3)_2Cl_{10}$	Tolbachik, Russia
Starovaite	$KCu_5O(VO_4)_3$	Tolbachik, Russia
Stoiberite	$Cu_5O_2(VO_4)_2$	Isalco, Salvador
Vergasovaite	$Cu_3O(MoO_4)(SO_4)$	Tolbachik, Russia
Yaroshevskite	$Cu_9O_2(VO_4)_4Cl_2$	Tolbachik, Russia

analogues have been prepared in the course of chemical transport reactions. Inspired by these studies, Magarill *et al.* (2000) applied this approach to Hg oxysalts with 'additional' O atoms and constructed a hierarchy of structures (including those of many minerals) based upon units consisting of polymerized (OHg_4) tetrahedra.

The description of crystal structures of minerals with 'additional' anions (such as O^{2-} and N^{3-}) in terms of anion-centred tetrahedra is justified in the cases in which the bonds to these anions from cations are stronger than other bonds formed by these cations in the structure. For minerals, the most common cations that form oxocentred OM_4 tetrahedra are Cu^{2+} and Pb^{2+} (see reviews by Siidra *et al.* (2008) and Krivovichev *et al.* (2013) and references therein). Rather exotic examples are the structures of mosesite, $[Hg_2N]Cl(H_2O)$, and kleinite, $[Hg_2N]Cl(H_2O)_n$, that are based on the cristobalite- and tridymite-like $[NHg_2]^+$ frameworks (Fig. 6). It is of interest that, in contrast to silicates, the frameworks bear a positive charge with the framework cavities filled by Cl^- anions and water molecules.

Topological diversity of structural units based upon anion-centred polyhedra display a much broader range of variations compared to the structural units of 'traditional' crystal chemistry of cation-centred polyhedra. For instance, edge sharing between adjacent polyhedra appears to be much more common, due to the relatively low charge of central ions that reduces the repulsion forces across shared edges (see Section 2.2.4).

Krivovichev (2008) demonstrated that some minerals containing "additional" anions may be considered as based upon structural units of linked anion-centred octahedra. Similar synthetic inorganic compounds are known as antiperovskites or inverse perovskites that have a perovskite structure but with cations replaced by anions and vice versa. The ideal antiperovskite structure as adopted by some ternary carbides and nitrides with a general formula A_3BX. The examples of simplest antiperovskites with $Pm\bar{3}m$ symmetry are Mg_3NAs, Mg_3NSb and the new superconductor Ni_3CMg (Mitchell, 2002). Another example of a cubic antiperovskite is the structure of Na_3OCl (Hippler *et al.* 1990), which is based on the framework of corner-sharing $[ONa_6]$ oxocentred octahedra with cavities occupied by large Cl^- anions.

Krivovichev (2008) described the crystal structures of some silicate, sulfate and phosphate minerals with a general formula $A_mX_n(TO_4)_k$ ($A = Na^+, K^+, Ca^{2+}$; $X = O^{2-}, F^-, Cl^-$; $T = Si^{4+}, S^{6+}, P^{5+}$) in terms of antiperovskite structural units (Table 5). The simplest antiperovskite unit is a cubic framework of corner-sharing octahedra. It is

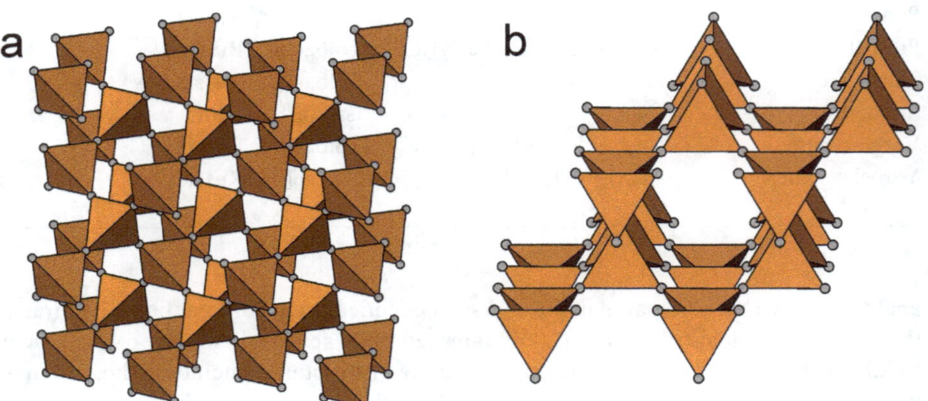

Figure 6. Frameworks of corner-sharing (OHg_4) tetrahedra in the crystal structures of kleinite (a) and mosesite (b). Note that the framework topologies are of cristobalite- and tridymite-types, respectively.

Table 5. Antiperovskite units in the crystal structures of minerals*.

X:*A*	*X*	*A*	Mineral name	Chemical formula
1:3-3*C*	F, Cl	Na	Sulfohalite	[$\mathbf{FClNa_6}$]$(SO_4)_2$
1:3-2*H*	F	Na,Ca	Nacaphite	[$\mathbf{FNa_2Ca}$](PO_4)
1:3-5*H*	F, Cl	Na	Galeite	[$\mathbf{F_4ClNa_{15}}$]$(SO_4)_5$
1:3-7*H*	F, Cl	Na	Schairerite	[$\mathbf{F_6ClNa_{21}}$]$(SO_4)_7$
1:3-9*R*	F	Na	Kogarkoite	[$\mathbf{FNa_3}$](SO_4)
	O	Ca	Hatrurite	[$\mathbf{OCa_3}$](SiO_4)
1:4	F	Na,Ca	Arctite	Ba[$\mathbf{F_3Na_5Ca_7}$]$(PO_4)_6$
1:3	F	Na,Ca	Polyphite	[$\mathbf{F_6Ca_3Na_{15}}$]$Na_2Mg(Ti,Mn)_4O_2$ $(Si_2O_7)_2(PO_4)_3$

* Composition of anion-centred units is given in bold.

realized in the structure of sulfohalite, $Na_6FCl(SO_4)_2$ (Pabst, 1934), where the framework consists of alternating [FNa_6] and [$ClNa_6$] octahedra (Fig. 7). Tetrahedral (SO_4) groups reside in the framework cavities. Due to the F-Cl ordering, the *a* parameter is doubled in comparison with the 'usual' cubic perovskite unit cell. Thus, sulfohalite is actually an ordered double antiperovskite with the anti-elpasolite structure and the $Fm\bar{3}m$ space group. The structure of nacaphite, $Na_2CaF(PO_4)$ (Krivovichev *et al.*, 2007) (Fig. 8a), is a hexagonal antiperovskite with the

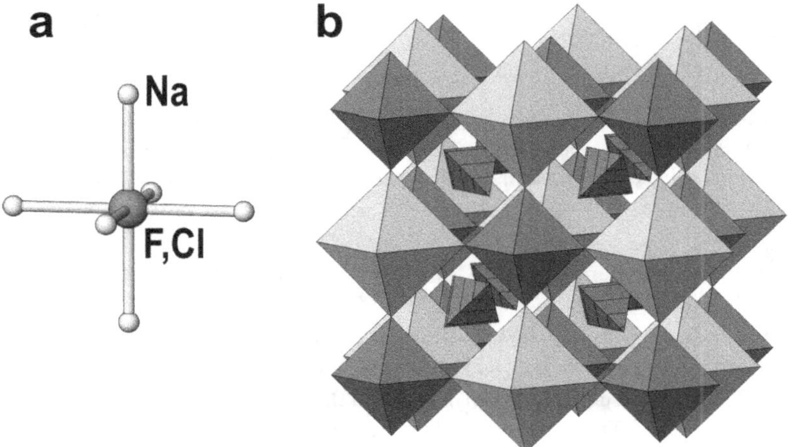

Figure. 7. Anion-centred octahedron (a) in the structure of sulfohalite (b). Note that the structure (b) is shown as a combination of anion- (octahedra) and cation- (tetrahedra) centred coordination polyhedra. Anion-centred [FNa_6] and [$ClNa_6$] octahedra are dark and light, respectively; sulfate tetrahedra are lined.

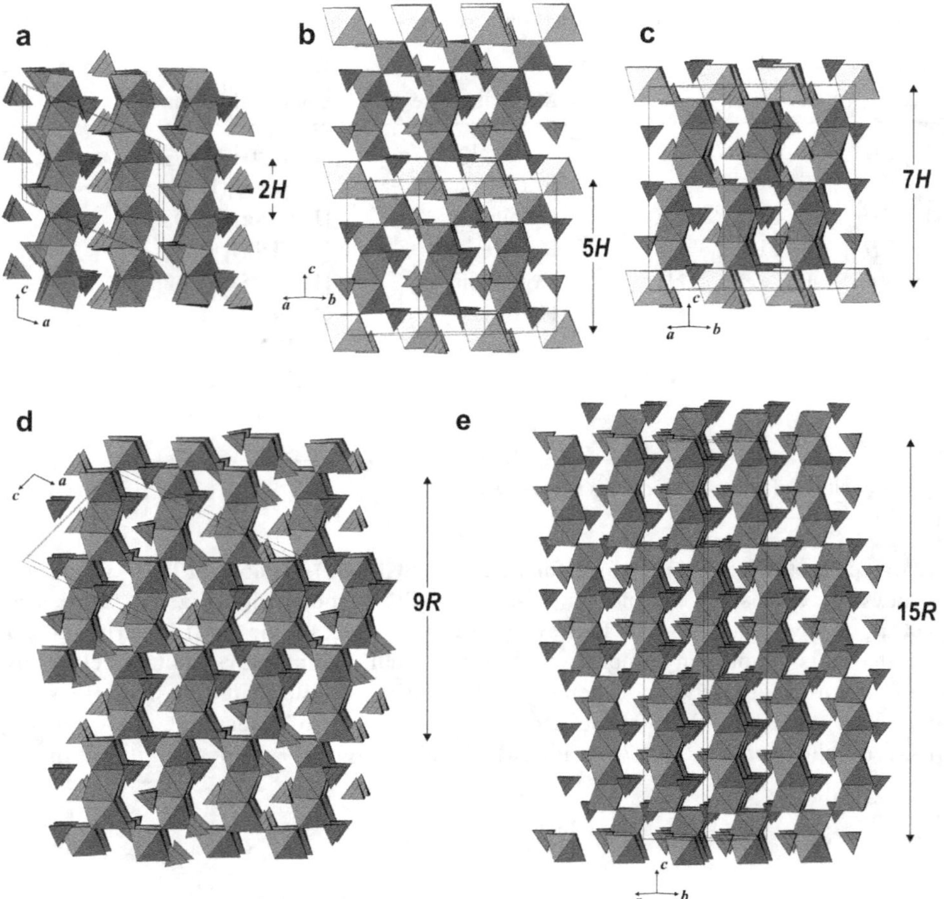

Figure 8. Crystal structures of minerals with antiperovskite structure shown as a combination of anion-centred octahedra and cation-centred tetrahedra: (a) nacaphite; (b) galeite; (c) schairerite; (d) kogarkoite; (e) rhombohedral polymorph of nacaphite. $[FNa_6]$ and $[ClNa_6]$ octahedra are dark and light, respectively; TO_4 tetrahedra (T = P, S) are lined.

2*H*-polytype arrangement of the $[FNa_4Ca_2]^{7+}$ octahedra that share faces to produce $[FNa_2Ca]^{3-}$ chains parallel to the *c* axis. The 5*H* and 7*H* polytypes are observed in the structures of galeite, $Na_{15}F_4Cl(SO_4)_5$ (Fanfani *et al.*, 1975a) (Fig. 8b), and schairerite, $Na_{21}F_6Cl(SO_4)_7$ (Fanfani *et al.*, 1975b) (Fig. 8c), respectively. Again, the arrangement of Cl^- and F^- anions is completely ordered, and the frameworks consist of $[FNa_6]^{5+}$ and $[ClNa_6]^{5+}$ octahedra. The fluorine-centred octahedra usually share faces, whereas the chlorine-centred ones do not. Figure 8d shows the structure of kogarkoite, $Na_3F(SO_4)$ (Fanfani *et al.*, 1980), which is similar to the structure of hatrurite, $Ca_3O(SiO_4)$ (Dunstetter *et al.*, 2006), an important phase known in the cement industry as a 'tricalcium silicate'. Here $[XA_6]$ octahedra share faces to form triplets linked further into a three-dimensional framework by sharing corners. This arrangement corresponds to a 9*R* hexagonal antiperovskite. Closely related to the 9*R* polytype is the 15*R* polytype shown in Fig. 8e. Here the octahedral framework consists of pentaplets of $[XA_6]$ octahedra. This structure is observed for the rhombohedral polymorph of nacaphite, $Na_2CaF(PO_4)$, reported by Sokolova *et al.* (1999).

An interesting case is exemplified by the crystal structure of polyphite, $[Ca_3Na_{15}F_6]Na_2Mg(Ti,Mn)_4O_2(Si_2O_7)_2(PO_4)_3$, a complex silicate-phosphate (Sokolova *et al.*, 1987) (Fig. 9). It has a modular structure (see Section 3.2) and is composed of alternating 2*H* antiperovskite (nacaphite-like) and *HOH* heterophyllo-silicate modules.

2.2.4. Schlegel diagrams and their use in crystal chemistry

Schlegel diagrams are projections of the edge networks of three-dimensional convex polyhedra onto a two-dimensional plane. As far as we know, it was P.B. Moore who first used Schlegel diagrams in the discussion of crystal structures of minerals, in his 1970 paper on the crystal chemistry of basic iron phosphates (Moore, 1970b). He analysed geometrical parameters of so-called *h* clusters (clusters of three face-sharing FeO_6 octahedra) in the crystal structures of dufrenite, rockbridgeite and beraunite (Fig. 10), using Schlegel diagrams, which "afford clear and concise means of illustrating distances for comparative purposes" (Moore, 1970b). Moore (1974) also used Schlegel diagrams of octahedra with shaded shared edges to describe topology of local linkage of octahedra in octahedral complexes. The approach was re-invented by Hoppe and Köhler (1988) for discussions of polyhedral linkages and geometrical parameters in solid-state chemistry. Krivovichev (1997) and Krivovichev *et al.* (1997) suggested using Schlegel diagrams to describe connectivity of polyhedra in polyhedral units. The example of such a description is shown in Fig. 11, which shows the $[O_9Pb_{14}]^{10+}$ layer of oxocentred OPb_4 tetrahedra in the crystal structure of kombatite, $Pb_{14}O_9(VO_4)_2Cl_4$ (Cooper and Hawthorne, 1994).

For the tetrahedron, the Schlegel diagram is simply the view from above onto a regular tetrahedron placed on one of its triangular bases. The following ways of representing the connectivity elements on the connectivity diagram (a modified Schlegel diagram) have been adopted by Krivovichev (1997): (1) the corner designated by a circle links the given tetrahedron to another; (2) if the corner links the given tetrahedron to more than one other tetrahedron, then the number of such tetrahedra is indicated next to the corner; and (3) the edge identified by a semi-bold line is common to two polyhedra.

The connectivity diagrams are equivalent if the relative disposition of the linkage corners and/or edges in them is the same (or is mirror-symmetrical). Wells (1970) suggested that, in polyhedral complexes, polyhedra with the same relative disposition of the shared elements be referred to as topologically equivalent. This means that tetrahedra with equivalent connectivity diagrams are referred to as 'topologically equivalent' tetrahedra in the complex.

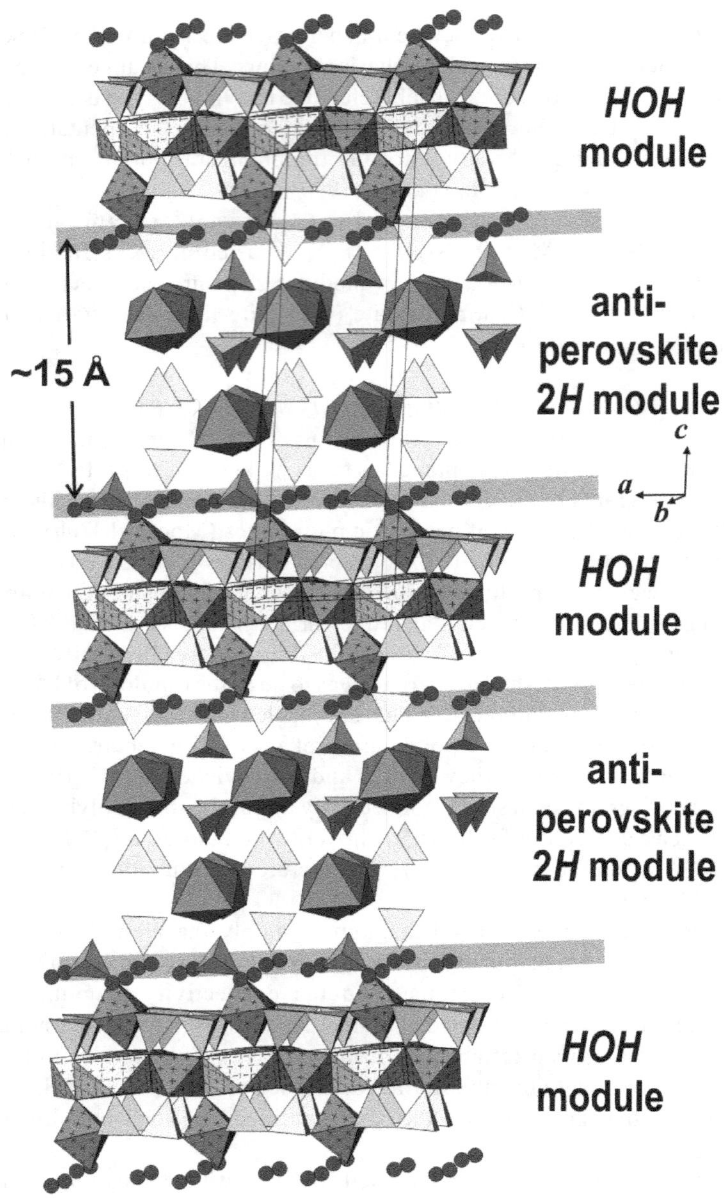

Figure 9. Polyphite, $[Ca_3Na_{15}F_6]Na_2Mg(Ti,Mn)_4O_2(Si_2O_7)_2(PO_4)_3$, has a modular structure composed of alternating *2H* antiperovskite (nacaphite-like) and *HOH* heterophyllosilicate modules. Note that the *HOH* modules consist solely of cation-centred polyhedra, whereas the antiperovskite modules are combinations of F-centred octahedra and cation-centred tetrahedra. The octahedra share faces to form one-dimensional chains running approximately perpendicular to the plane of the figure.

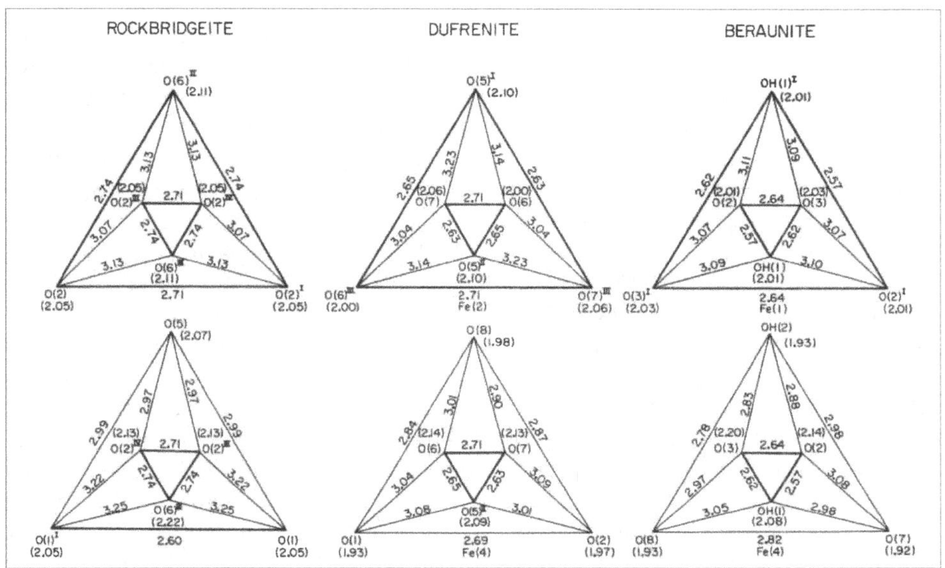

Figure 10. Schlegel diagrams with geometrical parameters of octahedra in the *h*-clusters in the crystal structures of rockbridgeite, dufrenite and beraunite (after Moore, 1970b).

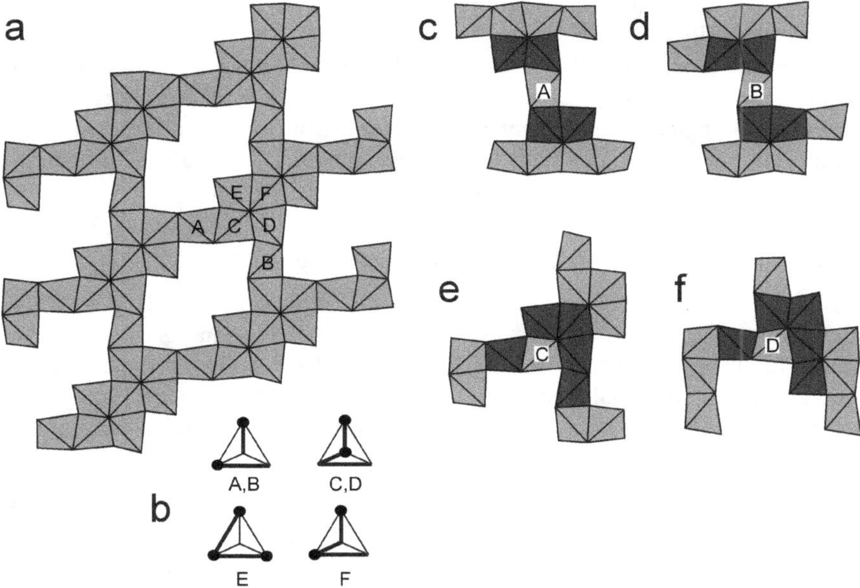

Figure 11. The $[O_9Pb_{14}]^{10+}$ layer of (OPb_4) tetrahedra from the structure of kombatite, $Pb_{14}O_9(VO_4)_2Cl_4$ (a), connectivity diagrams for its tetrahedra (b) and first coronas of the A (c), B (d), C (e), and D (f) tetrahedra.

The description of the disposition of the shared elements in the tetrahedron (with the aid of the connectivity diagram) is still insufficient to define the position of the tetrahedron in the complex. In the same structure, topologically equivalent tetrahedra may have different environments comprising tetrahedra with which they have common elements. Within the $[O_9Pb_{14}]^{10+}$ layer shown in Fig. 11a, there are six crystallographically distinct tetrahedral, denoted as A, B, C, D, E and F. However, there are only four different topological types of tetrahedra, as the A and B, and C and D tetrahedra have the same connectivity diagrams (Fig. 11b). To distinguish between global topological functions played by topologically equivalent tetrahedra in the complex, one has to analyse their configuration. In order to do this, Krivovichev (1997) suggested using the concept of corona introduced for description of topology of space tilings (Engel, 1986). The first corona, C^1(T) of the tetrahedron T is the tetrahedron itself plus all tetrahedra with which it has shared elements. The second corona constitutes the first corona plus all tetrahedra with which tetrahedra of the first corona share common elements. By analogy, *n*-corona, C^n(T) is defined as C^{n-1}(T) plus all tetrahedra with which tetrahedra of C^{n-1}(T) share common elements. Two tetrahedra are 'configurationally equivalent' if their *n*-coronas are equivalent for any *n*. For the topologically equivalent A and B tetrahedra of the complex shown in Fig. 11a, the 1-coronas are identical, but their 2-coronas are different (Fig. 11c,d), which indicates that these tetrahedra are configurationally non-equivalent. The same applies to the C and D tetrahedra (Fig. 11e,f). Thus, the the $[O_9Pb_{14}]^{10+}$ layer shown in Fig. 11a consists of six different configurational types (which, in this case, coincide with crystallographically different tetrahedra).

In general, there is a hierarchical relationship between crystallographic, configurational and topological equivalence. Configurationally equivalent tetrahedra are always topologically equivalent, but topologically equivalent tetrahedra may be configurationally non-equivalent. Crystallographically equivalent tetrahedra are always topologically and configurationally equivalent.

2.3. Fundamental building blocks and related concepts

According to Hawthorne (1994), the fundamental building block (FBB) is "a tightly-bonded unit within the structure that can be envisaged as the inorganic analogue of a molecule in an organic structure. The FBB is usually a homo- or heteropolyhedral cluster of coordination polyhedra with the strongest bond-valence linkages in the structure." The FBBs are finite and may polymerize to form extended structural units.

FBBs in borate minerals. The description of mineral structures in terms of FBBs appeared to be especially efficient for borate minerals, where FBBs behave as relatively rigid units with almost invariable internal bond lengths and angles (Filatov and Bubnova, 2000; Bubnova and Filatov, 2013). The borate FBBs are bipolyhedral, as they consist of different combinations of BO_3 triangles and BO_4 tetrahedra. Burns *et al.* (1995) proposed a descriptor for a borate FBB in the form *A*:*B*, where $A = i\Delta j\square$ gives the numbers *i* and *j* of borate triangles (Δ) and tetrahedra ($\square$), respectively. The *B* part is written such that adjacent Δ and □ represent polyhedra that share corners, and the delimiters <> indicate that the polyhedra form rings. The linkage of rings by sharing one, two or three polyhedra is indicated by the symbols –, = or ≡, respectively (Grice *et al.*, 1999). As an example, Fig. 12 shows eight distinct FBBs that occur in the borate minerals with chain structures.

FBBs in octahedral-tetrahedral structural units. From their analysis of mixed anionic frameworks composed from octahedra and tetrahedra, Voronkov *et al.* (1975) derived eight types of basic 'microblocks', which polymerize to form extended units (Fig. 13). They also gave some illustrative examples of how these blocks link to form structural units. For instance, Fig. 14 shows octahedral-tetrahedral units formed by condensation of type-3 'microblocks'. Voronkov *et al.* (1975) made it clear that the 'microblocks' that they described are in fact pieces of anion close packings with cations in octahedral and tetrahedral interstices. Thus, in some sense, the 'microblocks' (as well as some FBBs) are modules excised from a parent (archetype: see Section 3.2.1) structure and, under certain circumstances, octahedral-tetrahedral frameworks can be considered as defect closest packings of anions. For instance, Baur and Fischer (2013) considered the crystal structure of pharmacosiderite, $K[Fe_4(OH)_4(AsO_4)_3](6-7H_2O)$, as an anion- and cation-deficient MgO- (or NaCl) type arrangement, which can itself be described as

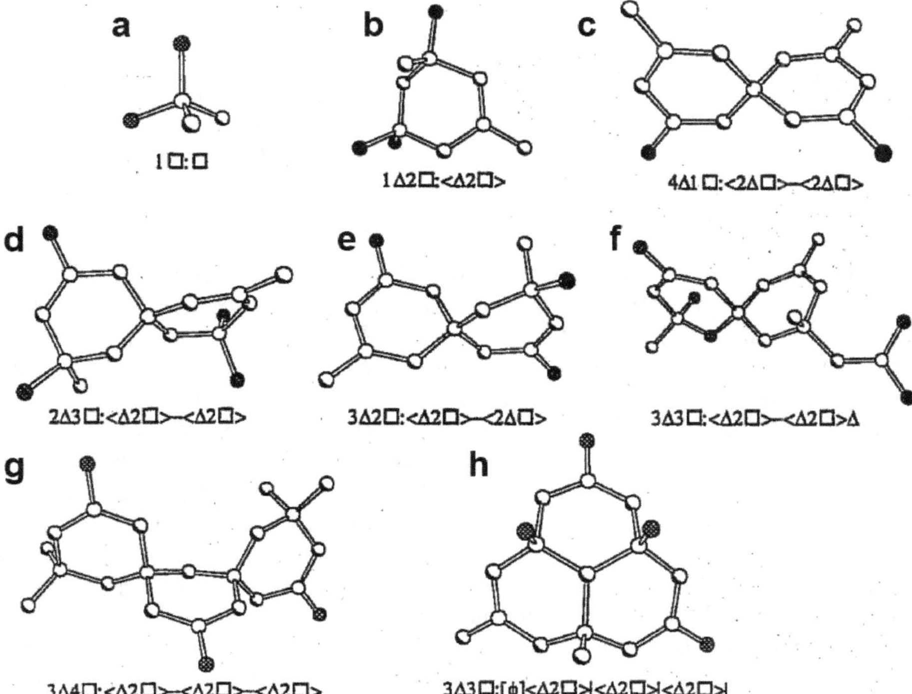

Figure 12. FBBs in the crystal structures of chain-borate minerals (after Grice *et al.*, 1999).

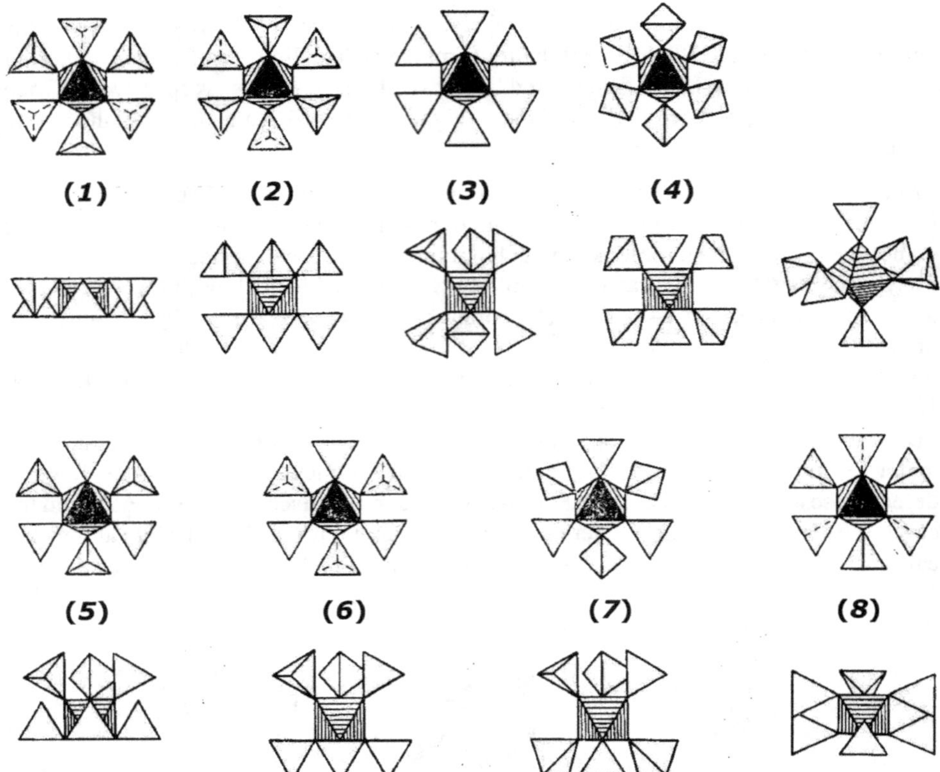

Figure 13. Eight 'microblocks' proposed by Voronkov *et al.* (1975) as structure-forming units in mixed anionic octahedral-tetrahedral radicals.

cubic close packing of anions with cations filling all octahedral interstices. An alternative description of the pharmacosiderite framework is in terms of FBBs that are octahedral tetramers linked by tetrahedra in three dimensions (Fig. 15). Similar FBBs are known in natural titanosilicates such as ivanyukite-group minerals, $A_n[Ti_4O_2(OH)_2(SiO_4)_3](H_2O)_m$ (A = K, Na, Cu; n = 1, 2; m = 6–9; Yakovenchuk *et al.*, 2009), and sitinakite, $KNa_2[Ti_4O_5(OH)(SiO_4)_2](H_2O)_4$ (Sokolova *et al.*, 1989), and have also been observed in the crystal structure of phosphovanadylite-group minerals, $M[V^{4+}_4(PO_4)_2O_4(OH)_4](H_2O)_{12}$ (M = Ca, Ba; Medrano *et al.*, 1998; Kampf *et al.*, 2013). The topology of linkage of the tetrameric FBBs in the structures of pharmacosiderite-type minerals, sitinakite and phosphovandylite is different. The persistence of identical FBBs in the structure types with different topologies provides more evidence for their stability and possible existence as pre-nucleation clusters in aqueous solutions from which corresponding crystalline solids have been crystallized.

Another interesting octahedral-tetrahedral FBB is that shown in Fig. 16a,b. It consists of four octahedra forming a 'butterfly-shaped' arrangement and surrounded by

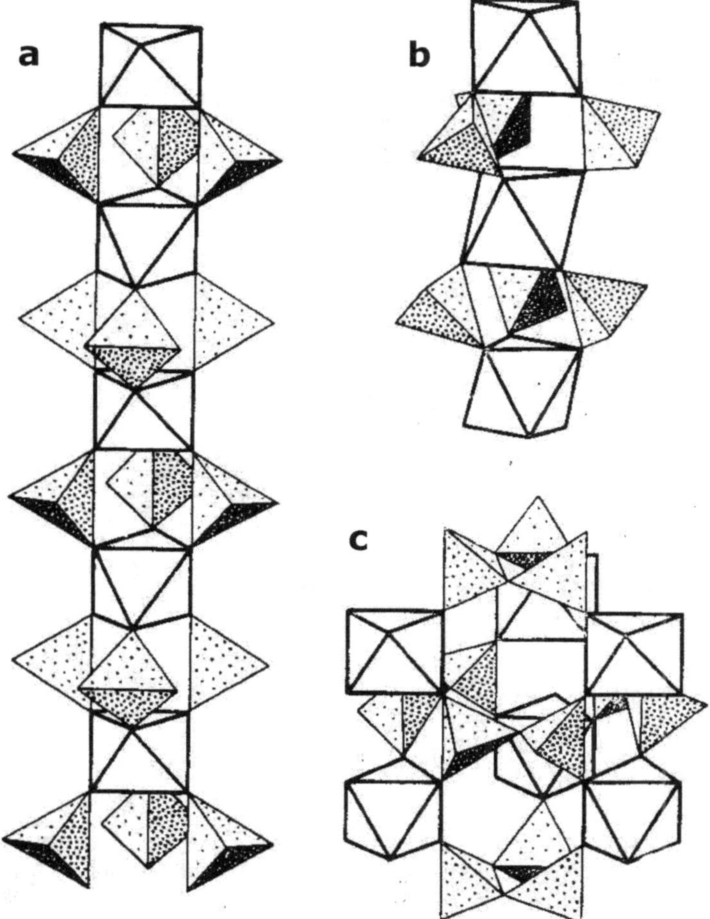

Figure 14. MARs based on the 'microblock' of the type (3) (Fig. 13): (a) chain in the structure of $Na_3Sc(SO_4)_3(H_2O)_5$; (b) finite unit in the crystal structure of coquimbite; (c) fragment of the three-dimensional framework in the structure of wadeite (after Voronkov *et al.*, 1975).

six tetrahedra. This FBB was observed in several synthetic metal phosphates (Férey, 1995, 1998, 2001) and is also known from the crystal structures of isotypic leucophosphite, $K_2[Fe_4(OH)_2(H_2O)_2(PO_4)_4](H_2O)_2$ (Moore, 1972), spheniscidite, $(NH_4)[Fe_2(OH)(H_2O)(PO_4)_2](H_2O)$ (Yakubovich and Dadachov, 1992), and tinsleyite, $K_2[Al_4(OH)_2(H_2O)_2(PO_4)_4](H_2O)_2$ (Dick, 1999). In the structures of these minerals, FBBs are linked by sharing corners to form frameworks with rather simple bcu (body-centred cubic) topology (Fig. 16c,d).

FBBs in sulfides. The idea of FBBs is not restricted to oxysalts, but can be applied also to other chemical classes of minerals. For instance, the crystal structures of pentlandite,

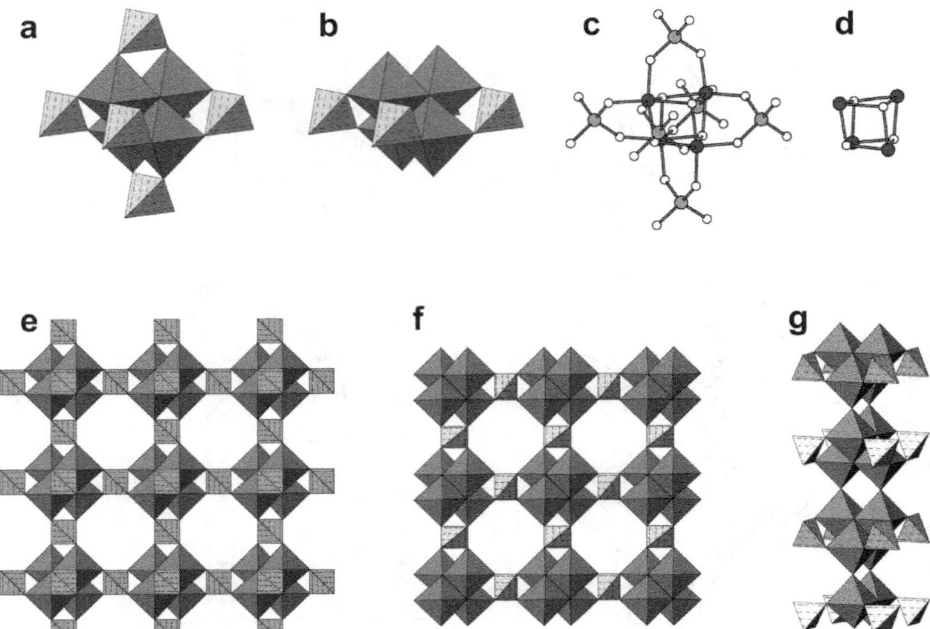

Figure 15. Octahedral-tetrahedral clusters in pharmacosiderite-related frameworks (a, b, c) and $M_4(OH)_4^{8+}$ polycation consisting of four M^{3+} cations and four OH^- groups at the core of these clusters (d). The structures of the minerals of the pharmacosiderite group are based on the 3D framework of the pcu regular net topology (e). The The $[M_4(O,OH)_4(XO_4)_4]$ FBBs (b) in the structures of $A_2[Ti_2O_3(SiO_4)](H_2O)_n$ (A = Na, H) are linked into the 3D framework (e) consisting of chains shown in (f).

$(Fe,Ni)_9S_6$, djerfisherite, $K_6NaFe_{25}S_{26}Cl$, bartonite, $K_3Fe_{10}S_{14}$, and related minerals are based on the microporous frameworks formed by condensation of $[Fe_8S_{14}]$ cubic clusters consisting of eight edge-sharing FeS_4 tetrahedra (Fig. 17). The clusters share edges and/or corners to form three different arrangements as shown in Fig. 18. The assembly of the cubic clusters into frameworks can be modelled using cellular automata (see Krivovichev (2004a, 2005) for more details).

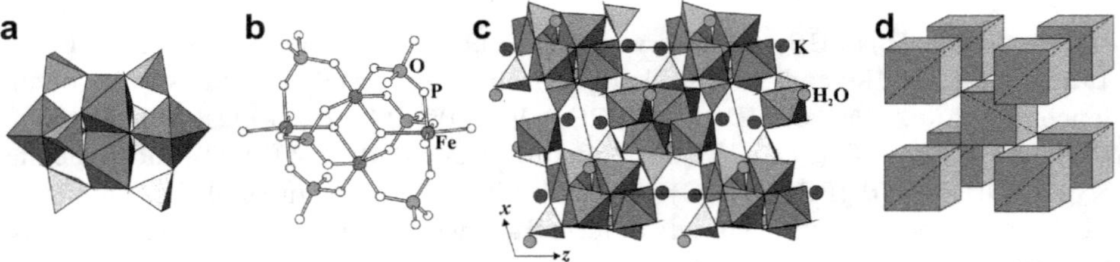

Figure 16. "Butterfly-shaped" cluster (FBB) consisting of four octahedra and six tetrahedra shown in polyhedral (a) and ball-and-stick (b) aspects. The structure of leucophosphite, $K_2[Fe_4(OH)_2(H_2O)_2(PO_4)_4](H_2O)_2$ (c) and the scheme demonstrating arrangement of FBBs in leucophosphite (d).

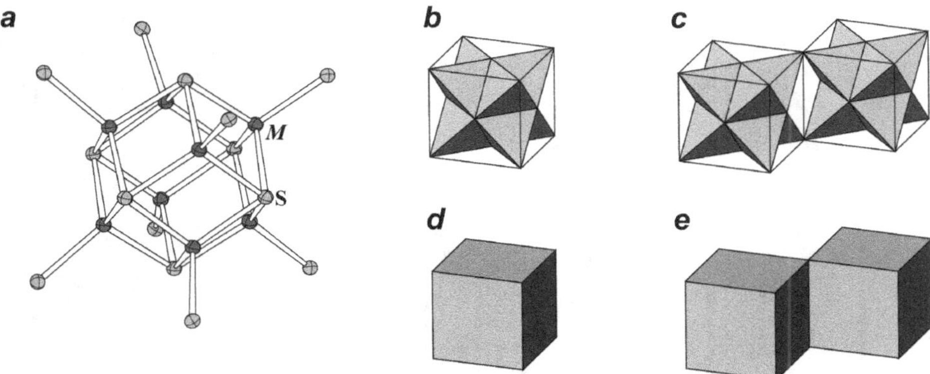

Figure 17. The M_8S_{14} cluster (M = Fe, Cu, Ni, Co) that occurs in transition metal sulfides of the pentlandite-djerfisherite-bartonite series (a) and its representation as a group of eight MS_4 tetrahedra (b) and as a cube (d). Linkage of two clusters (c) can be represented as linkage of two cubes via an edge (e).

Related concepts in synthetic inorganic chemistry. Recent years witnessed an explosion of publications on real mechanisms of crystallization of complex solids (primarily, inorganic oxysalts) through the assembly of pre-nucleation clusters consisting of cation-centred polyhedra. These clusters can be considered as FBBs for the respective structural units. In chemical literature, they are usually identified as secondary building blocks (SBUs) as opposed to coordination polyhedra, which are primary building units (PBUs) (Ferey, 1995, 1998; Serre *et al.*, 2003a,b, *etc.*). One of the most common SBUs in the process of crystallization of silicate and aluminosilicate zeolites is a double four-membered ring (D4R) (Kinrade *et al.*, 1998; Peister *et al.*, 2006), whereas, for metal phosphates, single four-membered rings (S4Rs) are considered as more common (Rao *et al.*, 2001).

2.4. Networks and graphs

2.4.1. General notes

The use of nets and graphs in crystal chemistry was pioneered by A.F. Wells who used nets to describe the topology of interpolyhedral and interatomic linkages in complex inorganic structures (Wells, 1954, 1970). One of the many advantages of this approach is that structures with the same overall topology may differ essentially in terms of space groups and unit-cell dimensions. Therefore, their description as graphs provides a unique and efficient way to reveal global similarities and relations between crystal structures, which may be hidden in the parameters of local geometrical arrangements. Smith and Rinaldi (1962) implicitly and Smith (1968) explicitly used nets to derive novel types of tetrahedral frameworks (in the 1962 paper they also introduced the famous **U** and **D** notations for tetrahedra pointing up and down, respectively) (Fig. 19), whereas Moore (1974) employed graph theory to derive octahedral clusters with various connectivities. Nowadays, the use of nets and graphs in structural chemistry is a well established and fashionable field (especially in chemistry of zeolites and

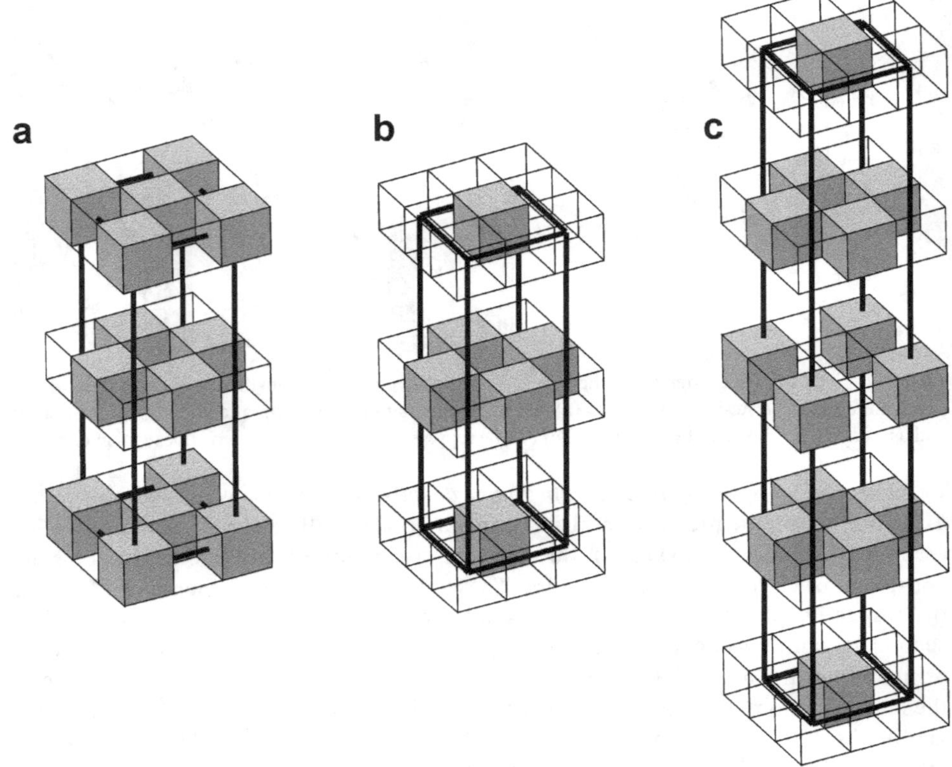

Figure 18. Pentlandite (a), djerfisherite (b), and bartonite (c) frameworks separated into layers of cubes which symbolize cubic FBBs and expanded in the direction perpendicular to the layers.

coordination polymers), owing to the active exploitation of different computer packages such as *TOPOS* (Blatov *et al.*, 2000). A review of the history of this whole field is beyond the scope of the present chapter and, below, we will consider only applications of nets and graphs in structural mineralogy.

2.4.2. Three-dimensional nets in mineral structures

Most attention in mineralogy has been paid to the topology of 4-connected three-dimensional nets due to their occurence in a range of important rock-forming framework silicates. The first papers by Smith and Rinaldi (1962) and Smith (1968) were followed by numerous theoretical works that led to the derivation of hundreds of novel topologies, most of which have never been found in minerals. Neverthless, the utilization of 4-connected nets allowed establishment of structural relationships between different mineral structures. Investigations of natural zeolites (Armbruster and Gunter, 2001) led to the development of a whole industry of molecular sieves that found applications in many areas of modern technology.

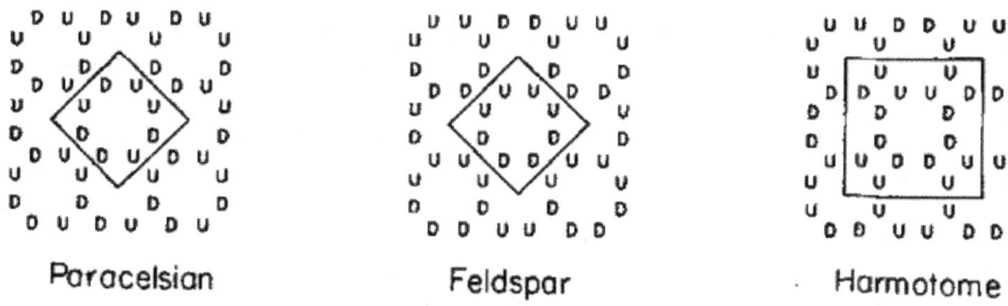

Figure 19. Description of tetrahedral frameworks in paracelsian, feldspar and harmotome with tetrahedra and their orientation symbolized by the U and D symbols after Smith and Rinaldi (1962).

McCusker *et al.* (2003), McCusker (2005) and Liebau (2003) compiled some recommendations for the description of ordered microporous and mesoporous materials on the basis of different methods and approaches devised in the second half of the 20th century. In particular, Liebau (2003) listed the following types of building units that can be separated in a framework structure:

(1) basic building units, **BBU**, usually coordination polyhedra of cations;
(2) composite building units, **CBU**, which consist of a finite or infinite number of **BBU**s; **CBU**s include ring, chains and polyhedral building units (**PBU**s);
(3) periodic building units, **PerBU**, which are periodic in one (chains) or two (layers) dimensions; these units were introduced by the Structure Commission of the International Zeolite Association in order to describe disordered zeolite frameworks systematically;
(4) fundamental building units, **FBU**s, from which a framework can be constructed completely by successive linkage by sharing peripheral atoms; obviously, the FBU is the same as FBB as outlined in Section 2.3.

The list of different notations, formulae and concepts introduced by McCusker *et al.* (2003) and Liebau (2003) is quite long and it is hard to believe that it will find its applications in the everyday work of a mineralogist or materials scientist. However, it provides some important ideas and insights that may help to systematize current knowledge about porous frameworks. For now, almost exhaustive lists of microporous zeolite-like frameworks can be found in the reference volumes by Baerlocher *et al.* (2007), Smith (2000), Baur and Fischer (2000, 2002), and Fischer and Baur (2006, 2009, 2013, 2014).

Despite the dominance of materials science and solid-state chemistry perspectives in modern research on porous materials, mineralogy continues to supply new data to the development of the field, *e.g.* by the recent discoveries of new 4-connected nets in the cancrinite-sodalite family of microporous aluminosilicates (Cámara *et al.*, 2010, 2012; Bonaccorsi *et al.*, 2012). This family of mineral species provides an excellent example of different descriptions of tetrahedral frameworks.

The cancrinite-sodalite supergroup consists of feldspathoids with the Al:Si ratio of 1:1 (Bonaccorsi and Merlino, 2005), which are not considered as zeolites in the mineralogical classification (Coombs *et al.*, 1998). However, topologies of their tetrahedral frameworks are included in the structure database of zeolite framework types of the International Zeolite Association (IZA).

One of the possible descriptions of the crystal structures of the cancrinite-sodalite supergroup minerals is that which employs layers of 6-membered rings of tetrahedra. Each ring is linked to three rings in the preceding layer and to three rings in the suceeding layer (Bonaccorsi and Merlino, 2005). By analogy with closest sphere packings, rings in a layer may have three different positions indicated by the letters A, B and C. If rings in the layer are in the A positions, rings in the next layer should be either in B or C positions. Thus, each framework can be described as a sequence of symbols that describe positions of 6-membered rings of tetrahedra in the successive construction of its unit cell. The simplest frameworks are the 2-layer cancrinite framework (CAN) with the AB sequence, and the 3-layer sodalite framework (SOD) with the ABC sequence. The maximum number of layers within a unit cell observed for the framework of the supergroup so far is 36, found in kircherite (Cámara *et al.*, 2012).

Considering a single layer of 6-membered rings as a **CBU** (or **PerBU**), the cancrinite-sodalite supergroup can be described as a series with structures controlled by the successive arrangement of two-dimensional layers. Table 6 provides a list of minerals of the supergroup that correspond to unique tetrahedral framework types (the full list of minerals of the supergroup can be found in Bonaccorsi and Merlino (2005) and current mineralogical literature.

Another way to look at the framework construction in the sodalite-cancrinite supergroup is to describe the frameworks as arrays of **PBU**s. There are five different polyhedral units (cages) present in the frameworks of the supergroup (Bonaccorsi and Merlino, 2005) and they are shown in Fig. 20. The simplest CAN ($n = 2$) and SOD ($n = 3$) frameworks are based upon *can* and *sod* cages, respectively) (Fig. 21). The 4-layer LOS and 8-layer AFG frameworks (Fig. 21) are based upon cages of two different types (*can*

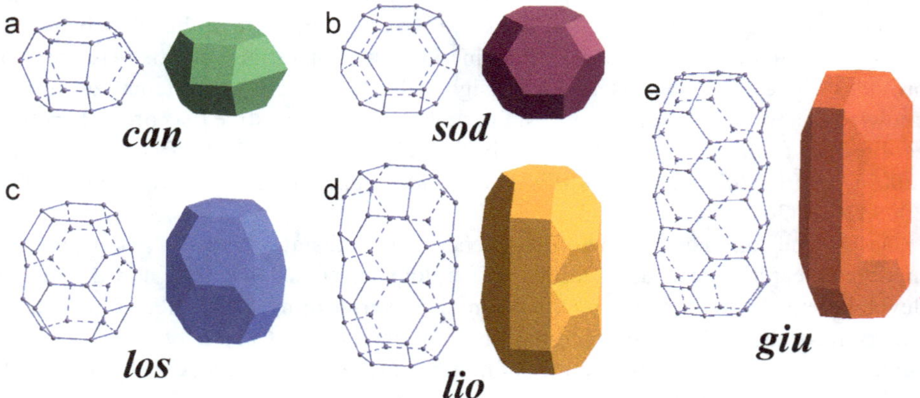

Figure 20. Polyhedral building units present in the frameworks of the cancrinite-sodalite-supergroup minerals.

Table 6. Structural data on topology of framework aluminosilicates of the cancrinite-sodalite supergroup.

Number of layers	Mineral name	Chemical formula	Layer sequence	Ref.	IZA code
2	Cancrinite	$Na_2(Ca,Na)_6(CO_3)(H_2O)_2[Si_6Al_6O_{24}]$	AB	1	CAN
3	Sodalite	$Na_8[Al_6Si_6O_{24}]Cl_2$	ABC	2	SOD
4	Bystrite	$Ca(Na,K)_7[Al_6Si_6O_{24}]S_3H_2O$	ABAC	3	LOS
6	Liottite	$Na_{10}K_6Ca_8[Si_{18}Al_{18}O_{72}](SO_4)_5Cl_{3.5}F_{0.5}$	ABABAC	4	LIO
8	Afghanite	$Na_{17}K_5Ca_{10}[Si_{24}Al_{24}O_{96}](SO_4)_6Cl_6$	ABABACAC	5	AFG
10	Franzinite	$Na_{21}K_9Ca_{10}[Si_{30}Al_{30}O_{120}](SO_4)_{10}(H_2O)_2$	ABCABACABC	6	FRA
12	Tounkite	$Na_{31}Ca_{16}K[Si_{36}Al_{36}O_{144}](SO_4)_{10}Cl_8$	ABABACACABAC	7	TOL
12	Marinellite	$Na_{31}K_{11}Ca_6[Si_6Al_6O_{24}]_6(SO_4)_8Cl_2(H_2O)_6$	ABCBCBACBCBC	8	MAR
14	Farneseite	$Na_{36}K_9Ca_8[Al_{42}Si_{42}O_{168}](SO_4)_{11}Cl(H_2O)_3$	ABCABABACBACAC	9	FAR
16	Giuseppettite	$Na_{43}K_{16}Ca_5[Si_{48}Al_{48}O_{192}](SO_4)_{10}Cl_2(H_2O)_5$	ABABABACBABABABC	10	GIU
28	Sacrofanite	$Na_{61}K_{19}Ca_{32}[Si_{84}Al_{84}O_{336}](SO_4)_{26}Cl_2F_6(H_2O)_2$	ABCABACACABACBAC BACABABACABC	11	–
33*/11**	Fantappièite	$(Na_{82.5}Ca_{33}K_{16.5})[Al_{99}Si_{99}O_{396}](SO_4)_{33}(H_2O)_6$	ACBACABACBACBACB CACBACBACBABCBACB	12	–
36*/12**	Kircherite	$Na_{90}Ca_{36}K_{18}(Al_{108}Si_{108}O_{432})(SO_4)_{36}(H_2O)_6$	ACABCABCABCACBCAB CABCABCBABCABCABCAB	13	–

* *R*-centering is ignored.

** *R*-centering is taken into account.

References: (1) Hassan and Grundy, 1991; (2) Loens and Schulz, 1967; (3) Pobedimskaya *et al.*, 1991; (4) Ballirano *et al.*, 1996; (5) Ballirano *et al.*, 1997; (6) Ballirano *et al.*, 2000; (7) Rozenberg *et al.*, 2004; (8) Bonaccorsi and Orlandi, 2003; (9) Cámara *et al.*, 2005; (10) Bonaccorsi, 2004; (11) Bonaccorsi *et al.*, 2012; (12) Cámara *et al.*, 2010; (13) Cámara *et al.*, 2012.

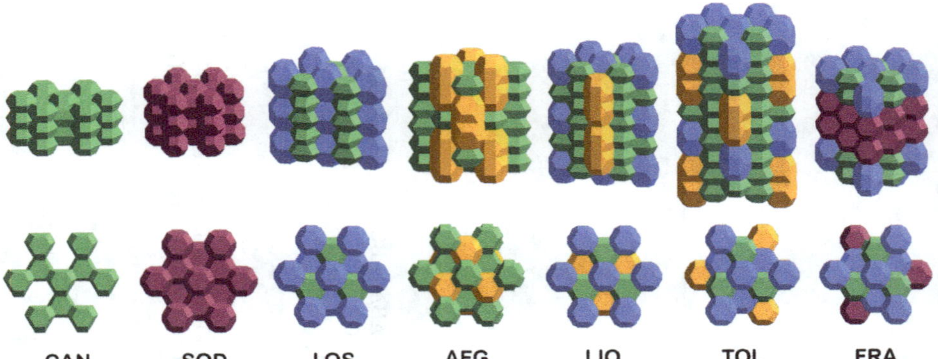

Figure 21. Frameworks of the cancrinite-sodalite-supergroup minerals as built up from **PBU**s: CAN, SOD, LOS, AFG, LIO, TOL and FRA frameworks.

and *los*, and *can* and *lio*, respectively). The 6-layer LIO framework (Fig. 21) consists of cages of three different types. Both LIO and TOL frameworks are based upon the *can*, *los* and *lio* cages, but their arrangement in the latter is more complex than in the former (Fig. 21). The MAR and FAR frameworks contain cages of the same types (*lio*, *sod* and *can*), but, again, with different arrangements (Fig. 22). The MAR framework can be considered as based upon two different columns of cages, ...-*lio-lio*-... and ...*sod-can-can-sod-can*-..., whereas the FAR framework contains the columns ...*sod-can-sod-lio*-... and ...*sod-can-can-sod-can-can*-... The FRA and kircherite frameworks are also based upon

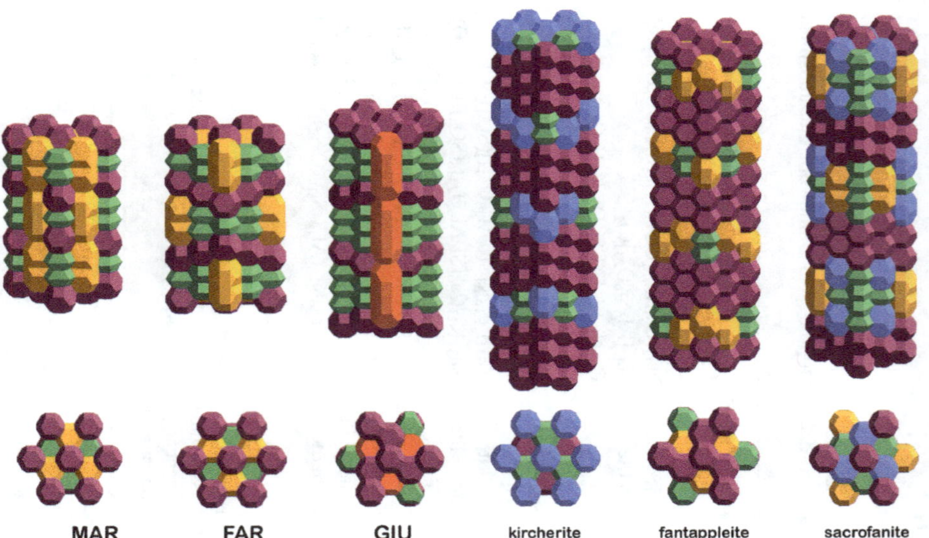

Figure 22. Frameworks of the cancrinite-sodalite-supergroup minerals as built up from **PBU**s: MAR, FAR, GIU frameworks and framework present in kircherite, fantappieite and sacrofanite.

the same cages (Fig. 21, 22). The sacrofanite framework (Fig. 22) is the most complex in the supergroup.

The extension of the nodal representation to non-tetrahedral (*e.g.* octahedral-tetrahedral) frameworks in minerals poses serious difficulties, due to the enormous increase in the number of topological possibilities and variations. Krivovichev (2005) suggested subdividing complex graphs that serve as underlying topologies in octahedral-tetrahedral frameworks according to the convenience of their description into those based upon PBUs, those based upon one-dimensional units and those based upon two-dimensional layers.

Two examples of octahedral-tetrahedral frameworks based on PBUs are shown in Figs. 23 and 24. In the $[M_2(O,OH)_2(Si_4O_{12})]$ octahedral-tetrahedral framework in labuntsovite-group minerals (Fig. 23a), $M\phi_6$ octahedra (M = Ti, Nb; ϕ = O, OH) share corners to produce single chains that are interlinked by Si_4O_{12} four-membered silicate rings. The resulting framework has channels oriented perpendicular to the octahedral chains and occupied by low-valence cations and H_2O molecules. The three-dimensional net of the framework (where white and black notes symbolize Si and M polyhedra, respectively) is shown in Fig. 23b. It can be considered as consisting of two types of polyhedral units shown in Figs 23c and d. Figure 24a shows the crystal structure of minerals of the shcherbakovite-batisite family that are based on an octahedral-tetrahedral titanosilicate framework with the composition $[Ti_2O_2(Si_4O_{12})]^{4-}$. The three-dimensional net that corresponds to this framework and its PBUs are shown in Figs 24b,c,d.

Among frameworks based on one-dimensional units, the most interesting are those that can be described as consisting of tubular chains. Figure 25a shows the structure of benitoite, $BaTiSi_3O_9$, projected along the *c* axis (Fischer, 1969). The benitoite structure type is common for many minerals and inorganic compounds, including bazirite, $BaZrSi_3O_9$, pabstite, $BaSnSi_3O_9$ (Hawthorne, 1987), $BaSi^{VI}Si_3^{IV}O_9$ (Finger *et al.*, 1995), *etc.* The structure represents a framework of isolated MO_6 octahedra and Si_3O_9 silicate rings. Nodal representation of the framework is shown in Fig. 25b. Analysis of its topology allows subdivision into two polyhedral units as shown in Figs 25c and d. However, the benitoite 3D net cannot be constructed solely from these units as it contains the 1D tubular unit shown in Fig. 25e. This unit cannot be unequivocally separated into polyhedral units and thus should be considered as an independent

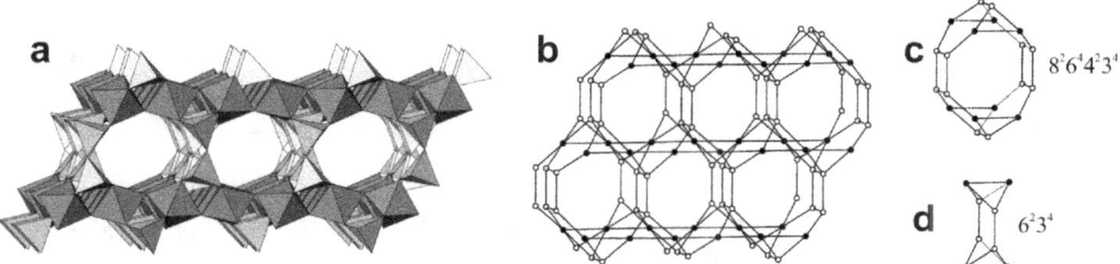

Figure 23. Octahedral-tetrahedral framework in the structures of labuntsovite-group minerals (a) and its nodal representation. The labuntsovite 3D net consists of two types of **PBU**s (c, d).

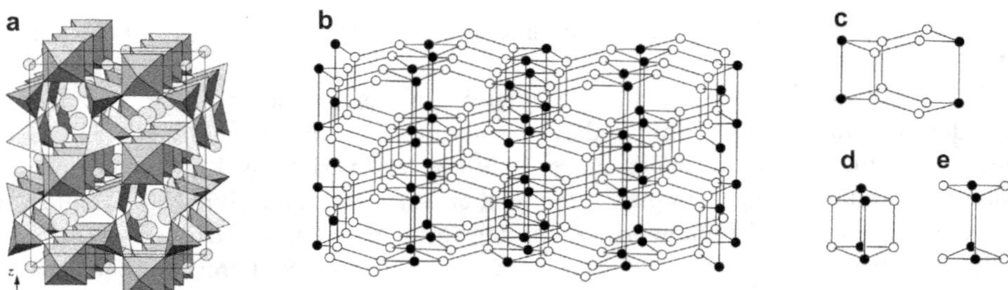

Figure 24. Octahedral-tetrahedral framework in minerals of the shcherbakovite-batisite series (a) and its nodal representation (b). The 3D black-and-white net can be constructed by assembling three different **PBU**s (c, d, e).

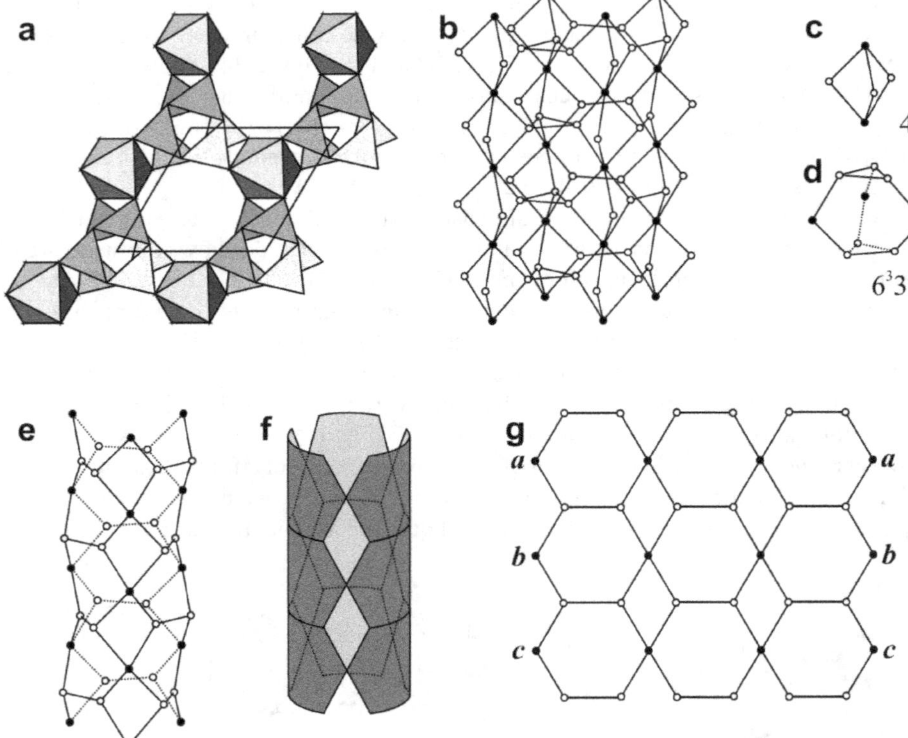

Figure 25. Octahedral-tetrahedral framework in the structure of benitoite, $BaTiSi_3O_9$ projected along the *c* axis (a) and its nodal representation (b). The benitoite net can be represented as consisting of polyhedral voids of two types (c and d) and tubular units (e, f). The tubular unit can be constructed by folding and gluing the tape-like black-and-white graph shown in (g). See text for details.

building unit of the benitoite net. A different kind of tubular unit is observed in the octahedral-tetrahedral framework of hilairite type, $Na_2ZrSi_3O_9(H_2O)_3$, (Ilyushin *et al.*, 1981) (Fig. 26). This framework consists of corner sharing of isolated ZrO_6 octahedra and single Si_6O_{18} chains. The hilairite net is assembled from tubular units composed of 8- and 3-membered rings. Unfolding of the tubular unit provides the tape shown in Fig. 26d. To make the tubular unit, one has to fold the tape and to join the points identified by the same letters. In contrast to benitoite and vlasovite, equivalent points are not opposite each other but, instead, are in diagonal orientation. Their joining produces a 'chiral' unit that has a 'helical' structure. It is noteworthy that the structures of the hilairite-group minerals (sazykinaite-(Y), hilairite, calciohilairite, komkovite and pyatenkoite-(Y)) have the same space group, *R*32, that contains only rotational symmetry elements. The tubular unit shown in Fig. 25e can be considered as a black-and-white graph on the surface of a cylinder. It consists of 6- and 4-membered rings. The tubular unit can be constructed using a procedure which is used to describe the topology of carbon nanotubes and which is known as "folding and gluing" (Kirby, 1997). First, one has to construct a black-and-white graph in the form of a one-dimensional ribbon. In order to obtain the tubular unit, the ribbon is folded and opposite sides are glued by joining equivalent points to make a cylinder. The idealized model for a tubular unit in benitoite and its prototape are shown in Figs 25f and g, respectively.

The analysis of 3D nets in gittinsite, $CaZrSi_2O_7$ (Roelofsen-Ahl and Peterson, 1989) and $SrZrSi_2O_7$ (Huntelaar *et al.*, 1994) reveals that both nets are based upon the same 2D net shown in Fig. 27a. It consists of three- and seven-membered rings. The structures differ in the way the 2D nets are linked to each other. In gittinsite, the adjacent 2D nets have the same orientation of triangles (Fig. 27b), whereas, in $SrZrSi_2O_7$, the adjacent 2D nets are related to each other by rotation by 180° around an axis vertical to the plane of the nets (the triangles in the adjacent nets have opposite orientation; Fig. 27c). More examples of nodal description of heteropolyhedral frameworks can be found in Krivovichev (2005, 2009).

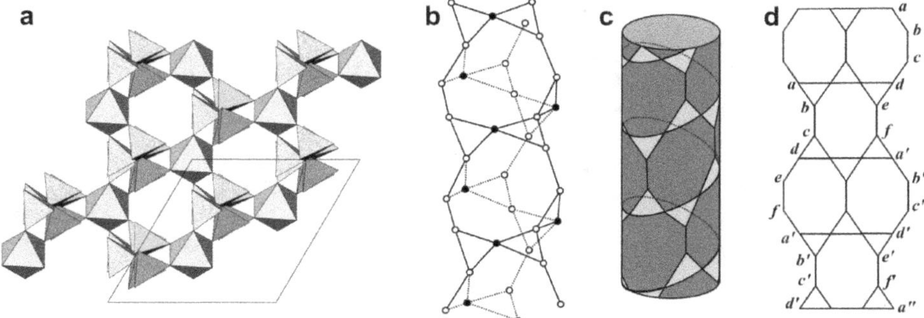

Figure 26. Octahedral-tetrahedral framework in the structure of hilairite $Na_2ZrSi_3O_9(H_2O)_3$ (a) consists of tubular units (b,c) composed from eight- and three-membered rings. Unfolding of the tubular unit provides the ribbon shown in (d). Note that the tubular unit is chiral.

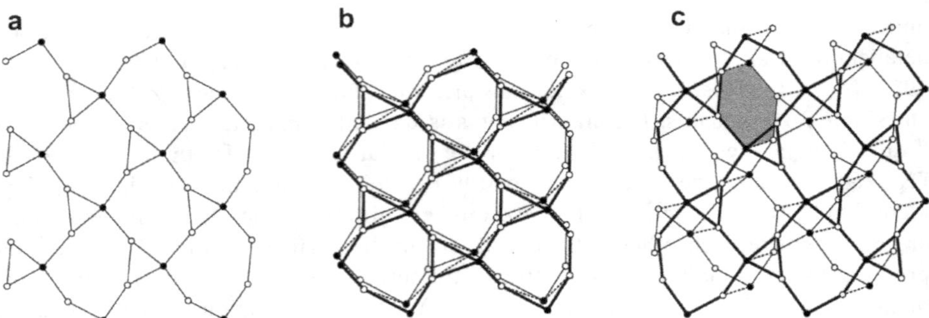

Figure 27. The 2D net consisting of three- and seven-membered rings (a) is a basis for the 3D nets in gittinsite, $CaZrSi_2O_7$ (b) and $SrZrSi_2O_7$ (c). In gittinsite, the adjacent 2D nets have the same orientation of triangles (b), whereas, in $SrZrSi_2O_7$, the adjacent 2D nets are related to each other by rotation by 180° around axis vertical to the plane of the nets (c).

2.4.3. Low-dimensional networks in mineral structures

The topology of low-dimensional networks can be described conveniently using graphs, especially in cases in which coordination polyhedra are linked to each other by sharing corners. Nodal description of tetrahedral layers and chains in silicates with branched anions has been used by Liebau (1978, 1985) (*cf.* Fig. 28, taken from Liebau, 1978), and many other examples are scattered throughout the literature. Moore (1974) and Hawthorne (1983) applied graphs and combinatorics to the enumeration of polyhedral clusters of different kinds and formulated the concepts of graphical

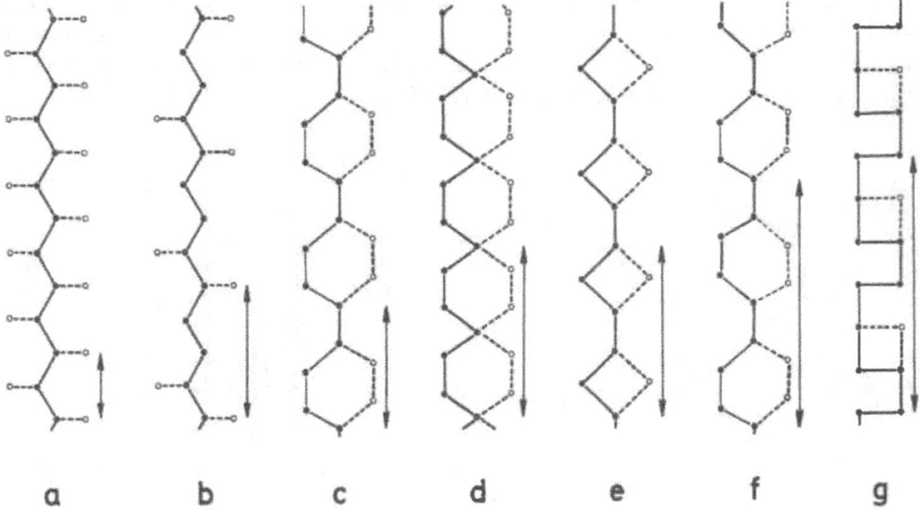

Figure 28. Nodal representations of silicate chain anions in the crystal structures of astrophyllite (a), aenigmatite (b), deerite and howieite (c), lemoynite (d), vlasovite (e), pellyite (f), and nordite (g), according to Liebau (1978).

(topological) and geometrical isomerism (see below). Graphs of heteropolyhedral structural units have been used for classification purposes by Rastsvetaeva and Pushcharovsky (1989), Hawthorne *et al.* (2000), Huminicki and Hawthorne (2002), Hawthorne and Huminicki (2005), Burns (2005), Krivovichev (2004c, 2005, 2009), *etc.*

2.4.4. Isomerism of structural units in minerals and its description

The concept of topological and geometrical isomerism was introduced into structural mineralogy by Moore (1970a), who also used the term "combinatorial polymorphism" (Moore, 1975), which did not find many applications. Hawthorne (1983) defined graphical (or topological isomers) as clusters that have the same specific chemical formula but different graphs. Geometrical isomers are clusters that have the same graphs but are different from the viewpoint of local polyhedral linkage configurations. Figure 29 shows a series of octahedral-tetrahedral clusters that have the same composition $M_2(TO_4)_2\varphi_N$. Note that black and white vertices denote tetrahedra and octahedra, respectively. The notations 001110 and 101010 describe the topology of linkage between the vertices as independent elements of the adjacency matrices of the graphs. For instance, the adjacency matrix of the first graph (001110) has the form:

	1	2	3	4
1	–	0	0	1
2	0	–	1	1
3	0	1	–	0
4	1	1	0	–

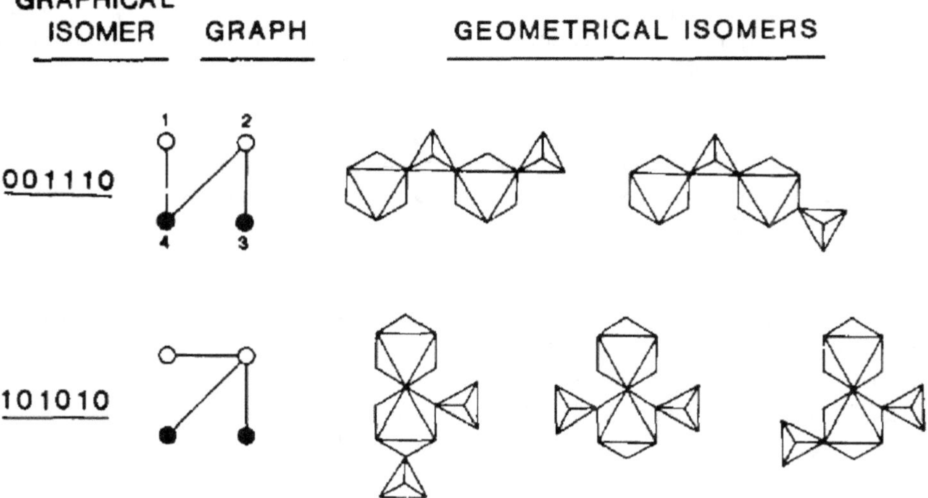

Figure 29. Octahedral-tetrahedral clusters for two graphical and topological isomers of the $M_2(TO_4)_2\varphi_N$ composition (after Hawthorne, 1983).

Here the matrix elements correspond to the numbers of atoms shared between two adjacent polyhedra (*i.e.* 0 in the case of no linkage and 1 in the case of corner linkage). The 001110 and 101010 graphs are graphical isomers. However, the same graph may be realized as different geometrical isomers due to the possibility of *cis*- and *trans*-linkage between octahedra and tetrahedra. The 001110 graph has two geometrical isomers, whereas the 101010 graph has three geometrical isomers. Therefore, one of the mechanisms that gives rise to the formation of geometrical isomers is due to the *cis*- or *trans*- orientation of shared corners within octahedra.

In 1975, P.B. Moore described another kind of geometrical isomerism observed for the octahedral-tetrahedral sheets in the crystal structures of laueite, pseudolaueite, stewartite and metavauxite (Moore, 1975). Figure 30a shows the $[Fe_2(PO_4)_2(OH)_2(H_2O)_2]^{2-}$ octahedral-tetrahedral sheet observed in the structure of laueite, $MnFe_2(PO_4)_2(OH)_2(H_2O)_8$ (Moore, 1965). Within this sheet, $Fe\varphi_6$ octahedra (φ = O, OH, H_2O) share *trans*-vertices to form chains which are further interlinked by PO_4 tetrahedra. Each tetrahedron is 3-connected (*i.e.* is linked to three octahedra), whereas $Fe\varphi_6$ octahedra are either 6-connected (linked to two octahedra and four tetrahedra) or 4-connected (linked to two octahedra and two tetrahedra).

Figure 31a shows the $[Fe_2(OH)_2(H_2O)_2(PO_4)_2]^{2-}$ octahedral-tetrahedral sheet observed in the structure of pseudolaueite, a polymorph of laueite (Baur, 1969). This sheet is identical chemically to the sheet in laueite. However, its topological structure is different. Its black-and-white graph is different from that of laueite in that all its black vertices are 5-connected, whereas, in the laueite graph, they are either 4- or 6-connected. Therefore, the sheets in laueite and pseudolaueite should be considered as 'topological' isomers.

However, laueite has another polymorph, stewartite (Moore and Araki, 1974). It is also based upon the $[Fe_2(OH)_2(H_2O)_2(PO_4)_2]^{2-}$ shown in Fig. 30b. Its black-and-white graph is isomorphous with the laueite graph. However, detailed examination of orientations of tetrahedra within the sheets reveal that the sheets are geometrically different and therefore should be regarded as 'geometrical' isomers. The geometrical isomerism is caused by the presence of 3-connected PO_4 tetrahedra with the fourth corner oriented either up or down relative to the plane of the sheet. To distinguish between the two sheets shown in Figs 30a and b, Krivovichev (2004b) proposed use of the concept of orientation matrix that lists the orientations of tetrahedra along the vertices of the graphs (here the **u** and **d** symbols correspond to the 'up' and 'down' orientations of tetrahedra, respectively, whereas the '□' symbol indicates tetrahedral vacancy). Following this approach, the orientation matrix of the laueite sheet can be written as (**ud**□□)(□**ud**□)(□□**ud**)(**d**□□**u**), whereas that of the stewartite sheet can be written as (**u**□□**ud**□□**d**)(**ud**□□**du**□□) (□**du**□□**ud**□)(□□**ud**□□**du**)(**d**□□**du**□□**u**)(**du**□□**ud**□□)(□**ud**□□**du**□)(□□**du**□□**ud**) (Fig. 30). In the case of stewartite, the matrix is much more complex than that for laueite and has dimensions of 8×8.

The pseudolaueite graph is typical for the 2D sheets in the structures of pseudolaueite and metavauxite, which contain two different geometrical isomers of the same

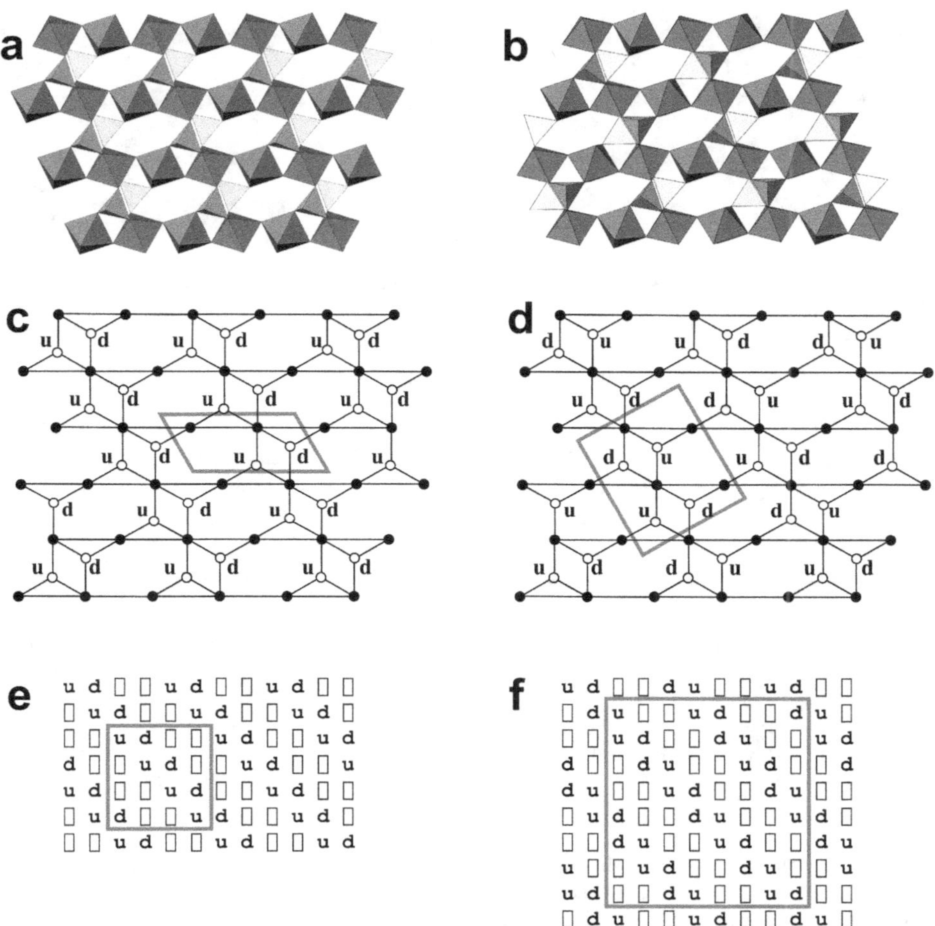

Figure 30. Octahedral-tetrahedral sheets in the structures of laueite (a) and stewartite (b), their 2D black-and-white graphs with orientations of non-shared vertices of tetrahedra written near the white circles (c and d for laueite and stewartite sheets, respectively) (legend: black circles = octahedra; white circles = tetrahedra). Martices of tetrahedra orientations for the sheets shown in a and b are given in e and f, respectively.

topology. This is demonstrated clearly by Fig. 31. According to the method of description used, the orientation matrices of tetrahedra for the sheets shown in Figs 31a and b are (**ud**□□)(□□**du**) and (**ud**□□)(□□**ud**), respectively.

2.5. Tilings

Tiling of n-dimensional Euclidean space is a countable number of n-dimensional bodies ('tiles') that cover space without gaps and overlaps. Here we shall consider only

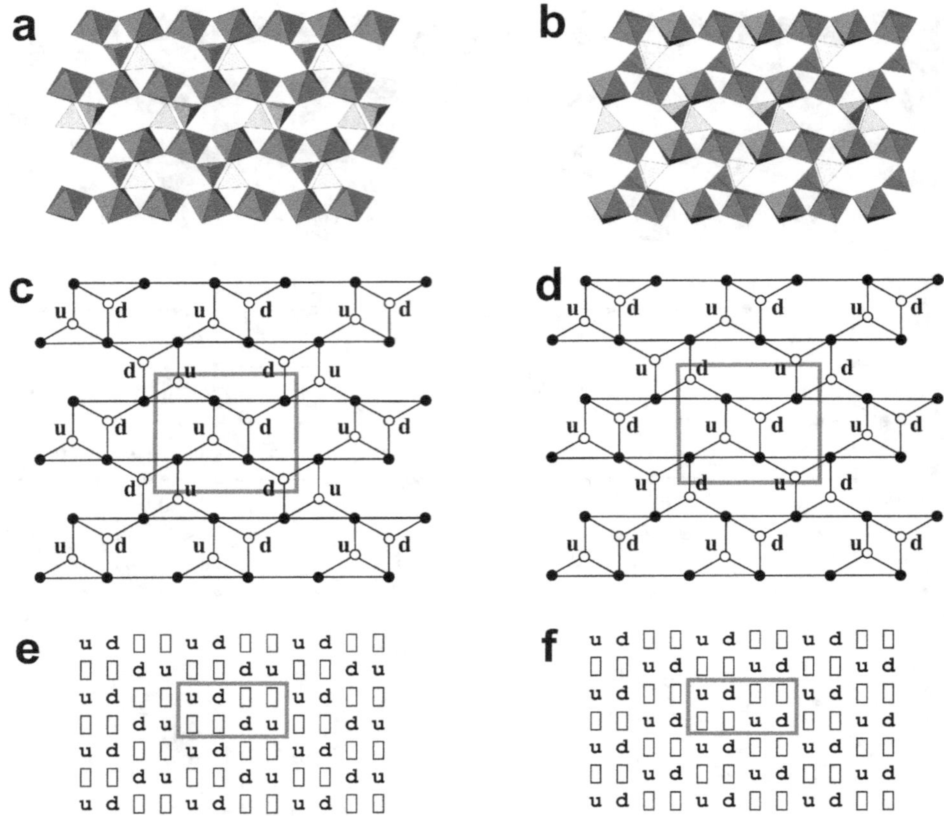

Figure 31. Octahedral-tetrahedral sheets in the structures of pseudolaueite (a) and metavauxite (b), their 2D black-and-white graphs with orientations of non-shared vertices of tetrahedra written near the white circles (c and d, respectively) (legend as in Fig. 30). Orientation matrices of tetrahedra for the sheets shown in a and b are given in e and f, respectively.

tilings with polyhedral tiles. Tiling is called 'face-to-face' if adjacent polyhedral tiles share vertices, whole edges, whole faces or have no points in common. If a tiling has a finite number n of topologically and geometrically different tiles, it is called n-hedral. If $n = 1$, tiling is isohedral; if $n = 2$, tiling is dihedral; if $n = 3$, tiling is trihedral, *etc.*

A special class of isohedral face-to-face tiling of 3D space is that in which all tiles are in the same orientation. Such tiles are called parallelohedra. For 3D Euclidean space, there are exactly five types of parallelohedra derived by E.S. Fedorov: cube, hexagonal prism, truncated octahedron, rhombic dodecahedron and elongated rhombic dodecahedron. Tiles of isohedral face-to-face tiling of 3D space with not necessarily parallel orientations of tiles are called stereohedra. The complete list of stereohedra for a 3D Euclidean space is still unknown.

In structural mineralogy, 2D and 3D tilings were used extensively in descriptions of complex 2- and 3-dimensional atomic arrangements.

2.5.1. 2D tilings: anion topologies

In order to describe dense sheet structures in uranyl oxides and oxysalts, Burns *et al.* (1996) developed the method of anion topologies. Figure 32a shows polyhedral representation of the uranyl silicate sheet in the crystal structure of uranophane. In this mineral, (UO_7) pentagonal bipyramids share their equatorial edges to form chains that are further linked by (SiO_4) tetrahedra. According to Burns *et al.* (1996), the anion topology of this sheet (Fig. 32a,b) can be constructed as follows: (1) each anion that is not bonded to at least two cations within the sheet, and that is not an equatorial anion of a bipyramid or pyramid within the sheet, is removed from further consideration

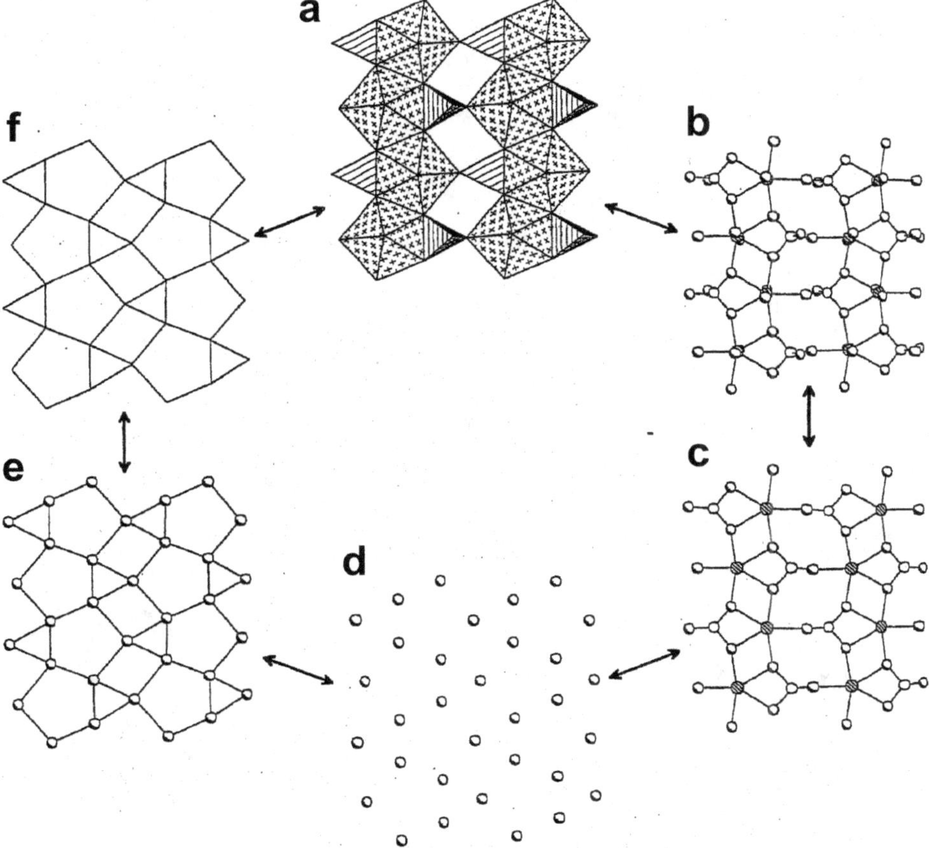

Figure 32. Construction of anion topology (after Burns *et al.*, 1996). See text for details.

(Fig. 32c); (2) cations are removed, along with all cation–anion bonds, leaving an array of unconnected anions (Fig. 32d); (3) anions are joined by lines, with only those anions that may be realistically considered as part of the same coordination polyhedron being connected (Fig. 32e); (4) anions are removed from further consideration, leaving only a series of lines that represent the anion topology.

The anion topology emphasizes the arrangement of anions in the plane of the sheet and, as a consequence, is most suitable for the planar 2D units. Uranyl oxysalts with dense sheets are especially appropriate for this method due to the strong tendency of uranyl-centred bipyramids to polymerize by sharing equatorial edges (Burns *et al.*, 1997), and a structural hierarchy of uranium minerals based on anion topologies was developed by Burns *et al.* (1996) (see Burns (2005) for an update). However, some non-uranium oxysalts can also be treated with this approach (see Krivovichev (2009) for the relevant examples).

The anion topology is obviously a tiling of the 2-D Euclidean plane into convex polygons. The tiling is of the 'edge-to-edge' type, which means that two polygons either do not have common points or share common corners or whole edges. In order to classify anion topologies, one can use their cyclic symbols defined as $n_1^{m1}n_2^{m2}n_3^{m3}$..., where n is the number of corners in a given polygon (3 for a triangle, 4 for a square, 5 for a pentagon, *etc.*), and m is the proportional number of these polygons in the anion topology. For example, the uranophane topology shown in Fig. 32e consists of pentagons, squares and triangles in the proportion 1:1:1. Thus, the cyclic symbol for this anion topology is $5^1 4^1 3^1$.

Some anion topologies are remarkable in their ability to accommodate different cation populations (Burns *et al.*, 1996; Burns, 1999, 2005). For instance, phosphuranylite topology (Fig. 33b) consists of hexagons, pentagons, squares and triangles (its cyclic symbol is $6^1 5^2 4^2 3^2$). Two sheets shown in Figs 33a and c can both be described using this anion topology but its population by cations is different in the two structures. In phosphuranylite itself (Fig. 33a), hexagons and pentagons are

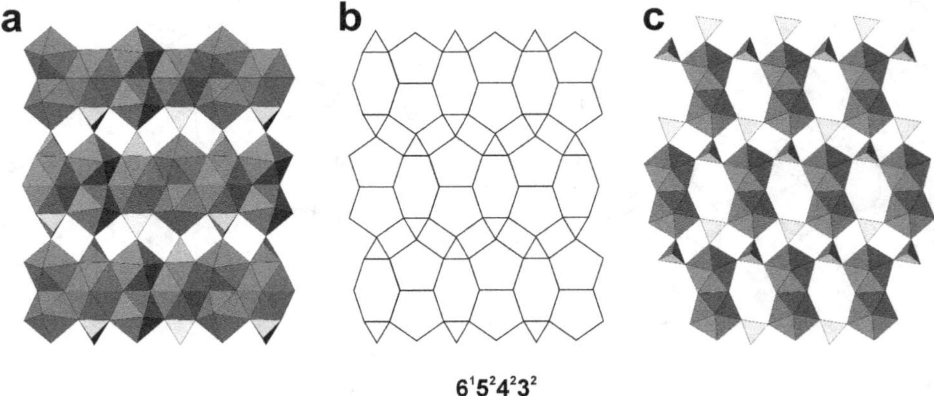

Figure 33. Phosphuranylite anion topology (b) may have different cation populations (a, c).

populated by uranyl cations, whereas squares are empty. In the 2-D unit shown in Fig. 33c, only pentagons and triangles are occupied, whereas triangles and hexagons are empty.

That the uranophane and phosphuranylite topologies are amongst the most commonly occuring in minerals and synthetic inorganic oxysalts is probably due to their large geometrical and energetic stabilities. Both topologies contain triangles that can be occupied by tetrahedral anions. In this case, the triangle comprises a triangular base of a tetrahedron with three tetrahedral corners being within the plane of the sheet. The fourth corner does not participate in the intra-sheet polymerization of polyhedra and may be oriented either up or down relative to the plane of the sheet. The possibility of the 'up' and 'down' orientations results in the appearance of different orientational geometrical isomers. Locock and Burns (2003a,b) considered geometrical isomerism in phosphuranylite- and uranophane-related structures and Krivovichev (2009) re-analysed and described them in terms of orientation matrices (see Section 2.4.4).

There is a number of 1-D structural units that can be considered as being derived from corresponding anion topologies. Figure 34a shows how dense chains of edge-sharing polyhedra can be produced by cutting the uranophane anion topology into polygonal

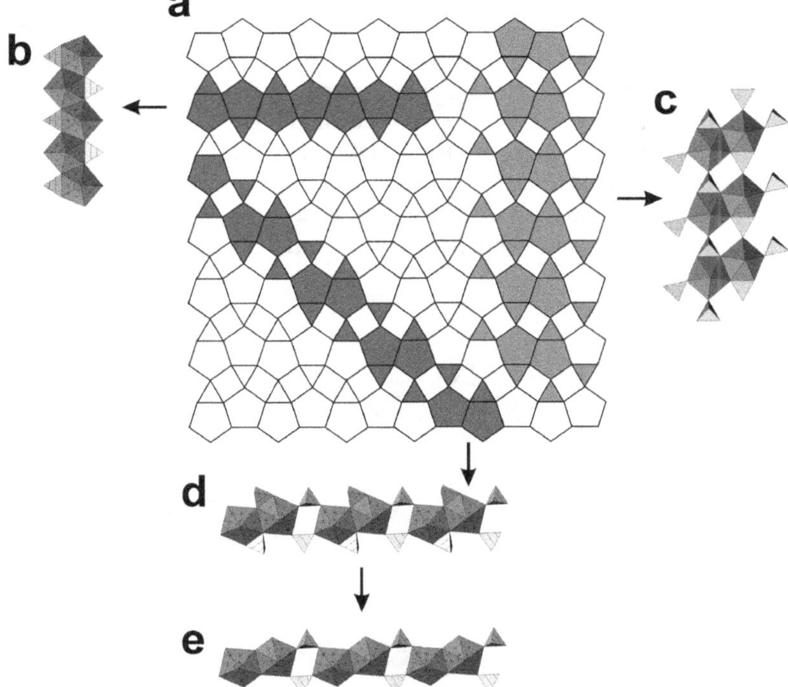

Figure 34. Production of heteropolyhedral chains in inorganic oxysalts through excision of 1-D regions from the uranophane anion topology. See text for details.

chains. Cutting the topology parallel to the direction of the chains of edge-sharing pentagons generates the $[NpO_2(CrO_4)]$ chain in $Cs[NpO_2(CrO_4)](H_2O)_2$ (Fig. 34b). Excision of the chain in the perpendicular direction results in the $[(UO_2)(TO_4)_2]$ chain (T = P, As) (Fig. 34c) that has been observed in the structures of parsonsite, $Pb_2[(UO_2)(PO_4)_2](H_2O)_n$ (Burns, 2000) and its As-analogue hallimondite (Locock *et al.* 2005). Finally, cutting the topology in the diagonal direction produces chains observed in $Cs_3[NpO_2(SO_4)_2](H_2O)_2$ (Fig. 34d), and in $K_2[(UO_2)F_2(SO_4)](H_2O)$ and $[N_2C_6H_{16}][UO_2F_2(SO_4)]$ (Fig. 34e).

2.5.2. 3D tilings

For obvious reasons, 3D tilings have been of much use in terms of the description of complex framework structures, especially those in tetrahedral framework silicates and aluminosilicates (see Figs 21 and 22, where tilings are used to describe conveniently the complexity of three-dimensional networks in the crystal structures of minerals of the cancrinite-sodalite supergoup).

Figure 35 shows a dihedral tiling that simplifies the construction of the three-dimensional black-and-white network of the labuntsovite type (Fig. 23). It consists of two polyhedra which are not topologically equivalent to the polyhedral units shown in Fig. 23 and can be obtained from the latter by addition of edges linking two black vertices (oriented vertically in Fig. 23). The polyhedra of the tiling are arranged in the following way. First, polyhedra of the type shown in Fig. 35b are arranged in layers by sharing faces on their sides (Fig. 35c). These layers have polyhedral hollows that are suited perfectly to the polyhedra of the type shown in Fig. 35a (Fig. 35d). Whole tiling is obtained by assembling layers of larger tiles one under another and by filling gaps with the smaller tiles (Fig. 35e).

Another example is a dihedral tiling that provides a simplified description of the 3D black-and-white scherbakovite-batisite net. We recall that the net can be constructed by

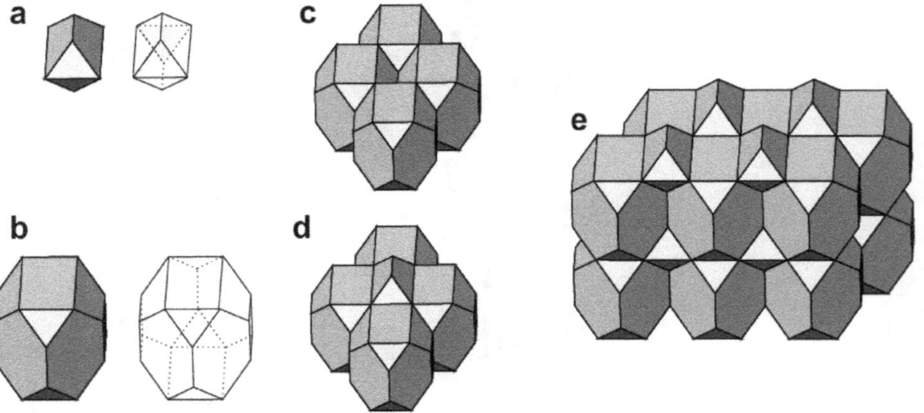

Figure 35. Dihedral tiling of 3D space that corresponds to the labuntsovite net. See text for details.

assembling the three different polyhedral building units shown in Fig. 24c, d and e. Using simple topological and geometrical operations, the **PBU** shown in the Fig. 24c unit can be transformed into a hexagonal prism, whereas those shown in Fig. 24d and e can be transformed into square prisms. Thus the 3D net shown in Fig. 24b can be considered to be a modified version of the tiling of 3D space into hexagonal and square prisms as shown in Fig. 36.

3. Concepts

3.1. Hierarchy

3.1.1. Types of hierarchy

The term 'structural hierarchy' was introduced into structural mineralogy by Moore (1970a), who defined it as "a general scheme which, through the appropriate algorithms, ties together a collection of arrangements". Most of the known structural classifications of minerals imply the construction of structural hierarchy schemes based upon specific criteria such as dimensionality that specify the respective hierarchical levels (see, *e.g.* systematics of silicates by Liebau (1985) discussed above).

In general, the modern science (or the modern philosophy of science) distinguishes between two different types of hierarchy (Salthe, 2012). The first type is a subsumption or specification hierarchy such as the Linnaean hierarchy of biological species. This form of hierarchy has its levels as nested subclasses and, in mineralogy, has a usual form of structural-chemical systematics of minerals (*e.g.* that developed by Strunz and Nickel, 2001). An example of specification hierarchical analysis applied to minerals is

{ minerals { silicates { framework silicates { zeolites { paulingite }}}}}.

Here the highest level is represented by minerals in general, *i.e.* crystalline compounds of natural origin, and the lowest level is a particular mineral, paulingite. The term 'structural hierarchy' used by Moore (1970a) relates to the kind of specification hierarchy that is based upon particular features of structural arrangements (dimensionality, connectivity and topology of structural units, for example). Graphically, this type of hierarchical relation is depicted as cladograms and nested trees (see, *e.g.* classification of mixed anionic radicals by Sandomirskiy and Belov (1984) in Fig. 4. Hawthorne (2014) provided a detailed review of the concept of structural hierarchy with illustrative examples taken from various classes of minerals and inorganic compounds.

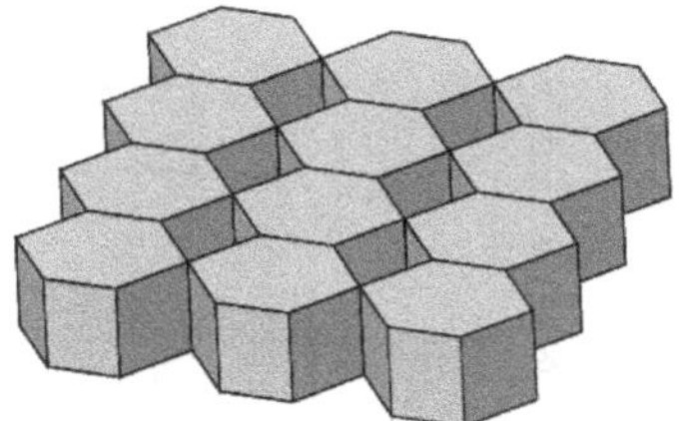

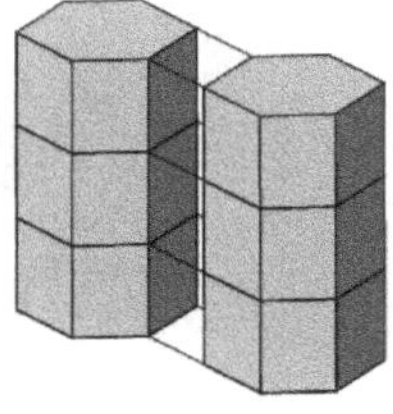

Figure 36. Dihedral tiling of 3D space into hexagonal and square prisms that corresponds to the shcherbakovite-batisite 3D net.

Another kind of hierarchy is a compositional hierarchy, also known as a scale or inclusion hierarchy. It considers a specific system as consisting of subsystems (components) nested within the system (whole). The example is an electron inside an atom inside a coordination polyhedron inside a structural unit inside a unit cell inside a crystal:

{ crystal { unit cell { structural unit { coordination polyhedron { atom { electron }}}}}}.

This kind of a hierarchy is implied in any structural classification of minerals and also has a high relevance for the related concept of modularity (see below). The quantitative basis for the compositional hierarchical analysis was established by Hawthorne (1983) by means of the bond-valence theory. The approach used by Hawthorne (1983) is consistent with the idea of a nearly decomposable system developed by Simon (1962) and below we compare these paradigms in more detail.

3.1.2. Crystal structure as a nearly decomposable system

In his famous paper 'The Architecture of Complexity', Simon (1962) suggested the concept of a near decomposability of complex systems, which are the systems "in which the interactions among the subsystems are weak, but not negligible." The system can be decomposed into subsystems by the strengths of interactions of subsubsystems within them. An example is provided by the graph shown in Fig. 37. It consists of fifteen vertices (components) with interactions depicted as edges of different kinds. The strength of the interaction along the solid edge is set to 1.00, whereas the strength of the interaction along the dashed line is set to 0.25. The total scheme of interactions within the graph can be described by the matrix shown in Fig. 37b, a nearly decomposable matrix, where black and grey '+' signs correspond to the interaction strengths of 1.00 and 0.25, respectively (it is assumed that the interaction of a vertex with itself is equal to 1.00 as well). The concept of near decomposability is based on the

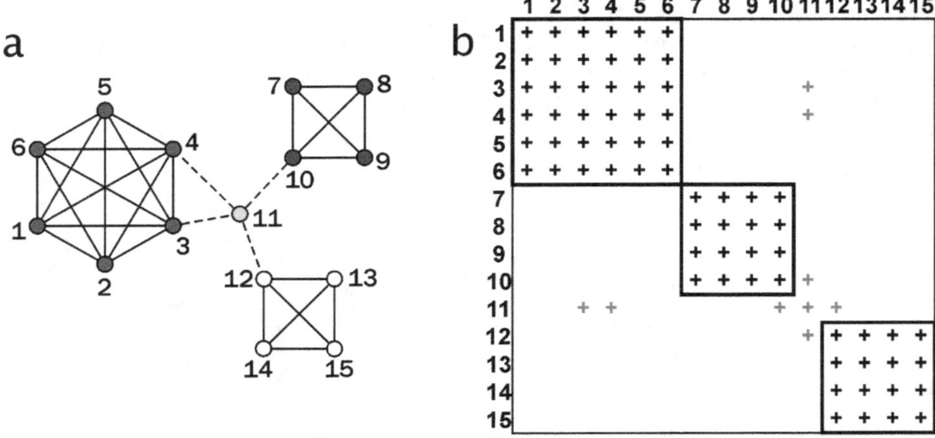

Figure 37. An example of nearly decomposable system (a) and its adjacency matrix (b).

choice of some number, ε, as the borderline parameter that separates strong interactions from weak interactions. By setting ε = 0.50, the graph shown in Fig. 37a can be decomposed into three weakly interacting subgraphs which can be considered as its subsystems. The graph therefore can be considered as a hierarchical system with three levels of a compositional hierarchy. The first level consists of all vertices, the second consists of the subgraphs [1,2,3,4,5,6], [7,8,9,10], [12,13,14,15], and [11], and the third level is the whole graph itself.

Since the first crystal-structure determinations of minerals, it has always been assumed intuitively that minerals are hierarchical structures with atoms representing the lowest level of hierarchy. However, it seems that Hawthorne (1983) was the first to establish a solid quantitative basis for the decomposition of the crystal structure of a mineral (or any inorganic compound in general) into subsystems. For this, he employed the bond-valence theory that provides numerical estimates of the strengths of interactions between atoms using empricial bond length–bond strength relationships (see also Section 2.2.2). In order to subdivide a crystal structure into tightly bonded subunits and to infer its hierarchical architecture, one has to analyse the adjacency matrix that has as its elements bond valences of respective bonds. For most ionic crystals, it is enough to construct such a matrix in the form known in structural mineralogy as a bond-valence table, *i.e.* the table that provides bond valences for the bonds between cations and anions. As an example, let us consider the bond-valence table for the crystal structure of symesite, $Pb_{10}(SO_4)O_7Cl_4(H_2O)$, a complex lead oxysulfate chloride hydrate described by Welch *et al.* (2000). Its bond-valence table shown in Fig. 38a has been constructed by calculations of the bond valences, s_{ij}, from the corresponding bond lengths, d_{ij}, between the ith and jth atoms using the equation:

$$s_{ij} = \exp[(r_o - d_{ij})/b] \tag{8}$$

where r_o and b are empirical parameters defined for a particular cation-anion pair. In our example, the $\{r_o, b\}$ parameters for the $Pb^{2+}-O^{2-}$, $S^{6+}-O^{2-}$, and $Pb^{2+}-Cl^{-}$ bonds have been taken as {1.963, 0.49}, {1.625; 0.37} and {2.53, 0.37}, respectively (Brese and O'Keeffe, 1991; Krivovichev and Brown, 2001). The analysis of the interaction matrix shown in Fig. 38a indicates that the strongest bonds (of 1.50 v.u. = valence units) in the structure are those between the S and O1–O4 atoms. The next strongest interactions are between the Pb atoms and the O5–O11 atoms, as their bond valences are in the range 0.33–0.62 v.u. (there are two Pb–Cl bonds with bond valences larger than 0.33 v.u., but they can be ignored for the moment). By setting ε = 0.33 v.u., the structure can be decomposed into subsystems highlighted in light and dark grey colours.

The next step in the construction of a structural hierarchy is to follow Hawthorne's proposal and to subdivide the structure into structural subunits or PBUs. The S atom is bonded solely to the O1–O4 atoms, whereas the latter are also weakly bonded to the Pb atoms. It is therefore natural to subdivide the SO_4 unit as the most tightly bonded subsystem in the structure. By analogy, the O5–O11 atoms are bonded solely to the Pb atoms, whereas the latter form additional weak bonds to other O atoms and to Cl atoms. It is also important that the O5–O11 atoms each form four bonds to Pb atoms, which

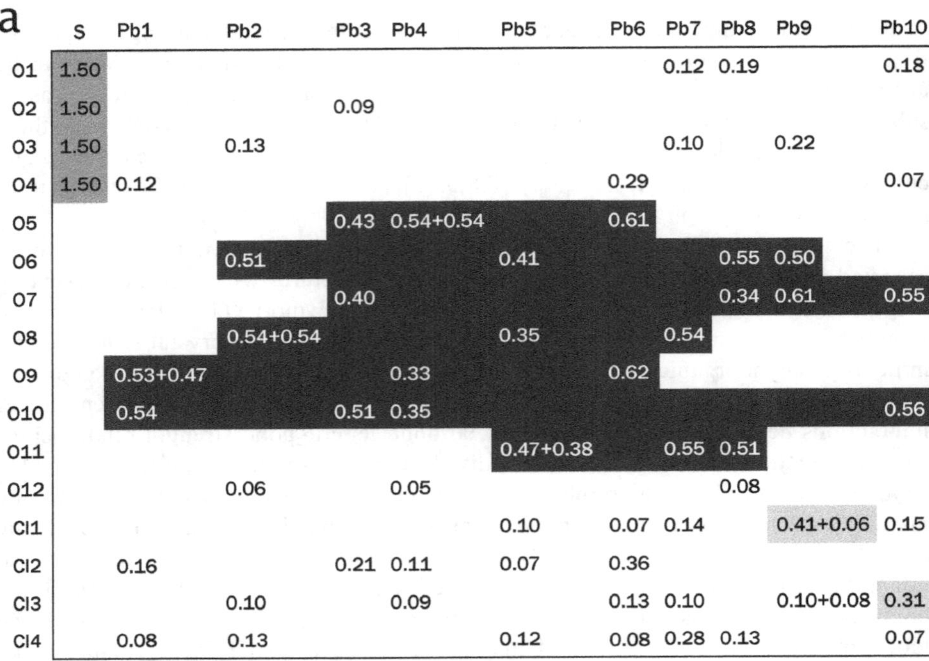

a

	S	Pb1	Pb2	Pb3	Pb4	Pb5	Pb6	Pb7	Pb8	Pb9	Pb10
O1	1.50							0.12	0.19		0.18
O2	1.50			0.09							
O3	1.50		0.13					0.10		0.22	
O4	1.50	0.12					0.29				0.07
O5				0.43	0.54+0.54		0.61				
O6			0.51			0.41			0.55	0.50	
O7				0.40					0.34	0.61	0.55
O8			0.54+0.54			0.35		0.54			
O9		0.53+0.47			0.33		0.62				
O10		0.54		0.51	0.35						0.56
O11						0.47+0.38		0.55	0.51		
O12			0.06		0.05				0.08		
Cl1						0.10	0.07	0.14		0.41+0.06	0.15
Cl2		0.16		0.21	0.11	0.07	0.36				
Cl3			0.10		0.09		0.13	0.10		0.10+0.08	0.31
Cl4		0.08	0.13			0.12	0.08	0.28	0.13		0.07

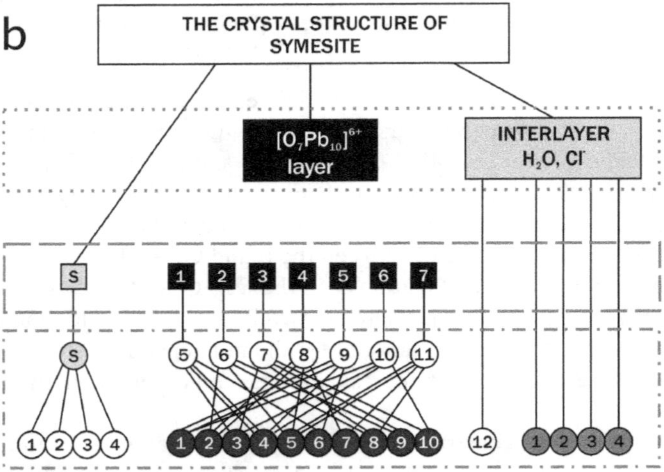

Figure 38. Bond-valence table (a) and compositional hierarchy scheme (b) for the crystal structure of symesite, $Pb_{10}(SO_4)O_7Cl_4(H_2O)$.

allows us to identify O-centred OPb_4 subunits as the configurations next in strength after the sulfate tetrahedra. The structure can thus be decomposed into the following subsystems: SO_4 tetrahedra, OPb_4 tetrahedra, H_2O groups (formed by the O12 atom), and Cl atoms. The situation is complicated by the fact that, whereas SO_4 tetrahedra are not linked to each other, the OPb_4 tetrahedra polymerize to form extended $[O_7Pb_{10}]^{6+}$ layers.

The compositional hierarchy scheme for the crystal structure of symesite can be constructed as shown in Fig. 38b. Here the circles symbolize atoms, whereas squares symbolize tetrahedra. The lowest hierarchical level (indicated by a dash-and-dotted line) consists of atoms that interact to form coordination polyhedra which constitute the second level (indicated by a dashed line). The OPb_4 tetrahedra interact further to form the $[O_7Pb_{10}]^{6+}$ layers which, along with the interlayer H_2O molecules and Cl atoms represent the third hierarchy level (shown by a dotted line). The fourth hierarchy level is the structure as a whole.

The example of symesite demonstrates that crystal structures of minerals can be considered as nearly decomposable hierarchical systems if quantified from the viewpoint of bond-valence theory following the hypothesis proposed by Hawthorne (1983).

Makovicky (1997) and Ferraris *et al.* (2004) discussed hierarchical description of crystal structures of minerals in terms of the following configurational levels:

- primary configurations: coordination polyhedra; secondary configurations: groups consisting of coordination polyhedra by sharing common ligands (finite clusters, chains, layers, *etc.*);
- tertiary configurations: dimensionally higher units obtained by the linkage of secondary configurations *via* bonds weaker than those involved at the previous level (*e.g.* linkage of chains *via* weak secondary bonds (*e.g.* hydrogen bonds) into layers);
- quaternary configurations emerge when tertiary configurations are combined with coordination polyhedra of other elements; according to Ferraris *et al.* (2004), this is the level of modules as used in a modular description (see below);
- quinary configurations: packing of quaternary or ternary configurations in a large-scale structural pattern; this is the level of crystal structures.

As an example of such a hierarchical description, Ferraris *et al.* (2004) discussed biopyriboles (pyroxenes, amphiboles, *etc.*). The primary units are SiO_4 tetrahedra that are linked by sharing O atoms to form idealized silicate chains as secondary configurations. The distortion of the silicate chains due to their adjustments to the structural environment results in the formation of real chains at the tertiary level (note that this contradicts the above-mentioned definition which considers tertiary configurations to be weakly bonded secondary units). The fourth-order (quaternary) configurations are rods consisting of two tetrahedral chains with a chain of edge-sharing octahedra as a core (so-called I-beams according to Thompson, 1978). Finally, the quinary level comprises the crystal structure as a whole.

The example of biopyriboles discussed by Ferraris *et al.* (2004) demonstrates that, despite the fundamental existence of hierarchical structures in minerals, the subdivision of a crystal structure into hierarchical levels depends upon the immediate research tasks and may vary from structure to structure and from researcher to researcher.

3.1.3. Hierarchy and complexity

Simon (1962) noticed that "hierarchy is one of the structural schemes that the architect of complexity uses". Figure 39 shows a general scheme of a hierarchical organization of a crystal structure of a mineral that combines the general ideas on structural description and classification discussed in the sections above. In general, five hierarchical levels can be identified (from bottom to top): (1) atoms as primary basic units (PBUs); (2) coordination polyhedra as basic building units (BBUs); (3) fundamental building blocks (FBBs) or composite building units (CBUs); (4) structural units and interstitial complexes; (5) structure as a whole. It is obvious that not all minerals are organized in such a way as to possess all five possible levels of hierarchy, and the number of hierarchical levels or a hierarchical depth (HD), can be considered as a measure of structural complexity. Another measure is a hierarchical span that reflects the size and diversity of organized entities on different hierarchy levels.

For instance, the crystal structure of native copper contains only two hierarchical levels (Fig. 40a) and its hierarchy is very shallow, indicating its structural simplicity. The hierarchical depths of the crystal structures of quartz (Fig. 40b) and albite (Fig. 40c) are three and four, respectively, which agrees well with the intuitive feelings of their structural complexity. A high complexity of the crystal structure of charoite can

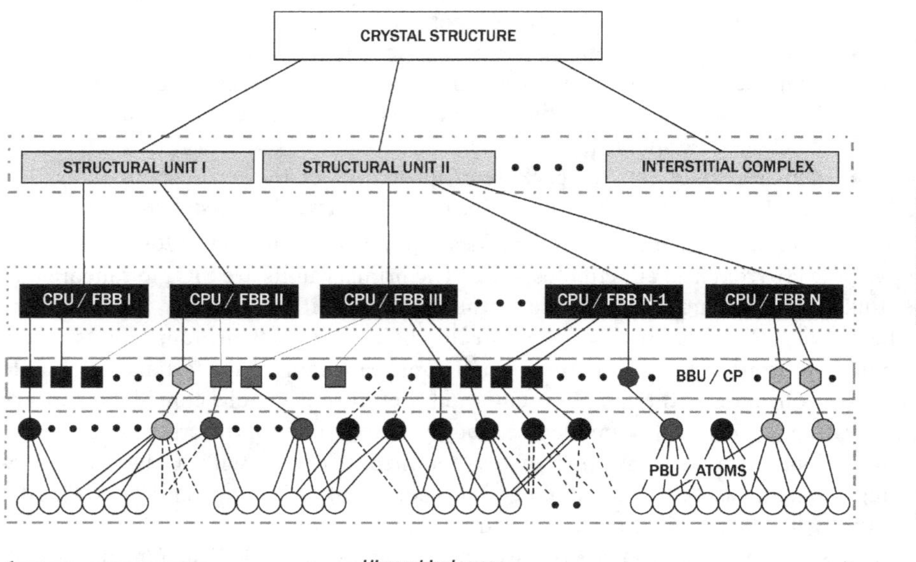

Figure 39. General scheme of hierarchical organization of an inorganic crystal structure.

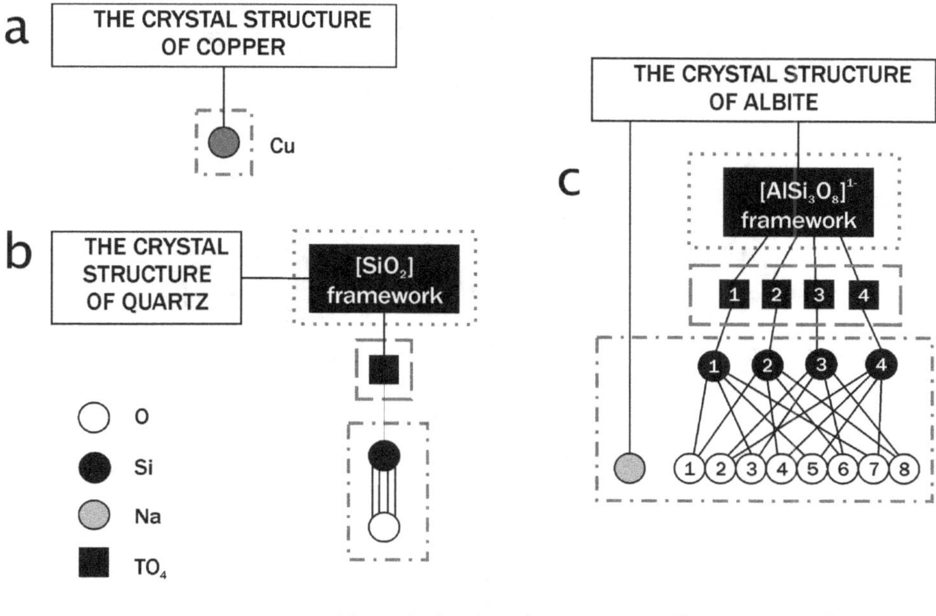

Figure 40. Examples of compositional hierarchy schemes of the crystal structures of minerals: native copper (a), quartz (b) and albite (c).

be seen from its structural hierarchy scheme (Fig. 41). The hierarchical organization of charoite contains five hierarchical levels (HD = 5) and its hierarchical span is spectacular. Even on the third level, it contains five different CBUs, three of which belong to three topologically different types of one-dimensional silicate anions.

However, the difficulty with quantifying complexity in terms of hierarchical depth and span lies in the sometimes subjective subdivision of hierarchy levels as was shown when discussing the hierarchical approach suggested by Ferraris *et al.* (2004). As was pointed out by Salthe (2012), "only some users of hierarchical forms would insist that particular levels actually exist as such in natural systems", as hierarchy is just a method of "cutting (a seemingly continuous) Nature at its joints." However, analysis of a hierarchical organization of the crystal structures of minerals agrees well with the quantitative analysis of their complexity using Shannon information theory (see below).

3.2. Modularity

Modularity is an inherent property of nearly decomposable complex systems and as such this concept has been applied extensively to a wide range of biological, technological and social systems (Schlosser and Wagner, 2004; Callebaut and Raskin-Gutman, 2005, *etc.*). That inorganic crystal structures have a modular construction has

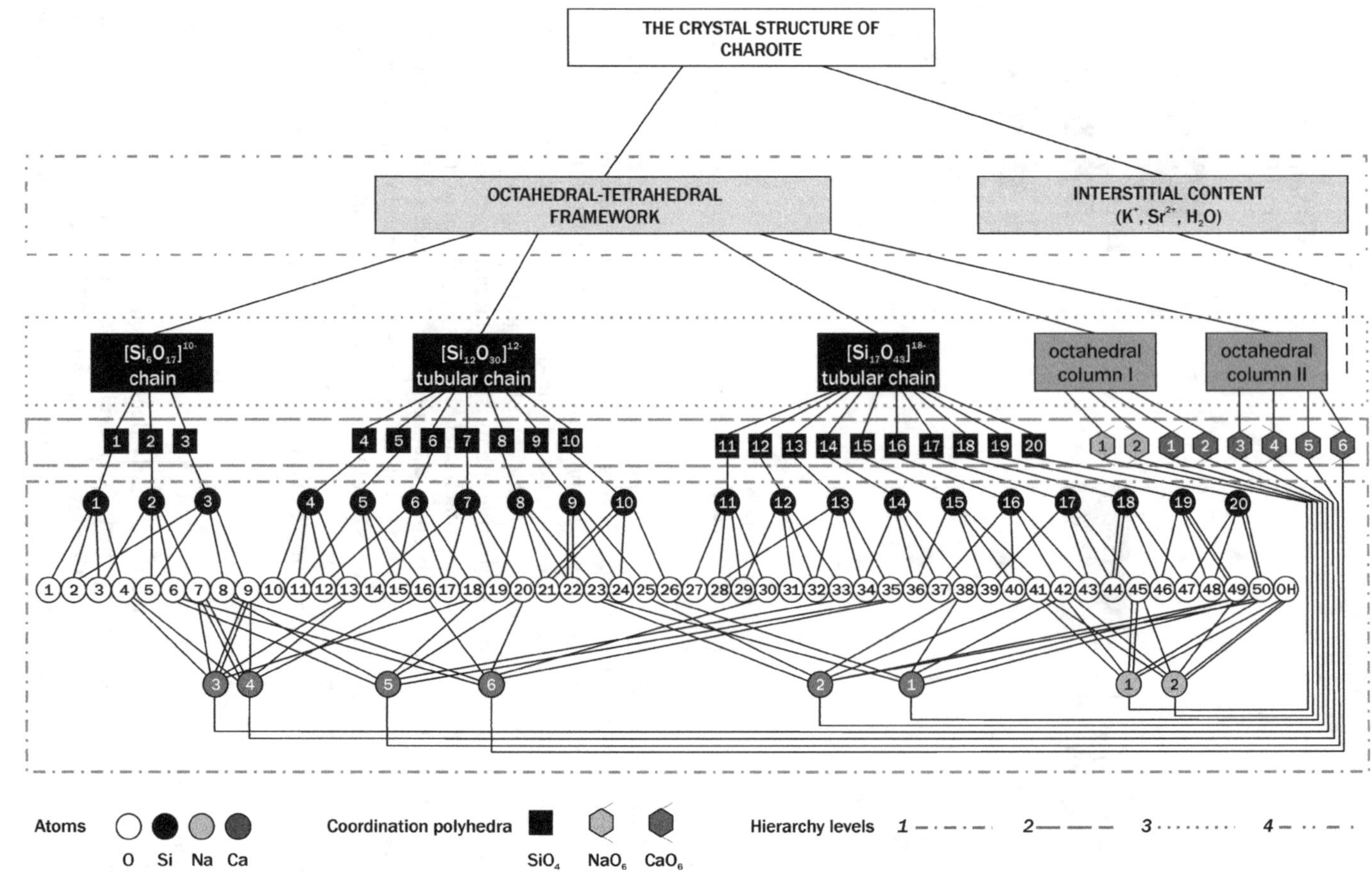

Figure 41. Compositional hierarchy scheme for the crystal structure of charoite.

long been recognized (see Merlino, 1997 and Ferraris *et al.*, 2004 for historical reviews of the subject) and modularity can be viewed as one of the central concepts in modern structural mineralogy. Here we consider only the basic definitions and ideas on modularity of minerals adopted from Makovicky (1997, 2006) and Ferraris *et al.* (2004). The reader is referred to these and other papers by G. Ferraris, E. Makovicky and S. Merlino as major contributors to the field for more details on the modularity approach in mineralogy.

3.2.1. Basic definitions

According to Makovicky (1997), "...*modular structures* are composed of moduli/fragments (blocks, rods or layers) of archetypal structures that are recombined in various orientations by the action of structure building operators".

In turn, archetypal structures are "...those simpler structures that display/encompass all fundamental bonding and geometric properties of the structure portions in the interior of building blocks/moduli" (Makovicky, 1997). In general, the term 'archetype' has many meanings in modern science and is usually defined as a prototype from which other things are copied, patterned or emulated (Krivovichev, 2014b). The examples of archetypal structures in minerals are PbS, SnS, ZnS, ReO_3, CaF_2, *etc.*

Modular structures of minerals may be monoarchetypal (*e.g.* humite series) or diarchetypal (*e.g.* sapphirine-type minerals in which layers of the pyroxene archetype combine with those from the spinel archetype (Merlino and Zvyagin, 1998; Zvyagin and Merlino, 2003)) (Ferraris *et al.*, 2004).

Skowron and Brown (1994) considered the crystal structures of stibnite, Sb_2S_3, robinsonite, $Pb_4Sb_6S_{13}$, boulangerite, $Pb_5Sb_4S_{11}$, zinckenite, $PbSb_2S_4$, and cosalite, $Pb_2Bi_2S_5$, as modular derivatives of the crystal structure of galena, PbS. The galena structure is first broken into (100) sheets as shown in Fig. 42a. Then the vacancies (shown by squares in Fig. 42b) are introduced, breaking the sheets into one-dimensional ribbons (Fig. 42c,d). The ribbons with different widths combine with

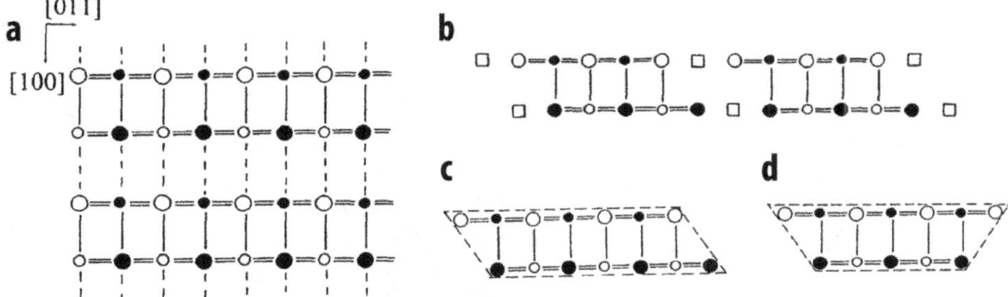

Figure 42. The crystal structure of galena viewed along [110] and split into sheets (a; the large and small circles represent S^{2-} and Pb^{2+}, respectively; open and filled circles lie at different levels), ordering of the cation vacancies splits the sheets into ribbons (b; the squares represent cation vacancies), the ribbons viewed parallel to their extension (c, d). After Scowron and Brown (1994).

each other to form the structures of the minerals under consideration (Fig. 43). The relation between the ribbons can be described in terms of structure-building operators.

3.2.2. Structure-building operators

Structure-building operators provide the description of mutual relations between the modules in a crystal structure. Ferraris *et al.* (2004) list the following basic structure-building operators: mirror-reflection twinning, glide-reflection twinning, axial twinning, cyclic twinning, non-commensurability, antiphase boundaries, crystallographic shear and unit-cell intergrowths. The structure-building operations generally correspond to the changes of chemical composition on the boundaries of the modules. For instance, Cooper and Hawthorne (1996) described the crystal structure of rapidcreekite, $Ca_2(SO_4)(CO_3)(H_2O)_4$, as a derivative of the structure of gypsum, $Ca(SO_4)(H_2O)_2$ (Fig. 44). The replacement of half of the sulfate groups in gypsum by the carbonate triangles results in the mirror-twinning (or unit-cell twinning) of gypsum modules across the carbonate groups and the formation of a new structure type.

3.2.3. Homologous series

The concept of a homologous series was introduced by Magnelí (1953) and later re-invented by Thompson (1978) as a polysomatic series. Following Makovicky (1997), we distinguish between the accretional and variable-fit homologues series.

Accretional series are those, in which the type, general shape of building blocks and structure-building operators are preserved, but the size of the blocks varies

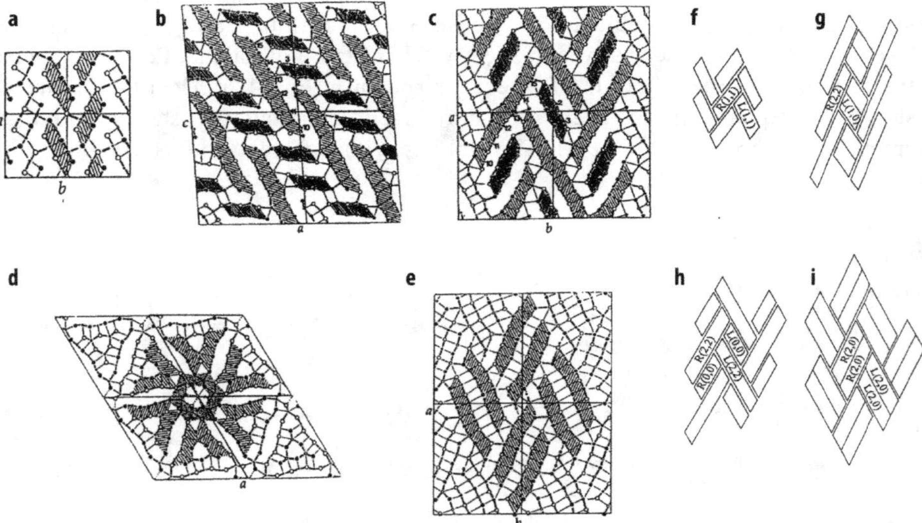

Figure 43. The crystal structures of stibnite (a), robinsonite (b), buolangerite (c), zinckenite (d) and cosalite (e), and schemes of arrangements of structural modules in stibnite (f), robinsonite (g), boulangerite (h) and cosalite (i). After Scowron and Brown (1994).

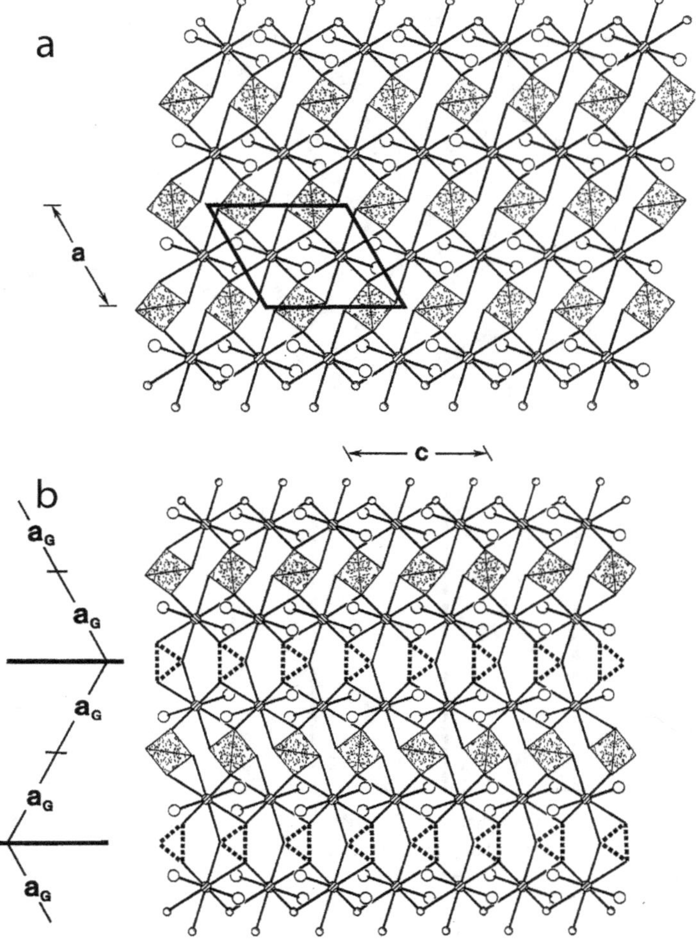

Figure 44. The crystal structure of rapidcreekite, $Ca_2(SO_4)(CO_3)(H_2O)_4$ (b), as generated by mirror-twinning of the crystal structure of gypsum, $Ca(SO_4)(H_2O)_2$ (a) (after Cooper and Hawthorne, 1996).

incrementally (Ferraris *et al.*, 2004). The important parameter is the order of a homologue, N, defined as a number of coordination polyhedra across a suitably defined diameter of the building block. An example of the accretional series is the lilianite series of structures, which are built from the blocks obtained from the archetype structure of galena by cutting along (311). Here the order is defined as a number of octahedra in the octahedral chains that run diagonally across an individual archetypal layer and is parallel to $[011]_{PbS}$ (Makovicky, 1997). Because the structures of the members of the series can be built from building blocks of two different orders, N_1 and N_2, each homologue is assigned a notation $^{N1,N2}L$. Figure 45a shows the crystal

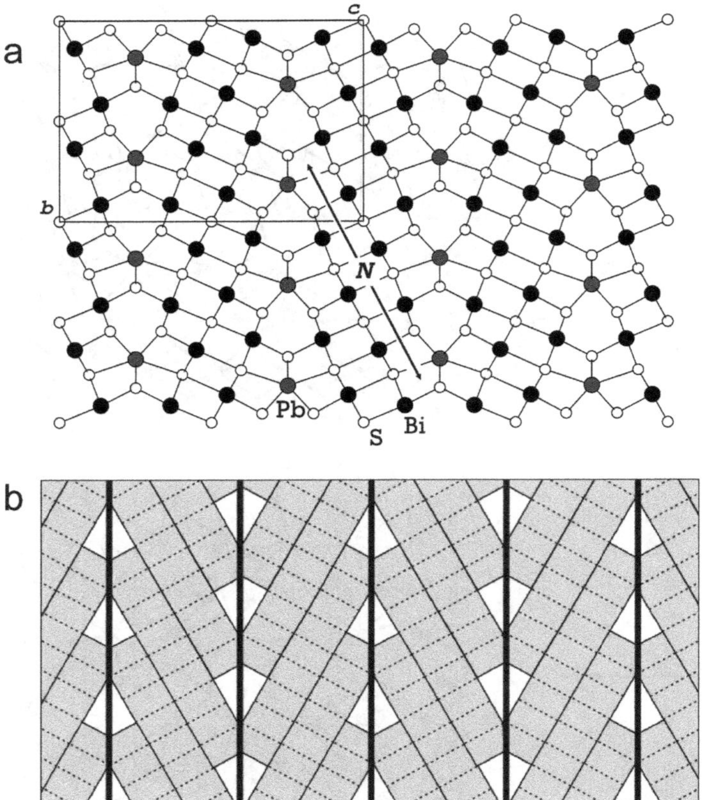

Figure 45. The crystal structure of gustavite, $PbAgBi_3S_6$, the ${}^{4,4}L$ homologues of the lilianite homologous series, projected along the *a* axis (a) and its schematic representation in terms of blocks excised from the structure of PbS (b).

structure of gustavite, $PbAgBi_3S_6$, which is composed of building blocks with the same order $N_1 = N_2 = 4$. The structure-building operator is a mirror-twinning indicated in Fig. 45b by a solid line. Thus, gustavite is the ${}^{4,4}L$ member of the lilianite accretional series. In contrast, the crystal structure of vikingite, $Pb_8Ag_5Bi_{13}S_{30}$, is the ${}^{4,7}L$ lilianite homologue, which is based upon slabs with $N_1 = 4$ and $N_2 = 7$ (Fig. 46).

Another example of the accretional homologues series are biopyriboles (Thompson, 1970). For more details on accretional series see: Makovicky (1997, 2006) and Ferraris *et al.* (2004).

Variable-fit homologous series occur in crystal structures composed of two alternating non-commensurate layers (Ferraris *et al.*, 2004), as the structures of cannizarite family (Makovicky, 2006). Their crystallographic description requires special tools of *n*-dimensional crystallography as described by Bindi and Chapuis (2017, this volume) in a companion chapter in this volume.

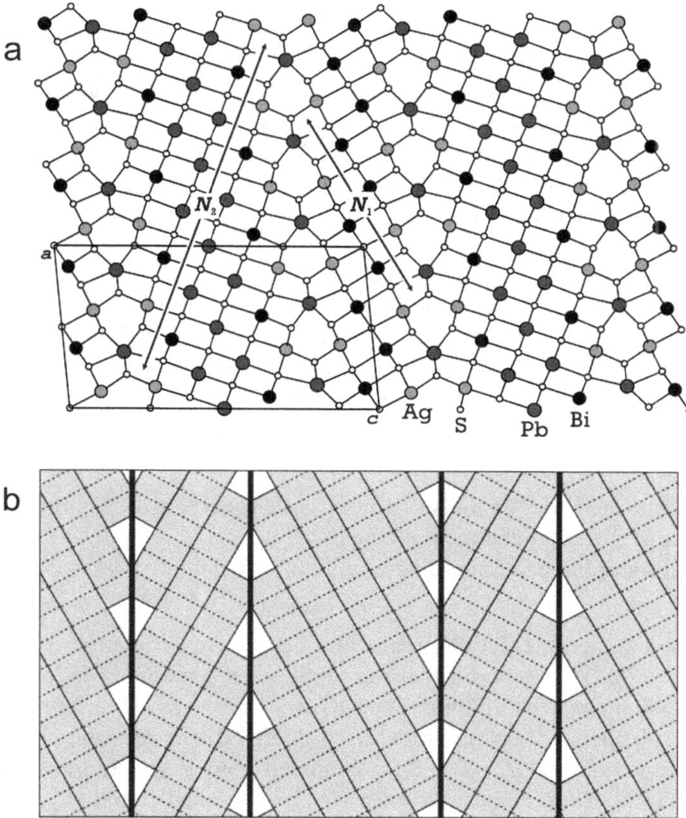

Figure 46. The crystal structure of vikingite, $Pb_8Ag_5Bi_{13}S_{30}$, the $^{4,7}L$ homologues of the lilianite homologous series, projected along the *b* axis (a) and its schematic representation in terms of blocks excised from the structure of PbS (b).

3.2.4. Polytypism

Polytypism is a special kind of modularity that involves different stacking modes of layers with the same topology and chemical composition (Ferraris *et al*., 2004). The formation of particular polytypes in Nature usually happens under particular geochemical and thermodynamic conditions and thus can be used to decipher those conditions from the crystal chemical analysis (see Merlino, 1997 for more details).

3.3. Complexity

3.3.1. Historical notes

The concept of complexity has a direct relevance to the description and interpretation of the crystal structures of minerals. Many mineralogists working with crystal structures of minerals often describe them or parts of them as 'complex' or 'complicated', which implies the existence of a certain intuitive feeling and appreciation of structural complexity. However, very little attention was paid to the quantitative aspects of complexity as applied to crystal structures in general.

The first attempt to evaluate structural complexity of crystals should be assigned to Linus Pauling's fifth ('parsimony') rule "determining the structure of complex ionic crystals" (Pauling, 1929): "the number of different types of constituents in a crystal tends to be small." Hawthorne (2006) noted that "this rule has not seen much use, apart from in teaching, but it certainly seems to hold reasonably well, and its underlying cause is certainly of interest."

Pauling (1929) did not provide a quantitative procedure for evaluating structural parsimony (he did not even mention this rule in the first edition of *The Nature of the Chemical Bond* (Pauling, 1939)), and this path was undertaken by Baur *et al.* (1983). These authors proposed two types of parsimony indices in order to describe topological and geometrical complexity of crystal structures.

The topological parsimony index, I_t, was defined by Baur *et al.* (1983) as

$$I_t = (t - e)/t \qquad (9)$$

where t is a number of topologically distinct atomic species in the asymmetric unit and e is the number of different chemical elements.

The crystallographic parsimony index, I_c, can be calculated as

$$I_c = (c - e)/c \qquad (10)$$

where c is the number of crystallographically distinct constituents.

Baur *et al.* (1983) pointed out that the topological parsimony index is of primary importance and suggested subdividing structures into lavish ($I_t > 0.66$), intermediate ($0.66 > I_t > 0.33$) and parsimonius ($I_t < 0.33$). However, the index has little discriminative power: *e.g.* $I_t = I_c = 0$ for cubic halite (NaCl), tetragonal rutile (TiO_2), and cubic $BaTiO_3$. Both indices are insensitive to chemical complexity and symmetry and reflect only relationships between the number of chemical elements and the number of topologically distinct atomic species. It is obvious that, for some complex minerals, it may happen that $e > t$, and therefore $I_t < 0$. Though parsimony is related generally to structural complexity, the parsimony indices proposed by Baur *et al.* (1983) cannot be used as complexity measures of crystal structures.

As outlined by Pauling (1929), Oganov and Valle (2009) proposed a number of parameters to investigate complexity of atomic arrangements in any atomic system, including crystals. In particular, they suggested using the parameter called quasi-entropy, S_{str}, defined as follows:

$$S_{str} = -\sum_{A} \tfrac{N'}{N} \langle \ln(1 - D_{A_i}A_j \rangle \qquad (11)$$

where N is the total number of atoms, N' is the number of atoms of the type A, and $D_{A_{iAj}}$ is the distance between fingerprint functions of the atoms A_i and A_j (fingerprint functions are similar to, but not identical to, a pair distribution function, see Oganov and Valle, 2009 for more details). Quasi-entropy provides a measure of distinction between the geometrical environments of the atoms of the same element in a crystal structure. It has a non-zero value only when there are two distinct crystallographic sites occupied by the same element. This means that the quasi-entropies of such structures as halite (NaCl), tetragonal rutile (TiO_2), and cubic $BaTiO_3$ are equal to zero, despite the obvious difference in their structural complexity. However, quasi-entropy can be of much use in investigating the evolution of geometrical complexity along the continuous changes of geometrical parameters, *e.g.* in the course of thermal expansion or compression.

An alternative approach to structural complexity was suggested by Lister *et al.* (2004, 2009), who used the number of atoms in an asymmetric unit as a measure of complexity for a commensurate structure. This method has many advantages for triclinic and monoclinic structures but, unfortunately, does not provide appropriate quantification of the complexity of high-symmetry structures such as paulingite, a complex natural zeolite which was identified by Mackay and Klinowski (1986) as one of the most complex inorganic structures.

Mackay (2001) and later Estevez-Rams and González-Férez (2009) proposed estimating the complexity of crystal structures on the basis of the complexity of the algorithms that can be used to generate them from an unordered array of atoms. In particular, Estevez-Rams and González-Férez (2009) proposed to calculate the Kolmogorov complexity $K\,(s|N)$ of the crystal structure, s, consisting of N atoms according to the following formula:

$$K\,(s|N) = |r^*| + |v^*| + |a^*| \tag{12}$$

where r^* is the set of generators of the space group, v^* are atom coordinates, and a^* is the algorithm used to compute all atomic positions from r^* and v^*. The notion $|s|$ means 'the length of the string s'. In this definition, $K\,(s|N)$ provides a universal measure of the process of generating crystal structures from lattice parameters, space-group generators and atomic coordinates. The measure can be applied to any crystal structure without regard to chemical composition and topology. However, there is no clear way to calculate the $K\,(s|N)$ value; in particular, Estevez-Rams and González-Férez (2009) provided no example of, or a clear procedure for, such a calculation, which makes this parameter non-applicable in its present form. In addition, structural parameters can be determined with different levels of precision, which would result in different amounts of information assigned to the same structure studied by different methods.

Different approaches to structural complexity of crystals were discussed by Steurer (2011). He pointed out the difference between algorithmic, symbolic and combinatorial measures of complexity. Symbolic complexity is "a function of system size, which is a function of the repeat period" for periodic structures, whereas combinatorial complexity "...can best account for higher symmetries of structural subunits (clusters). The less probable (symmetric) a configuration is, the higher is its complexity" (Steurer, 2011). In other words, symbolic complexity is a function of unit-cell volume or the number of atoms in the unit cell, whereas combinatorial complexity is a function of symmetry. Steurer (2011) draws special attention to establishing a universal measure of complexity which could be applied equally to normal periodic structures, incommensurately modulated structures and quasicrystals. He concludes that "no single complexity concept alone is able to reflect all facets of structural complexity".

3.3.2. Information-based measures of structural complexity

One of the approaches to the evaluation of complexity of natural objects is to quantify the amount of information they contain. The more information the object contains, the more complex it is. Krivovichev (2012, 2013a, 2014a) proposed calculating the structural information content of a crystal structure using the modified Shannon formula:

$$I_G = -\sum_{i=1}^{k} p_i \log_2 p_i \qquad \text{(bits/atom)} \tag{13}$$

where k is the number of crystallographic orbits (Wyckoff sites) and p_i is the random choice probability for an atom from the ith crystallographic orbit, that is:

$$p_i = m_i\ v \tag{14}$$

where m_i is a multiplicity of the crystallographic orbit relative to the reduced unit cell, and v is the number of atoms in the reduced unit cell (= number of vertices in the quotient graph).

For instance, the structure of lopezite, $K_2Cr_2O_7$, is triclinic, space group $P\bar{1}$ (Weakley *et al.*, 2004). All atoms in the structure occupy $2i$ sites: there are four K, four Cr and 14 O sites, which corresponds to the total number of 22 equivalence classes. The total number of atoms in the reduced unit cell (which in this case coincides with the crystallographic unit cell) is 44. Each equivalence class has the same random choice probability equal to 2/44 = 0.0455. The structural information content for lopezite is calculated as

$$I_G = -22 \times (2/44) \log_2 2/44 = 4.459 \text{ bits/atom.}$$

The total information content is determined according to the formula

$$I_{G,\text{total}} = v \times I_G \qquad \text{(bits/(unit cell = u.c.))} \tag{15}$$

For lopezite, $I_{G,\text{total}} = 44 \times 4.459 = 196.196$ bits.

The maximal structural information content $I_{G,\max}$ for a crystal structure with v atoms in the unit cell is achieved when all atoms are symmetrically non-equivalent. It can be calculated according to the following formula:

$$I_{G,\max} = \log_2 v \qquad \text{(bits/atom)} \tag{16}$$

In order for I_G to be independent of v, a normalized structural information content, $I_{G,\text{norm}}$, may be defined as

$$I_{G,\text{norm}} = I_G/I_{G,\max} = -\ [\sum_{i=1}^{k} p_i \log_2 p_i]/\log_2 v \tag{17}$$

For the structure of lopezite, $I_{G,\max} = \log_2 44 = 5.459$ bits/atom and $I_{G,\text{norm}} = 4.459/5.459 = 0.817$. The $I_{G,\text{norm}}$ parameter shows the degree of deviation of the crystal structure from triclinic symmetry and therefore is symmetry-sensitive. I_G is sensitive to the symmetry of the structure as well and thus describes its combinatorial complexity, whereas $I_{G,\text{total}}$ combines both the symmetry and the size of the structure and therefore combines combinatorial and symbolic complexities in the context discussed by Steurer (2011).

Another useful measure of structural complexity is the information density, which can be defined as the number of bits of information per cubic Å:

$$\rho_{inf} = I_{G,total}/V \text{ (bits/Å}^3\text{)} \quad (18)$$

For lopezite, $\rho_{inf} = 0.270$ bits/Å^3. For another modification of $K_2Cr_2O_7$ (Krivovichev *et al.*, 2000), $\rho_{inf} = 0.156$ bits/Å^3, which indicates its lower complexity compared to lopezite. It is important to note that information density should be calculated using the volume of the reduced unit cell.

For two structures with similar unit cells, but different symmetry, information density will be higher for the structure with lower symmetry (and higher complexity). For instance, cation ordering within the same unit cell would lead to the formation of a superstructure with greater information density than that of the original structure. Information density can be used as a measure of complexity for crystal structures with similar compositions. The fact that it is a function of both information content and unit-cell volume makes it dependent on the size of atoms and ions that constitute the respective crystal as well as upon temperature and pressure. Suanite, $Mg_2B_2O_5$, and natrosilite, $Na_2Si_2O_5$, have the same total structural information content (114.117 bits), but their information densities are drastically different (0.333 and 0.242 bits/Å^3, respectively), which is an obvious consequence of different ionic radii of the Mg^{2+} and Na^+ ions, on one hand, and of the B^{3+} and Si^{4+} ions, on the other.

It is necessary to distinguish between structural and topological complexity of a crystal structure. Nepheline, $Na_3K(AlSiO_4)_4$, and kalsilite, $KAlSiO_4$, are two important feldspathoid minerals with structures based on a three-dimensional framework of a high-tridymite type stuffed with alkali metal cations. The topological information content of the framework of tetrahedra (which corresponds to its modification with the highest symmetry) is very small (17.510 bits). However, Al-Si ordering, stuffing the framework with Na^+ and K^+ ions, as well as its complex distortions due to tiltings of tetrahedra, lead to the significant increase of the information content. The usual nepheline with the space group $P6_3$ (Foreman and Peacor, 1970) has $I_{G,total} = 193.134$ bits, whereas kalsilite 1*H* (Dollase and Freeborn, 1977) and kalsilite 1*T* (Cellai *et al.*, 1997) register 45.510 and 29.793 bits, respectively. Trikalsilite (Bonaccorsi *et al.*, 1988) and panunzite (a natural analogue of tetrakalsilite; Merlino *et al.*, 1985), are modifications of kalsilite, where a combination of atomic ordering and a complex system of distortions results in a dramatic increase in the $I_{G,total}$ values to 562.942 and 1182.496 bits, respectively. The structural complexity defined by equations 13, 15 and 18 may be essentially greater than topological complexity, which makes the latter more important in the studies of structural topology of minerals. In addition, information-based parameters of structural complexity are dependent on the proper crystal-structure determination, whereas topological information content is a measure of a topological complexity only.

3.3.3. Structural complexity of minerals: statistics and systematics

Krivovichev (2013a) provided statistical data on the complexity of minerals based on 19,531 structure reports extracted from the Inorganic Crystal Structure Database

(ICSD, version 2011-2) and analysed using *TOPOS*. From these, 3949 entries were selected as representatives of particular minerals with known crystal structures. Figure 47 shows the total structural information content of the minerals ($I_{G,\text{total}}$) plotted against the information content per atom (I_G) and the number of atoms in the reduced unit cell (v). The graphs show that neither the symmetry-sensitive I_G, nor the size-sensitive v parameters can be used as single complexity measures, whereas the $I_{G,\text{total}}$ value provides their combination sensitive to both size and symmetry. The data allowed calculation of the average information content for mineral structures as 228(6) bits per structure (unit cell) and 3.23(2) bits per atom. The classification of minerals according to their structural information content or complexity is presented in Table 7. Known minerals can be subdivided into very simple, simple, intermediate, complex and very complex. Most of the rock-forming minerals fall into first three categories. However, several exceptions are known. For instance, vesuvianite, a common mineral in skarns and metasomatic rocks, formed by interactions of ultrabasic and basic rocks with granite intrusions, is regarded as very complex. The structural complexity of vesuvianite is a consequence of its modularity: its structure consists of one-dimensional modules excised from the structure of grossular (Warren and Modell, 1931). Channels between the modules are occupied by a range of cation and anion sites, which results in complex crystal-chemical behaviour (Allen and Burnham, 1992; Gnos and Armbruster, 2006).

Inspection of Table 7 shows that the information-based estimates of complexity agree with the hierarchical complexity of mineral structures. Hierarchically shallow structures such as diamond, copper, galena, fluorite, *etc.* are quantified as very simple, whereas more information-rich arrangements emerge as a result of complexification due to the appearance of additional structure elements (*e.g.* octahedral-tetrahedral layers in alunite or I-beams in diopside). Complex and very complex structures correspond to hierarchically deep and broad arrangements such as those observed in vesuvianite- and eudialyte-group minerals.

Among minerals, the greatest complexity is possessed by zeolites and complex feldspatoids, misfit layered silicates, minerals with nanoscale clusters and tubules, complex sulfates oxysalts, and complex modular structures (for more details see:

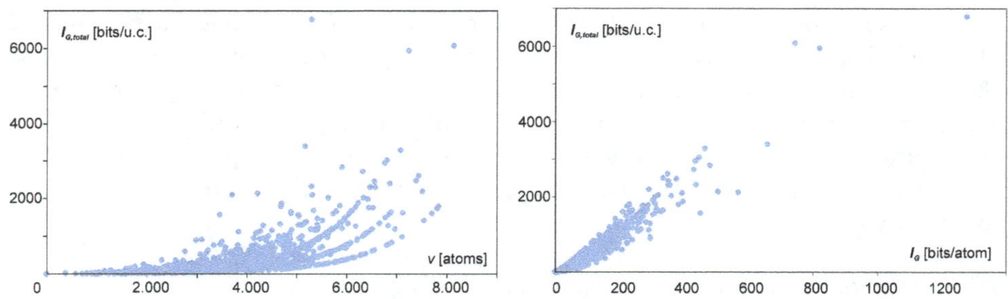

Figure 47. The total structural information content of minerals ($I_{G,\text{total}}$) plotted against the number of atoms in the reduced unit cell (v; above) and the information content per atom (I_G; below).

Table 7. Classification of minerals according to their complexity (after Krivovichev, 2013a).

Category	Total information content (bits)	Approximate number of mineral species	Examples
Very simple	0–20	600	Diamond, copper, halite, galena, uraninite, fluorite, quartz, corundum, ringwoodite, calcite, dolomite, zircon, goethite, lepidocrocite
Simple	20–100	1100	Alunite, jarosite, nepheline, kieserite, szomolnokite, kaolinite, olivine-group minerals, diopside, orthoclase, albite, biotite $1M$
Intermediate	100–500	1800	Enstatite, epidote, biotite $2M_1$, leucite, apatite, natrolite, talc $2M$, pyrope, grossular, beryl, muscovite $2M_1$, staurolite, actinolite, holmquistite, coesite, tourmaline, analcime, boracite
Complex	500–1000	300	Eudialyte, steenstrupine, coquimbite, sapphirine, alum, cymrite, aluminite
Very complex	>1000	100	Vesuvianite, paulingite, bouazzerite, ashcroftine-(Y), bementite, antigorite

Krivovichev, 2013a). The most complex mineral known to date is the zeolite, paulingite, $K_6Ca_{16}(Al_{38}Si_{130}O_{336})(H_2O)_{113}$ (Gordon *et al.*, 1966), which registers 6766.998 bits. Its structure (Fig. 48a,b) is a complex aluminosilicate framework of three different types of secondary building units (Fig. 48c) linked into two interpenetrating **pcu** (primitive cubic) nets (Fig. 48d,e) connected by additional T nodes. The problem of the complexity of paulingite was pointed out by Mackay and Klinowski (1986), and there are different approaches to understanding why the mineral is so complex. From the genetic point of view, the most reasonable explanation of zeolite complexity is in the crystal growth mechanism of zeolites from prenucleation nanoclusters already existing in the crystallization medium (Depla *et al.*, 2011, and references therein). It is of interest that, among zeolites, the paulingite framework, which has the symbolic notation PAU in the *Atlas of Zeolite Structure Types* (Baerlocher *et al.*, 2001), is the second in its topological complexity (4763.456 bits) after the framework SFV reported for the synthetic zeolite SSZ-57 by Baerlocher *et al.* (2011) (19557.629 bits) (see Krivovichev (2013b) for detailed discussion of zeolite complexity).

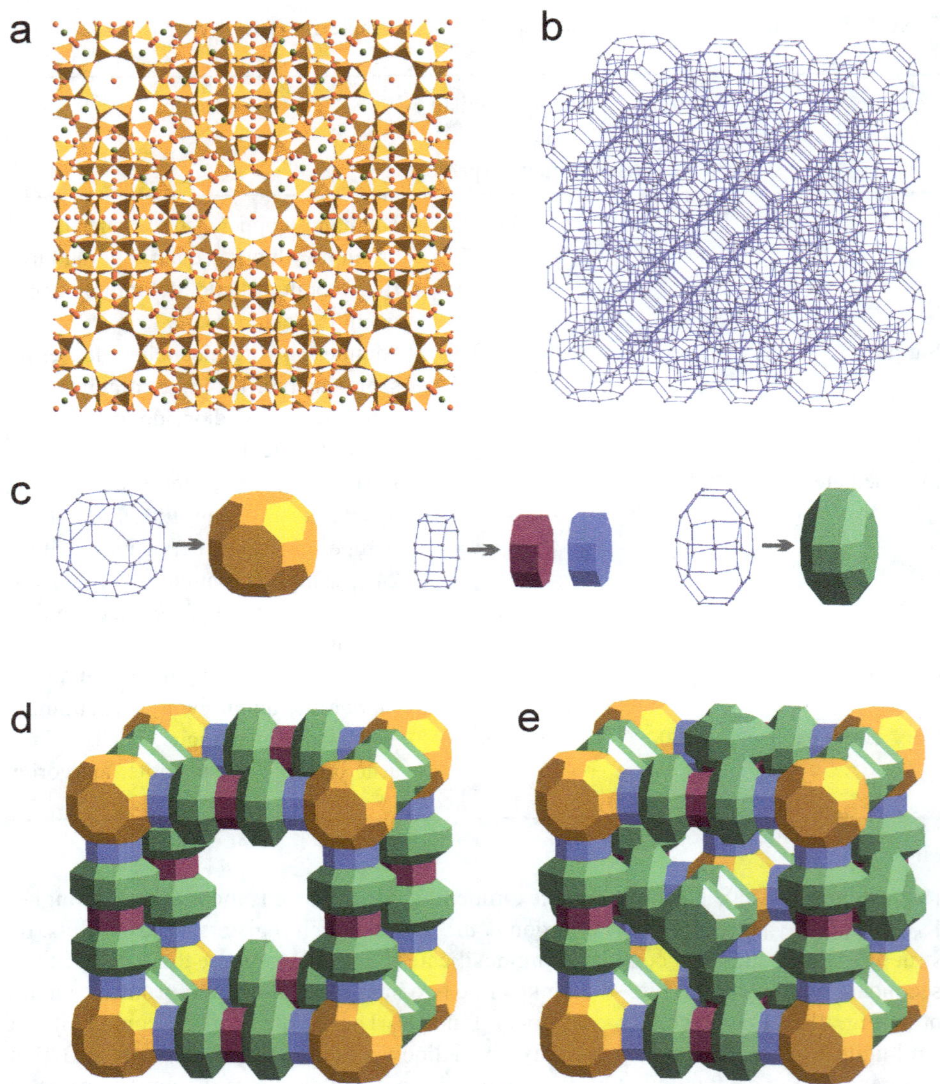

Figure 48. (a) The structure of paulingite viewed along the *a* axis and (b) the nodal representation of tetrahedral framework. The framework can be considered as consisting of cages (c) that share edges to form a three-dimensional framework (d). Two frameworks interpenetrate and are linked by additional nodes (not shown) to form a complete paulingite topology (e).

It is of interest that the most complex inorganic structure reported so far is that of the intermetallic compound $Al_{55.4}Cu_{5.4}Ta_{39.1}$ (*ACT*-71) (Conrad *et al*., 2009; Weber *et al.* 2009). Its total structural information content is equal to 48538.637 bits/u.c., which is about seven times greater than that of paulingite. The reduced unit cell contains v =

5814 atom sites and structural information content per atom (I_G) is 8.349 bits. The exceptional complexity of this structure is the result of a combination of different structural modules, which was described by Weber *et al.* (2009) and Conrad *et al.* (2009) as follows: the $Al_{12}Ta_{28}$ fullerene cluster shells (orange polyhedra in Fig. 49a consisting of 40 faces and 76 vertices) are linked by the Ta_{15} bifrusta (red polyhedra consisting of 12 faces and 15 vertices) into a porous framework with large cavities occupied by supertetrahedral clusters of two types. The first type (Fig. 49b,c) consists

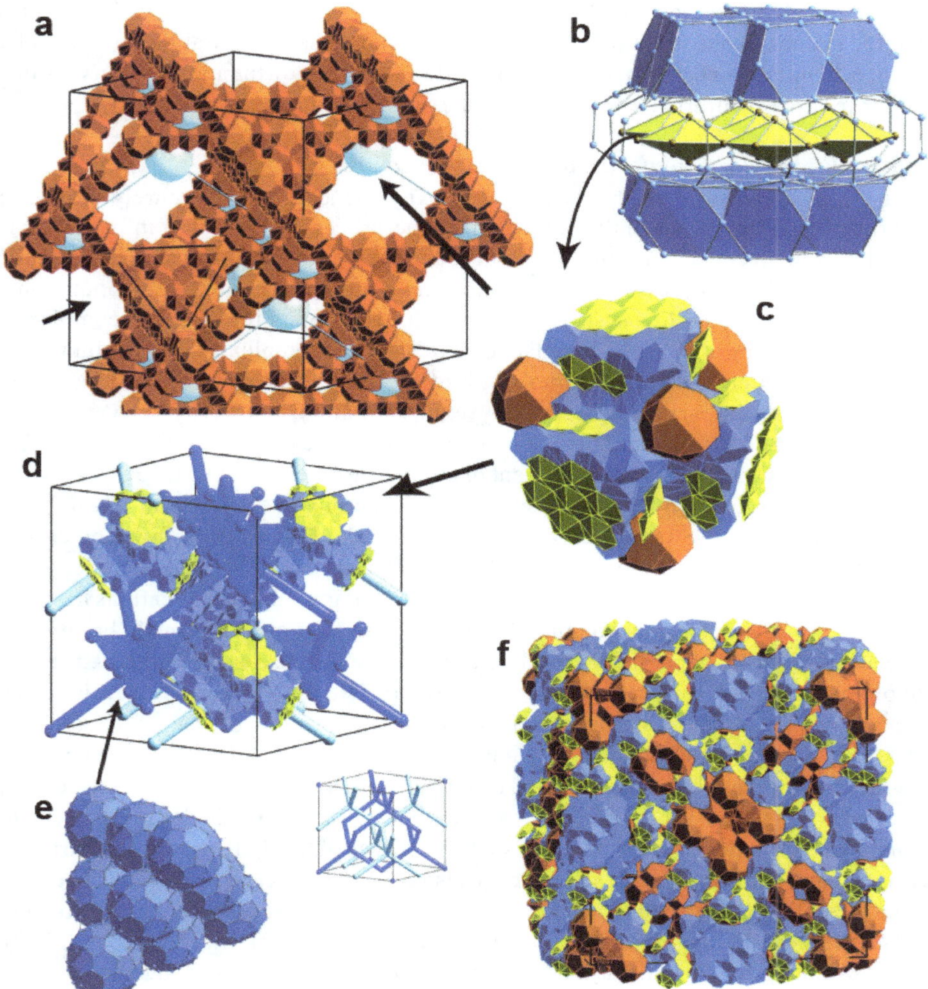

Figure 49. Structural building principle of $Al_{55.4}Cu_{5.4}Ta_{39.1}$ (ACT-71). After Conrad *et al.* (2009). See text for details.

of 146 Ta-centred Al_{12} Friauf polyhedra, Ta_8 hexagonal bipyramids, and $Al_{12}Ta_{28}$ fullerene shells. The second type is formed by ten $Al_{102}Ta_{57}$ fullerenes (Fig. 49e). The supertetrahedral clusters form two interpenetrating diamond-type networks (Fig. 49d) and, together with the fullerene-bifrustum framework (Fig. 49a), form the unit cell of extraordinary complexity (Fig. 49f).

The quantitative evaluation of structural complexity on the basis of Shannon information has some interesting applications regarding the structural evolution of minerals in the course of macro- and micro-scale processes, their formation during metastable crystallization, and the relations between chemical and structural complexity (Krivovichev, 2013a and references therein; Siidra *et al.*, 2014). It seems that the chemical and structural complexity of minerals has a natural limit dictated by the geochemical constraints, on one hand, and the ability of the inorganic crystalline matter to process information encoded in the atomic arrangements.

3.3.4. Algorithmic complexity of mineral structures

Shannon information describes static complexity, which addresses purely structural informational aspects and is independent of the processes by which information is encoded and decoded. Another type of complexity measures describes dynamic complexity, which addresses the question of how much dynamical or computational effort is required to build a system (Ilachiski, 2001). Among the dynamic complexity measures, one of the most popular is algorithmic complexity (or Kolmogorov complexity) which is defined as the length of the shortest programme such as, when executed, yields the particular state of a system. The natural and reality-relevant way to measure algorithmic complexity of a crystal structure is to investigate possible processes of its growth by the attachment of atomic particles to the surface of a growing crystal. In the science of crystal growth, one of the most accepted and widely used models is the Kossel-Stransky or Terrace-Ledge-Kink (TLK) model, in which crystal growth is viewed as happening due to the sequential attachment of cubic particles to the surface of a crystal. This model as well as the model of growth on dislocations has been useful in describing thermodynamic and kinetic aspects of crystal growth, but it is far from the real world in that the real particles are not ideal cubes and the crystal-structure topology is far from an ideal cubic lattice. More realistic models can be developed by the analysis of topologies of real structures and the manner of their generation by preferential attachment of atomic particles or building blocks. Mackay (1976) pointed out that crystal growth can be modelled using finite automata and, in particular, cellular automata. This idea was applied by Krivovichev (2004a) to modelling growth of structures consisting of cubic modules such as sulfides of the djerfisherite and bartonite groups. The cellular automata models have also been used to generate novel framework types in zeolites (Shevchenko and Krivovichev, 2008), to describe growth and disorder of heteropolyhedral layered structures in uranyl selenates (Krivovichev, 2010, 2012), and to discuss possible pathways of self-assembly in the structures of lovozerite-group minerals (Shevchenko *et al.*, 2010). It has been demonstrated that different periodic topologies may be generated by the same automaton depending upon the initial conditions (structure of 'nuclei'). In another report, Krivovichev *et al.* (2012) suggested using finite automata to describe generation of tetrahedral structures in $CaAl_2Si_2O_8$ polymorphs and to use complexity-of-state diagrams to evaluate algorithmic complexity of the resulting structures. This theme was developed further (Krivovichev, 2014b) by the consideration of a wider range of inorganic structures, including minerals, zeolites, inorganic oxysalts, and coordination polymers. It has been demonstrated that a large number of known and widespread topologies can be described as orthogonal networks, which allows construction of their topology-generating finite automata and further analysis of their complexity. The use of finite automata provides the link between orthogonal networks and formal languages, and the complexity of finite automata (or their state diagram) can be used to evaluate algorithmic complexity of the corresponding orthogonal frameworks. We refer the reader to a recent paper by Krivovichev (2014b) for more details on this approach.

4. Concluding remarks

The first hundred years of crystal-structure analysis have resulted in the accummulation of large amounts of data on crystal chemistry and crystal-structure topology of minerals. The future of mineralogy will depend largely on our ability to understand and to incorporate these data into a larger framework of knowledge that links internal crystal-chemical properties of minerals to their formation and behaviour in the course of global and local geological history. This synthetic knowledge is impossible without coherent and deep understanding of the internal atomic- and molecular-scale mechanisms that govern complexity and diversity in the mineral world. In this respect, efforts to interpret the atomic arrangements of minerals from different points of view reviewed in this chapter serve as a starting point for the ongoing research programmes in structural mineralogy and mineralogical crystallography.

Acknowledgements

The writing of this chapter was supported through the grants obtained from the Russian Foundation of Basic Research (#17-05-01027) and the Council of the President of the Russian Federation (NSh-10005.2016.5).

References

Allen, F.M. and Burnham, C.W. (1992) A comprehensive structure-model for vesuvianite: Symmetry variation and crystal growth. *The Canadian Mineralogist*, **30**, 1–18.

Armbruster, T. and Gunter, M. (2001) Crystal structures of natural zeolites. Pp. 1–67 in: *Natural Zeolites: Occurrence, Properties, Applications* (D.L. Bish and D.W. Ming, editors). Reviews in Mineralogy and Geochemistry, **45**. Mineralogical Society of America and Geochemical Society, Chantilly, Virginia, USA.

Baerlocher, Ch., Meier, W.M. and Olson, D.H. (2001) *Atlas of Zeolite Framework Types*. Elsevier, Amsterdam, London, New York, Oxford, Paris, Shannon, Tokyo.

Baerlocher, Ch., Weber, T., McCusker, L.B., Palatinus, L. and Zones, S.I. (2011) Unraveling the perplexing structure of the zeolite SSZ-57. *Science*, **333**, 1134–1137.

Bakakin, V.V. and Belov N.V. (1964) Crystal chemistry of titanates, zirconates, titano- and zirconosilicates. *Geokhimiya*, **1964(2)**, 91–101 (in Russian).

Ballirano, P., Merlino, S., Bonaccorsi, E. and Maras, A. (1996) The crystal structure of liottite, a six-layer member of the cancrinite group. *The Canadian Mineralogist*, **34**, 1021–1030.

Ballirano, P., Bonaccorsi, E., Maras, A. and Merlino, S. (1997) Crystal structure of afghanite, the eight-layer member of the cancrinite group: evidence for long-range Si,Al ordering. *European Journal of Mineralogy*, **9**, 21–30.

Ballirano, P., Bonaccorsi, E., Maras, A. and Merlino, S. (2000) The crystal structure of franzinite, the ten-layer mineral of the cancrinite group. *The Canadian Mineralogist*, **38**, 657–668.

Baur, W.H. (1969) A comparison of the crystal structures of pseudolaueite and laueite. *American Mineralogist*, **54**, 1312–1322.

Baur, W.H. (2014) One hundred years of inorganic crystal chemistry – a personal view. *Crystallography Reviews*, **20**, 64–116.

Baur, W. and Fischer, R.X. (2000) *Microporous and other Framework Materials with Zeolite-Type Structures. Subvol. B. Zeolite Structure Codes ABW to CZP. Landolt-Börnstein. Group IV.* Vol. **14**. Springer-Verlag Berlin Heidelberg, 459 pp.

Baur, W. and Fischer, R.X. (2002) *Microporous and other Framework Materials with Zeolite-Type Structures. Subvol. C. Zeolite-Type Crystal Structures and their Chemistry. Framework Type Codes DAC to LOV. Landolt-Börnstein. Group IV.* Vol. **14**. Springer-Verlag Berlin Heidelberg, 459 pp.

Baur, W. and Fischer, R.X. (2013) Gaps in cubic closest packing: from MgO via spinel to the pharmacosiderite crystal structure type. *Mineralogy and Petrology*, **107**, 153–162.

Baur, W.H., Tillmanns, E.T. and Hofmeister, W. (1983) Topological analysis of crystal structures. *Acta Crystallographica*, **B39**, 669–674.

Belov, N.V. (1947) *Structure of Ionic Crystals and Metallic Phases*. Izd. Akad. Nauk SSSR, Moscow (in Russian).

Bergerhoff, G. and Paeslack, J. (1968) Sauerstoff als Koordinationszentrum in Kristallstrukturen. *Zeitschrift für Kristallographie*, **126**, 112–123.

Bindi, L. and Chapuis, G. (2017) Aperiodic mineral structures. Pp. 213–254 in: *Mineralogical Crystallography* (J. Majzlan and J. Plášil, editors). EMU Notes in Mineralogy, **19**. European Mineralogical Union and Mineralogical Society of Great Britain & Ireland, London.

Blatov, V.A., Shevchenko, A.P. and Serezhkin, V.N. (2000) *TOPOS 3.2* – a new version of the program package for multipurpose crystal-chemical analysis. *Journal of Applied Crystallography*, **33**, 1193.

Bokii, G.B. and Gorogotskaya, L.I. (1969) On the crystal chemical classification of sulfates. *Zhurnal Strukturnoy Khimii*, **10**, 624–632 (in Russian).

Bonaccorsi, E. (2004) The crystal structure of giuseppettite, the 16-layer member of the cancrinite-sodalite group. *Microporous and Mesoporous Materials*, **73**, 129–136.

Bonaccorsi, E. and Merlino, S. (2005) Modular microporous minerals: Cancrinite-davyne group and CSH phases. *Reviews in Mineralogy and Geochemistry*, **57**, 241–290.

Bonaccorsi, E. and Orlandi, P. (2003) Marinellite, a new feldspathoid of the cancrinite-sodalite group. *European Journal of Mineralogy*, **15**, 1019–1027.

Bonaccorsi, E., Merlino, S. and Pasero, M. (1988) Trikalsilite: its structural relationships with nepheline and tetraskalsilite. *Neues Jahrbuch für Mineralogie, Monatshefte*, **1988**, 559–567.

Bonaccorsi, E., Ballirano, P. and Cámara, F. (2012) The crystal structure of sarcofanite, the 74 Å phase of the cancrinite-sodalite supergroup. *Microporous and Mesoporous Materials*, **147**, 318–326.

Borisov, S.V. (1982) Crystal-chemical characteristics of compounds with heavy and high-valent cations. *Journal of Structural Chemistry*, **23**, 414–418.

Borisov, S.V. (1986) Cation sublattices in inorganic compounds. *Journal of Structural Chemistry*, **27**, 486–488.

Borisov, S.V. (1992) Crystalline state. *Journal of Structural Chemistry*, **33**, 871–877.

Borisov, S.V. (1996) Comparative crystal chemistry of heavy metal fluorides and complex niobates and tantalates in terms of a new crystalline state concept. *Journal of Structural Chemistry*, **37**, 773–779.

Bragg, W.L. (1930) The structure of silicates. *Zeitschrift für Kristallographie*, **74**, 237–305.

Bragg, W.L. and Claringbull, G.F. (1965) *Crystal Structures of Minerals*. Bell, London, 409 pp.

Brese, N.E. and O'Keeffe, M. (1991) Bond-valence parameters for solids. *Acta Crystallographica*, **B47**, 192–197.

Brown, I.D. (1981) The bond valence method. An empirical approach to chemical structure and bonding. Pp. 1–30 in: *Structure and Bonding in Crystals*. Vol. 2 (M. O'Keeffe and A. Navrotsky, editors). Academic Press, New York.

Brown, I.D. (2002) *The Chemical Bond in Inorganic Chemistry. The Bond Valence Model*. Oxford University Press, Oxford, UK, 288 pp.

Brown, I.D. (2009) Recent developments in the methods and applications of the bond valence model. *Chemical Reviews*, **109**, 6858–6919.

Brown, I.D. and Shannon, R.D. (1973) Empirical bond-strength – bond-length curves for oxides. *Acta Crystallographica*, **A29**, 266–282.

Bubnova, R.S. and Filatov, S.K. (2013) High-temperature borate crystal chemistry. *Zeitschrift für Kristallographie*, **228**, 395–428.

Burns, P.C. (1999) The crystal chemistry of uranium. Pp. 23–90 in: *Uranium: Mineralogy, Geochemistry, and the Environment* (P.C. Burns and R. Finch, editors). Reviews in Mineralogy, **38**. Mineralogical Society of America, Chantilly, Virginia, USA.

Burns, P.C. (2000) A new uranyl phosphate chain in the structure of parsonsite. *American Mineralogist*, **85**, 801–805.

Burns, P.C. (2005) U^{6+} minerals and inorganic compounds: insights into an expanded structural hierarchy of crystal structures. *The Canadian Mineralogist*, **43**, 1839–1894.

Burns, P.C., Grice, J.D. and Hawthorne, F.C. (1995) Borate minerals. I. Polyhedral clusters and fundamental building blocks. *The Canadian Mineralogist*, **33**, 1131–1151.

Burns, P.C., Miller, M.L. and Ewing, R.C. (1996) U^{6+} minerals and inorganic phases: a comparison and hierarchy of structures. *The Canadian Mineralogist*, **34**, 845–880.

Burns, P.C., Ewing, R.C. and Hawthorne, F.C. (1997) The crystal chemistry of hexavalent uranium: Polyhedral geometries, bond-valence parameters, and polymerization of polyhedra. *The Canadian Mineralogist*, **35**, 1551–1570.

Callebaut, W. and Raskin-Gutman, D. (editors) (2005) *Modularity. Understanding the Development and Evolution of Natural Complex Systems*. The MIT Press, Cambridge, Massachusetts, USA, 459 pp.

Cámara, F., Bellatreccia, F., Della Ventura, G. and Mottana, A. (2005) Farneseite, a new mineral of the cancrinite-sodalite group with a 14-layer stacking sequence: occurrence and crystal structure. *European Journal of Mineralogy*, **17**, 839-846

Cámara, F., Bellatreccia, F., Della Ventura, G., Mottana, A., Bindi, L., Gunter, M.E. and Sebastiani, M. (2010) Fantappièite, a new mineral of the cancrinite-sodalite group with a 33-layer stacking sequence: occurrence and crystal structure. *American Mineralogist*, **95**, 472-480

Cámara, F., Bellatreccia, F., Della Ventura, G., Gunter, M.E., Sebastiani, M. and Cavallo, A. (2012) Kircherite, a new mineral of the cancrinite-sodalite group with a 36-layer stacking sequence: occurrence and crystal structure. *American Mineralogist*, **97**, 1494–1504.

Caro, P. E. (1968) OM_4 tetrahedra and the cationic groups $(MO)^{n+}$ in rare earth oxides and oxysalts. *Journal of Less-Common Metals*, **16**, 367–377.

Cellai, D., Bonazzi, P. and Carpenter, M.A. (1997) Natural kalsilite, $KAlSiO_4$, with $P31c$ symmetry: crystal structure and twinning. *American Mineralogist*, **82**, 276–279.

Chen, F., Ewing, R.C. and Clark, S.B. (1999) The Gibbs free energies and enthalpies of formation of U^{6+} phases: an empirical method of prediction. *American Mineralogist*, **84**, 650–664.

Chermak, J.A. and Rimstidt, J.D. (1989) Estimating the thermodynamic properties (ΔG^0_f and ΔH^0_f) of silicate minerals at 298 K from the sum of polyhedral contributions. *American Mineralogist*, **74**, 1023–1031.

Christ, C.L. and Clark, J.R. (1977) A crystal-chemical classification of borate structures with emphasis on hydrated borates. *Physics and Chemistry of Minerals*, **2**, 59–87.

Conrad, M., Harbrecht, B., Weber, Th., Jung, D.Y. and Steurer, W. (2009) Large, larger, largest – a family of cluster-based tantalum copper aluminides with giant unit cells. II. The cluster structure. *Acta Crystallographica*, **B65**, 318–325.

Cooper, M.A. and Hawthorne, F.C. (1994) The crystal structure of kombatite, $Pb_{14}(VO_4)_2O_9Cl_4$, a complex heteropolyhedral sheet mineral. *American Mineralogist*, **79**, 550–554.

Cooper, M.A. and Hawthorne, F.C. (1996) The crystal structure of rapidcreekite, $Ca_2(SO_4)(CO_3)(H_2O)_4$, and its relation to the structure of gypsum. *The Canadian Mineralogist*, **34**, 99–106.

Depla, A., Verheyen, E., Veyfeyken, A., Van Houteghem, M., Houthoofd, K., Van Speybroeck, V., Waroquier, M., Kirschhock, C.E.A. and Martens, J.A. (2011) UV-Raman and ^{29}Si NMR spectroscopy investigation of the nature of silicate oligomers formed by acid catalyzed hydrolysis and polycondensation of tetramethylorthosilicate. *The Journal of Physical Chemistry C*, **115**, 11077–11088.

Dick, S. (1999) Über die Struktur von synthetischem Tinsleyit $K(Al_2(PO_4)_2OH)(H_2O))\cdot(H_2O)$. *Zeitschrift für Naturforschung*, **B54**, 1385–1390.

Dollase, W.A. and Freeborn, W.P. (1977) The structure of $KAlSiO_4$ with $P6_3mc$ symmetry. *American Mineralogist*, **62**, 336–340.

Dunstetter, F., de Noiffontaine, M.-N. and Courtial, M. (2006) Polymorphism of tricalcium silicate, the major compound of Portland cement clinker. 1. Structural data: review and unified analysis. *Cement and Concrete Research*, **36**, 39–53.

Eby, R.K. and Hawthorne, F.C. (1993) Structural relations in copper minerals. I. Structural hierarchy. *Acta Crystallographica*, **B49**, 28–56.

Effenberger, H. (1985) $Cu_2O(SO_4)$, dolerophanite: refinement of the crystal structure with a comparison of $OCu(II)_4$ tetrahedra in inorganic compounds. *Monatshefte Chemie*, **116**, 927–931.

Engel, P. (1986) *Geometrical Crystallography. An Axiomatic Introduction to Crystallography*. D. Reidel, Dordrecht, The Netherlands.

Ercit, T.S. and Hawthorne, F.C. (1995) Murataite, a UB_{12} derivative structure with condensed Keggin molecules. *The Canadian Mineralogist*, **33**, 1223–1229.

Estevez-Rams, E. and González-Férez, R. (2009) On the concept of long-range order in solids: the use of algorithmic complexity. *Zeitschrift für Kristallographie*, **224**, 179–184.

Fanfani, L., Nunzi, A., Zanazzi, P.F. and Zanzari, A.R. (1975a) The crystal structure of galeite, $Na_{15}(SO_4)_5F_4Cl$. *Mineralogical Magazine*, **40**, 357–361.

Fanfani, L., Nunzi, A., Zanazzi, P.F., Zanzari, A.R. and Sabelli, C. (1975b) The crystal structure of schairerite and its relationship to sulphohalite. *Mineralogical Magazine*, **40**, 131–139.

Fanfani, L., Giuseppetti, G., Tadini, C. and Zanazzi, P.F. (1980) The crystal structure of kogarkoite, Na_3SO_4F. *Mineralogical Magazine*, **43**, 753–759.

Férey, G. (1995) Oxyfluorinated microporous compounds ULM-*n*: chemical parameters, structures and a proposed mechanism for their molecular tectonics. *Journal of Fluorine Chemistry*, **72**, 187–193.

Férey, G. (1998) The new microporous compounds and their design. *Comptes Rendus de l'Acàdemie des Sciences, Series IIc*, **1**, 1–13.

Férey, G. (2001) Microporous solids: from organically templated inorganic skeletons to hybrid frameworks. Ecumenism in chemistry. *Chemistry of Materials*, **13**, 3084–3098.

Ferraris, G., Makovicky, E. and Merlino, S. (2004) *Crystallography of Modular Materials*. Oxford University Press, Oxford, UK.

Filatov, S.K. and Bubnova, R.S. (2000) Borate crystal chemistry. *Physics and Chemistry of Glasses*, **41**, 216–224.

Filatov, S.K., Semenova, T.F. and Vergasova, L.P. (1992) Types of polymerization of $[OCu_4]^{6+}$ tetrahedra in compounds with 'additional' oxygen atoms. *Doklady Akademii Nauk SSSR*, **322**, 536–539 (in Russian).

Finger, L.W., Hazen, R.M. and Fursenko, B.A. (1995) Refinement of the crystal structure of $BaSi_4O_9$ in the benitoite form. *Journal of Physics and Chemistry of Solids*, **56**, 1389–1393.

Fischer, K. (1969) Verfeinerung der Kristallstruktur von Benitoit $BaTi(Si_3O_9)$. *Zeitschrift für Kristallographie*, **129**, 222–243.

Fischer, R.X. and Baur, W. (2006) *Microporous and other Framework Materials with Zeolite-Type Structures. Subvol. D. Zeolite-Type Crystal Structures and their Chemistry. Framework Type Codes LTA to RHO. Landolt-Börnstein. Group IV. Vol.* **14**. Springer-Verlag Berlin Heidelberg, 584 pp.

Fischer, R.X. and Baur, W. (2009) *Microporous and other Framework Materials with Zeolite-Type Structures. Subvol. E. Zeolite-Type Crystal Structures and their Chemistry. Framework Type Codes RON to STI. Landolt-Börnstein. Group IV. Vol.* **14**. Springer-Verlag Berlin Heidelberg, 584 pp.

Fischer, R.X. and Baur, W. (2013) *Microporous and other Framework Materials with Zeolite-Type Structures. Subvol. F. Zeolite-Type Crystal Structures and their Chemistry. Framework Type Codes STO to ZON. Landolt-Börnstein. Group IV. Vol.* **14**. Springer-Verlag Berlin Heidelberg, 311 pp.

Fischer, R.X. and Baur, W. (2014) *Microporous and other Framework Materials with Zeolite-Type Structures. Subvol. G. 41 New Framework Type Codes. Landolt-Börnstein. Group IV. Vol.* **14**. Springer-Verlag Berlin Heidelberg, 427 pp.

Foreman, N. and Peacor, D.R. (1970) Refinement of the nepheline structure at several temperatures. *Zeitschrift für Kristallographie*, **132**, 45–70.

Gnos, E. and Armbruster, T. (2006) Relationship among metamorphic grade, vesuvianite rod polytypism, and vesuvianite composition. *American Mineralogist*, **91**, 862–870.

Gordon, E.K., Samson, S. and Kamb, W.B. (1966) Crystal structure of the zeolite paulingite. *Science*, **154**, 1004–1007.

Grice, J.D., Burns, P.C. and Hawthorne, F.C. (1999) Borate minerals. II. A hierarchy of structures based upon borate fundamental building block. *The Canadian Mineralogist*, **37**, 731–762.

Guan, Y.S., Simonov, V.I. and Belov, N.V. (1963) The crystal structure of bafertisite $BaFe_2TiO[Si_2O_7](OH)_2$. *Doklady Akademii Nauk SSSR*, **149**, 1416–1419.

Hassan, I. and Grundy, H.D. (1991) The crystal structure of basic cancrinite, ideally $Na_8(Al_6Si_6O_{24})(OH)_2 3H_2O$. *The Canadian Mineralogist*, **29**, 377–383.

Hawthorne, F.C. (1983) Graphical enumeration of polyhedral clusters. *Acta Crystallographica*, **A39**, 724–736.

Hawthorne, F.C. (1984) The crystal structure of stenonite and the classification of the aluminofluoride minerals. *The Canadian Mineralogist*, **22**, 245–251.

Hawthorne, F.C. (1985) Towards a structural classification of minerals: the $^{vi}M^{iv}T_2O_n$ minerals. *American Mineralogist*, **70**, 455–473.

Hawthorne, F.C. (1986) Structural hierarchy in $^{iv}M_x^{iii}T_y\phi_z$ minerals. *The Canadian Mineralogist*, **24**, 625–642.

Hawthorne, F.C. (1987) The crystal chemistry of the benitioite group minerals and structural relations in (Si_3O_9) ring structures. *Neues Jahrbuch für Mineralogie, Monatshefte*, **1987**, 16–30.

Hawthorne, F.C. (1990) Structural hierarchy in $M^{[6]}T^{[4]}\phi_n$ minerals. *Zeitschrift für Kristallographie*, **192**, 1–52.

Hawthorne, F.C. (1994) Structural aspects of oxide and oxysalt crystals. *Acta Crystallographica*, **B50**, 481–510.

Hawthorne, F.C. (1998) Structure and chemistry of phosphate minerals. *Mineralogical Magazine*, **62**, 141–164.

Hawthorne, F.C. (editor) (2006) *Landmark Papers: Structure Topology*. The Mineralogical Society of Great Britain & Ireland, London.

Hawthorne, F.C. (2014) The structure hierarchy hypothesis. *Mineralogical Magazine*, **78**, 957–1027.

Hawthorne, F.C. and Huminicki, D.M.C. (2002) The crystal chemistry of beryllium. Pp. 333–403 in: *Beryllium: Mineralogy, Petrology, and Geochemistry* (E.S. Grew, editor). Reviews in Mineralogy and Geochemistry, **50**, Mineralogical Society of America and Geochemical Society, Chantilly, Virginia, USA.

Hawthorne, F.C., Burns, P.C. and Grice, J.D. (1996) The crystal chemistry of boron. Pp. 41–116 in: *Boron: Mineralogy, Petrology, and Geochemistry* (L.M. Anovitz and E.S. Grew, editors). Reviews in Mineralogy, **33**. Mineralogical Society of America, Chantilly, Virginia, USA.

Hawthorne, F.C., Krivovichev, S.V. and Burns, P.C. (2000) Crystal chemistry of sulfate minerals. Pp. 1–112 in: *Sulfate Minerals: Crystallography, Geochemistry, and Environmental Significance* (C.N. Alpers, J.L. Jambor, and D.K. Nordstrom, editors). Reviews in Mineralogy and Geochemistry, **40**. Mineralogical Society of America and Geochemical Society, Chantilly, Virginia, USA.

Hazen, R.M. (1985) Comparative crystal chemistry and the polyhedral approach. Pp. 317–345 in: *Microscopic to Macroscopic: Atomic Environments to Mineral Thermodynamics* (S.W. Kieffer and A. Navrotsky, editors). Reviews in Mineralogy, **14**, Mineralogical Society of America. Chantilly, Virginia, USA..

Hazen, R.M. (1988) A useful fiction: polyhedral modeling of mineral properties. *American Journal of Science*, **288A**, 248–269.

Hippler, K., Sitta, S., Vogt, P. and Sabrowsky, H. (1990) Structure of Na_3OCl. *Acta Crystallographica*, **C46**, 736–738.

Hoppe, R. and Köhler, J. (1988). SCHLEGEL projections and SCHLEGEL diagrams – new ways to describe and discuss solid state compounds. *Zeitschrift für Kristallographie*, **183**, 77–111.

Huminicki, D.M.C. and Hawthorne, F.C. (2002) The crystal chemistry of phosphate minerals. Pp, 123–254 in:

Phosphates (M.L. Kohn, J. Rakovan and J.M. Hughes, editors). Reviews in Mineralogy and Geochemistry, **48**. Mineralogical Society of America and Geochemical Society, Chantilly, Virginia, USA.

Huntelaar, M.E., Cordfunke, E.H.P., van Vlaanderen, P. and Ijdo, D.J.W. (1994) $SrZr(Si_2O_7)$. *Acta Crystallographica*, **C50**, 988–991.

Ilyushin, G.D. (2003) *Modeling of Self-Assembly Processes in Crystal-Forming Systems.* Editorial URSS, Moscow, 376 pp. (in Russian).

Ilyushin, G.D. (2012) Theory of cluster self-organization of crystal-forming systems: geometrical–topological modeling of nanocluster precursors with a hierarchical structure. *Structural Chemistry*, **23**, 997–1043.

Ilyushin, G.D., Voronkov, A.A., Nevsky, N.N., Ilyukhin, V.V. and Belov, N.V. (1981) Crystal structure of hilairite $Na_2ZrSi_3O_9{\cdot}3(H_2O)$. *Soviet Physics Doklady*, **26**, 916–917.

Kampf, A.R., Nash, P. and Loomis, T.A. (2013) Phosphovanadylite-Ca, $Ca[V^{4+}_4P_2O_8(OH)_8{\cdot}12H_2O$, the Ca analogue of phosphovanadylite-Ba. *American Mineralogist*, **98**, 439–443.

Kepler, J. (1611) *Strena Seu de Nive Sexangula.* Gottfried Tampach, Frankfurt, Germany.

Kinrade, S.D., Knight, C.T.G., Pole, D.L. and Syvitski R.T. (1998) Silicon-29 NMR studies of tetraalkylammonium silicate solutions. 1. Equilibria, ^{29}Si chemical shifts, and ^{29}Si relaxation. *Inorganic Chemistry*, **37**, 4272–4277.

Kirby, E.C. (1997) Recent work on toroidal and other exotic fulleren structures. Pp. 263–296 in: *From Chemical Topology to Three-Dimensional Geometry* (A.T. Balaban, editor). Plenum, New York.

Krivovichev, S.V. (1997) On the use of Schlegel diagrams for description and classification of mineral crystal structures. *Zapiski Vserossiiskogo Mineralogicheskogo Obschestva* (*Proceedings of the Russian Mineralogical Society*), **126(2)**, 37–46 (in Russian).

Krivovichev, S.V. (2004a) Crystal structures and cellular automata. *Acta Crystallographica*, **A60**, 257–262.

Krivovichev, S.V. (2004b) Topological and geometrical isomerism in minerals and inorganic compounds with laueite-type heteropolyhedral sheets. *Neues Jahbuch für Mineralogie, Monatshefte*, **2004**, 209–220.

Krivovichev, S.V. (2004c) Combinatorial topology of inorganic oxysalts: 0-, 1- and 2-dimensional units with corner-sharing between coordination polyhedra. *Crystallography Reviews*, **10**, 185–232.

Krivovichev, S.V. (2005) Topology of microporous structures. Pp. 17–68 in: *Micro- and Mesoporous Mineral Phases* (G. Ferraris and S. Merlino, editors). Reviews in Mineralogy and Geochemistry, **57**. Mineralogical Society of America and Geochemical Society, Chantilly, Virginia, USA.

Krivovichev, S.V. (2008) Minerals with antiperovskite structure: A review. *Zeitschrift für Kristallographie*, **223**, 109–113.

Krivovichev, S.V. (2009) *Structural Crystallography of Inorganic Oxysalts.* Oxford University Press, Oxford, UK.

Krivovichev, S.V. (2010) Actinyl compounds with hexavalent elements (S, Cr, Se, Mo) – Structural diversity, nanoscale chemistry, and cellular automata modeling. *European Journal of Inorganic Chemistry*, **2010**, 2594–2603.

Krivovichev, S.V. (2012) Topological complexity of crystal structures: quantitative approach. *Acta Crystallographica*, **A68**, 393–398.

Krivovichev, S.V. (2013a) Structural complexity of minerals: information storage and processing in the mineral world. *Mineralogical Magazine*, **77**, 275–326.

Krivovichev, S.V. (2013b) Structural and topological complexity of zeolites: an information-based approach. *Microporous and Mesoporous Materials*, **171**, 223–229.

Krivovichev, S.V. (2014a) Which inorganic structures are the most complex? *Angewandte Chemie International Edition*, **53**, 654–661.

Krivovichev, S.V. (2014b) On the algorithmic complexity of crystals. *Mineralogical Magazine*, **78**, 415–435.

Krivovichev, S.V. and Brown, I.D. (2001) Are the compressive effects of encapsulation an artefact of the bond valence parameters? *Zeitschrift für Kristallographie*, **216**, 245–247.

Krivovichev, S.V. and Filatov, S.K. (1999) Metal arrays in structural units based on anion-centered metal tetrahedra. *Acta Crystallographica*, **B55**, 664–676.

Krivovichev, S.V., Filatov, S.K. and Semenova, T.F. (1997) On the systematics of polyions of linked polyhedra. *Zeitschrift für Kristallographie*, **212**, 411–417.

Krivovichev, S.V., Kir'yanova, E.V., Filatov, S.K. and Burns, P.C. (2000) β-$K_2Cr_2O_7$. *Acta Crystallographica*,

C56, 629–630.

Krivovichev, S.V., Yakovenchuk, V.N., Ivanyuk, G.Yu., Pakhomovsky, Ya.A., Armbruster, T. and Selivanova, E.A. (2007) The crystal structure of nacaphite, $Na_2Ca(PO_4)F$: A re-investigation. *The Canadian Mineralogist*, **45**, 915–920.

Krivovichev, S.V., Shcherbakova, E.P. and Nishanbaev, T.P. (2012) Crystal structure of svyatoslavite and evolution of complexity during crystallization of the $CaAl_2Si_2O_8$ melt: A structural automata description. *The Canadian Mineralogist*, **50**, 585–592.

Krivovichev, S.V., Mentré, O., Siidra, O.I., Colmont, M. and Filatov, S.K. (2013) Anion-centered tetrahedra in inorganic compounds. *Chemical Reviews*, **113**, 6459–6535.

Lebedev, V.I. (1972) Foundations of new crystal chemistry. *International Geological Review*, **14**, 543–547.

Liebau, F. (1956) Bemerkungen zur Systematik der Kristallstrukturen von Silikaten mit hochkondensierten Anionen. *Zeitschrift für Physikalische Chemie*, **206**, 73–92.

Liebau, F. (1962) Die Systematik der Silikate. *Naturwissenschaften*, **21**, 481–491.

Liebau, F. (1978) Silicates with branched anions: a crystallochemically distinct class. *American Mineralogist*, **63**, 918–923.

Liebau, F. (1980) Classification of silicates. Pp. 1–24 in: *Orthosilicates* (R.G. Burns, editor). Reviews in Mineralogy, **5**. Mineralogical Society of America, Chantilly, Virginia, USA.

Liebau, F. (1985) *Structural Chemistry of Silicates. Structure, Bonding and Classification.* Springer-Verlag, Berlin, 347 pp.

Liebau, F. (2003) Ordered microporous and mesoporous materials with inorganic hosts: definition of terms, formula notation, and systematic classification. *Microporous and Mesoporous Materials*, **58**, 15–72.

Lima-de-Faria, J. (editor) (1994) *Structural Mineralogy. An Introduction.* Kluwer Academic Publishers, Dordrecht, The Netherlands, 353 pp.

Lima-de-Faria, J. (2001) *Structural Classification of Minerals. Vol. 1. Minerals with A, A_mB_n and $A_pB_qC_r$ General Chemical Formulas.* Kluwer Academic Publishers, Dordrecht, The Netherlands, 150 pp.

Lima-de-Faria, J. (2003) *Structural Classification of Minerals. Vol. 2. Minerals with $A_pB_qC_rD_s$ to $A_pB_qC_rD_sE_xF$ General Chemical Formulas.* Kluwer Academic Publishers, Dordrecht, The Netherlands, 152 pp.

Lima-de-Faria, J. (2004) *Structural Classification of Minerals. Vol. 3. Minerals with $A_pB_q...E_xF_y..._nA_q$ General Chemical Formulas.* Kluwer Academic Publishers, Dordrecht, The Netherlands, 112 pp.

Lima-de-Faria, J. (2012) The close packing in the classification of minerals. *European Journal of Mineralogy*, **24**, 163–169.

Lister, S.E., Radoslavljević Evans, I., Howard, J.A.K. Coelho, A. and Evans, J.S.O. (2004) $Mo_2P_4O_{15}$ – the most complex oxide structure solved by single crystal methods? *Chemical Communications*, **2004**, 2540–2541.

Lister, S.E., Radoslavljević Evans, I. and Evans, J.S.O. (2009) Complex superstructures of $Mo_2P_4O_{15}$. *Inorganic Chemistry*, **48**, 9271–9281.

Locock, A.J. and Burns, P.C. (2003a) The crystal structure of bergenite, a new geometrical isomer of the phosphuranylite group. *The Canadian Mineralogist*, **41**, 91–101.

Locock, A.J. and Burns, P.C. (2003b) Structures and syntheses of framework triuranyl diarsenate hydrates. *Journal of Solid State Chemistry*, **176**, 18–26.

Locock, A.J., Burns, P.C. and Flynn, T.M. (2005) The role of water in the structures of synthetic hallimondite, $Pb_2[(UO_2)(AsO_4)_2](H_2O)_n$ and synthetic parsonsite, $Pb_2[(UO_2)(PO_4)_2](H_2O)_n$, $0 < n < 0.5$. *American Mineralogist*, **90**, 240–246.

Loens, J. and Schulz, H. (1967) Strukturverfeinerung von Sodalith, $Na_8Si_6Al_6O_{24}Cl_2$. *Acta Crystallographica*, **23**, 434–436.

Machatschki, F. (1928) Zur Frage der Struktur und Konstitution der Felspate. (Gleichzeitig vorläufige Mitteilung über die Prinzipien des Baues der Silikate. *Centralblatt für Mineralogie, Geologie und Paläontologie. Abteilung A, Mineralogie und Petrographie*, **1928**, 97–104.

Mackay, A.L. (1976) Crystal symmetry. *Physics Bulletin*, **1976**, 495–496.

Mackay, A.L. (2001) On complexity. *Crystallography Reports*, **46**, 524–526.

Mackay, A.L. and Klinowski, J. (1986) Towards a grammar of inorganic structure. *Computers and Mathematics with Applications*, **B12**, 803–824.

Magarill, S.A., Romanenko, G.V., Pervukhina, N.V., Borisov, S.V. and Palchik, N.A. (2000) Oxocentered polycationic complexes – an alternative approach to crystal-chemical investigation of the structure of natural and synthetic mercury oxosalts. *Journal of Structural Chemistry*, **41**, 96–105.

Magnelí, A. (1953) Structures of the ReO_3 type with recurrent dislocations of atoms: 'homologous series' of molybdenum and tungsten oxides. *Acta Crystallographica*, **6**, 495–500.

Makovicky, E. (1997) Modularity – different types and approaches. Pp. 315–343 in: *Modular Aspects of Minerals* (S. Merlino, editor). European Mineralogical Union Notes in Mineralogy. Vol. **1**. Eötvös University Press, Budapest.

Makovicky, E. (2006) Crystal structures of sulfides and other chalcogenides.Pp. 7–125 in: *Sulfide Mineralogy and Geochemistry* (D.J. Vaughan, editor). Reviews in Mineralogy and Geochemistry, **61**, Mineralogical Society of America and Geochemical Society, Chantilly, Virginia, USA.

McCusker, L.B. (2005) IUPAC nomenclature for ordered microporous and mesoporous materials and its application to non-zeolite microporous mineral phase. Pp. 1–16 in: *Micro- and Mesoporous Mineral Phases* (G. Ferraris and S. Merlino, editors). Reviews in Mineralogy and Geochemistry, **57**, Mineralogical Society of America and Geochemical Society, Chantilly, Virginia, USA.

McCusker, L.B., Liebau, F. and Engelhardt, G. (2003) Nomenclature of structural and compositional characteristics of ordered microporous and mesoporous materials with inorganic hosts (IUPAC recommendations 2001). *Microporous and Mesoporous Materials*, **58**, 3–13.

Medrano, M.D., Evans, H.T., Jr., Wenk, H.-R. and Piper, D.Z. (1998) Phosphovanadylite: A new vanadium phosphate mineral with a zeolite-type structure. *American Mineralogist*, **83**, 889–895.

Merlino, S. (editor) (1997) *Modular Aspects of Minerals*. EMU Notes in Mineralogy, **1**. Eötvös University Press, Budapest.

Merlino, S. and Zvyagin, B.B. (1998) Modular features of sapphirine-type structures. *Zeitschrift für Kristallographie*, **213**, 513–521.

Merlino, S., Franco, E., Mattia, C.A., Pasero, M. and de Gennaro, M. (1985) The crystal structure of panunzite (natural tetrakalsilite). *Neues Jahrbuch für Mineralogie, Monatshefte*, **1985**, 322–328.

Mitchell, R.H. (2002) *Perovskites. Modern and Ancient*. Almaz Press, Thunder Bay, Ontario, Canada, 318 pp.

Moore, P.B. (1965) The crystal structure of laueite, $MnFe_2(OH)_2(PO_4)_2(H_2O)_6(H_2O)_2$. *American Mineralogist*, **50**, 1884–1892.

Moore, P.B. (1970a) Structural hierarchies among minerals containing octahedrally coordinating oxygen. I. Stereoisomerism among corner-sharing octahedral and tetrahedral chains. *Neues Jahrbuch für Mineralogie, Monatshefte*, **1970**, 163–173.

Moore, P.B. (1970b) Crystal chemistry of the basic iron phosphates. *American Mineralogist*, **55**, 135–169.

Moore, P.B. (1972) Octahedral tetramer in the crystal structure of leucophosphite, $K_2[Fe_4^{3+}(OH)_2(H_2O)_2(PO_4)_4](H_2O)_2$. *American Mineralogist*, **57**, 397–410.

Moore, P.B. (1974) Structural hierarchies among minerals containing octahedrally coordinating oxygen. II. Systematic retrieval and classification of edge-sharing clusters: an epistemological approach. *Neues Jahrbuch für Mineralogie, Abhandlungen*, **120**, 205–227.

Moore, P.B. (1975) Laueite, pseudolaueite, stewartite and metavauxite: a study in combinatorial polymorphism. *Neues Jahrbuch für Mineralogie, Abhandlungen*, **133**, 148–159.

Moore, P.B. (1994) Foreword. Pp. ix–x in: *Structural Mineralogy. An Introduction* (J. Lima-de-Faria, editor). Kluwer Academic Publishers, Dordrecht, The Netherlands, 353 pp.

Moore, P.B. and Araki, T. (1974) Stewartite, $Mn(II)Fe(III)_2(OH)_2(H_2O)_6(PO_4)_2(H_2O)_2$. Its atomic arrangement. *American Mineralogist*, **59**, 1272–1276.

Náray-Szabó, S. (1930) Ein auf der Kristallstruktur basierendes Silicatsystem. *Zeitschrift für Physikalische Chemie*, **B9**, 356–377.

Oganov, A.R. and Valle, M. (2009) How to quantify energy landscapes of solids. *The Journal of Chemical Physics*, **130**, 104504.

O'Keeffe, M. and Hyde, B.G. (1985) An alternative approach to non-molecular crystal structures with emphasis on the arrangement of cations. *Structure and Bonding*, **61**, 77–144.

Pabst, A. (1934) The crystal structure of sulphohalite. *Zeitschrift für Kristallographie*, **89**, 514–517.

Pabst, A. (1950) A structural classification of flualuminates. *American Mineralogist*, **35**, 149–165.

Pauling, L. (1929) The principles determining the structure of complex ionic crystals. *Journal of the American Chemical Society*, **51**, 1010–1026.

Pauling, L. (1939) *The Nature of the Chemical Bond*. Cornell University Press, Ithaka, New York, 429 pp.

Peister, S.A., Schrader, W. and Schuth, F. (2006) Monitoring temporal evolution of silicate species during hydrolysis and condensation of silicates using mass spectrometry. *Journal of the American Chemical Society*, **128**, 4310–4317.

Pobedimskaya, Ye.A., Terent'eva, L.Ye., Rastsvetaeva, R.K., Sapozhnikov, A.N., Kashaev, A.A. and Dorokhova, G.I. (1991) Crystal structure of bystrite. *Doklady Akademii Nauk SSSR*, **319**, 873–878 (in Russian).

Povarennykh, A.S. (1966) *Crystal Chemical Classification of Minerals* (in Russian). English edition: Plenum Press, New York, London (1972), 2 vols., 766 pp.

Pyatenko, Yu.A., Voronkov, A.A. and Pudovkina Z.V. (1976) *Mineralogical Crystal Chemistry of Titanium*. Nauka, Moscow, 156 pp. (in Russian).

Rao, C.N.R., Natarajan, S., Choudhury, A., Neeraj, S. and Ayi, A.A. (2001) Aufbau principle of complex open-framework structures of metal phosphates with different dimensionalities. *Accounts of Chemical Research*, **34**, 80–87.

Rastsvetaeva, R.K. and Pushcharovsky, D.Yu. (1989) Crystal chemistry of sulfates. *Itogi Nauki I Tekhniki, ser Kristallokhimiya*, **23** (Advances in Science and Technology, ser Crystal Chemistry) (in Russian).

Rastsvetaeva, R.K., Mukhtarova, N.N. and Ilyukhin, V.V. (1983) Crystal chemistry of sulfates. Types of mixed radicals in anhydrous sulfates of trivalent metals. *Mineralogicheskii Zhurnal*, **5(4)**, 14–21 (in Russian).

Roelofsen-Ahl, J.N. and Peterson, R.C. (1989) Gittinsite: a modification of the thortveitite structure. *The Canadian Mineralogist*, **27**, 703–708.

Rozenberg, K.A., Sapozhnikov, A.N., Rastsvetaeva, R.K., Bolotina, N.B. and Kashaev, A.A. (2004) Crystal structure of a new representative of the cancrinite group with a 12-layer stacking sequence of tetrahedral rings. *Crystallography Reports*, **49**, 635–642.

Sabelli, C. and Trosti-Ferroni, R. (1985) A structural classification of sulfate minerals. *Periodico di Mineralogia*, **54**, 1–46.

Salthe, S.N. (2012) Hierarchical structures. *Axiomathes*, **22**, 355–383.

Sandomirskiy, P.A. and Belov, N.V. (1984) *Crystal Chemistry of Mixed Anionic Radicals*. Nauka, Moscow, 205 pp. (in Russian).

Santamaría-Pérez, D. and Liebau, F. (2011) Structural relationships between intermetallic clathrates, porous tectosilicates and clathrate hydrates. *Structure and Bonding*, **138**, 1–29.

Schlosser, G. and Wagner, D.P. (editors) (2004) *Modularity in Development and Evolution*. University of Chicago Press, 601 pp.

Serre, C., Lorentz, C., Taulelle, F. and Férey, G. (2003a) Hydrothermal synthesis of nanoporous metalofluorophosphates. 2. In situ and ex situ ^{19}F and ^{31}P NMR of nano- and mesostructured titanium phosphates crystallogenesis. *Chemistry of Materials*, **15**, 2328–2337.

Serre, S., Taulelle, F. and Férey, G. (2003b) Rational design of porous titanophosphates. *Chemical Communications*, **2003**, 2755–2765.

Shafranovsky, I.I. (1978) *History of Crystallography from the Ancient Times till XIXth Century*. Nauka, Leningrad, 297 pp. (in Russian).

Shevchenko, V.Ya. and Krivovichev, S.V. (2008) Where are genes in paulingite? Mathematical principles of formation of inorganic materials on the atomic level. *Structural Chemistry*, **19**, 571–577.

Shevchenko, V.Ya, Krivovichev, S.V. and Mackay, A. (2010) Cellular automata and local order in the structural chemistry of the lovozerite group minerals. *Glass Physics and Chemistry*, **36**, 1–9.

Siidra, O.I., Krivovichev, S.V. and Filatov, S.K. (2008) Minerals and synthetic Pb(II) compounds with oxocentered tetrahedra: review and classification. *Zeitschrift für Kristallographie*, **223**, 114–126.

Siidra, O.I., Zenko, D.S. and Krivovichev, S.V. (2014) Structural complexity of lead silicates: Crystal structure of $Pb_{21}[Si_7O_{22}]_2[Si_4O_{13}]$ and its comparison to hyttsjöite. *American Mineralogist*, **99**, 817–823.

Simon, H.A. (1962) The architecture of complexity. *Proceedings of the American Philosophical Society*, **106**, 467–482.

Skowron, A. and Brown, I.D. (1994) Crystal chemistry and structures of lead-antimony sulfides. *Acta*

Crystallographica, **B50**, 524–538.

Smirnova, N.L., Akimova, N.V. and Belov, N.V. (1967) On the crystal chemistry of sulfates. *Zhurnal Strukturnoy Khimii*, **8**, 80-84 (in Russian).

Smirnova, N.L., Akimova, N.V. and Belov, N.V. (1968) On the crystal chemistry of sulfates. II. *Zhurnal Strukturnoy Khimii*, **9**, 850-853 (in Russian).

Smith, J.V. (1968) Further discussion of framework structures built from four- and eight-membered rings. *Mineralogical Magazine*, **36**, 640–642.

Smith, J.V. (2000) *Microporous and other Framework Materials with Zeolite-Type Structures. Subvol. A. Tetrahedral Frameworks of Zeolites, Clathrates and Related Materials. Landolt-Börnstein. Group IV. Vol.* **14**. Springer-Verlag Berlin Heidelberg, 459 pp. *Subvol. A. Zeolite Structure Codes ABW to CZP. Landolt-Börnstein. Group IV. Vol.* **14**. Springer-Verlag Berlin Heidelberg, 331 pp.

Smith, J.V. and Rinaldi, F. (1962) Framework structures formed from parallel four- and eight-membered rings. *Mineralogical Magazine*, **33**, 202–212.

Sokolova, E.V., Egorov-Tismenko, Yu.K. and Khomyakov, A.P. (1987) Crystal structure of $Na_{17}Ca_3Mg(Ti,Mn)_4(Si_2O_7)_2(PO_4)_3O_2F_6$, a new representative of the familiy of layered titanium silicates. *Doklady AN SSSR*, **294**, 357–362.

Sokolova, E.V., Rastsvetaeva, R.K., Andrianov, V.I., Egorov-Tismenko, Y.K. and Men'shikov, Y.P. (1989) The crystal structure of a new sodium titanosilicate. *Soviet Physics Doklady*, **34**, 583–585.

Sokolova, E.V., Kabalov, Yu.K., Ferraris, G., Schneider, J. and Khomyakov, A.P. (1999) Modular approach in solving the crystal structure of a synthetic dimorph of nacaphite, $Na_2Ca(PO_4)F$, from powder-diffraction data. *Canadian Mineralogist*, **37**, 83–90.

Steurer, W. (2011) Measures of complexity. *Acta Crystallographica*, **A67**, C184.

Strunz, H. and Nickel, E. (2001) *Strunz Mineralogical Tables. Chemical-Structural Mineral Classification System.* 9th edition. Schweizerbart Science Publishers, Stuttgart, Germany, 870 pp.

Thompson, J.B., Jr. (1970) Geometrical possibilities for amphibole structures: model biopyriboles. *American Mineralogist*, **55**, 292–293.

Thompson, J.B., Jr. (1978) Biopyriboles and polysomatic series. *American Mineralogist*, **63**, 239–249.

Thompson, R.M. and Downs, R.T. (2001) Quantifying distortion from ideal closest-packing in a crystal structure with analysis and application. *Acta Crystallographica*, **B57**, 119–127.

Thompson, R.M. and Downs, R.T. (2010) Packing systematics of the silica polymorphs: The role played by oxygen-oxygen nonbonded interactions in the compression of quartz. *American Mineralogist*, **95**, 104–111.

Thompson, R.M., Yang, H. and Downs, R.T. (2012) Packing systematics and structural relationships of the new copper molybdate markascherite and related minerals. *American Mineralogist*, **97**, 1977–1986.

Urusov, V.S. and Orlov, A.P. (1999) State-of-art and perspectives of the bond-valence model in inorganic crystal chemistry. *Crystallography Reports*, **44**, 686–709.

Vegas, A. (2000) Cations in inorganic solids. *Crystallography Reviews*, **7**, 189–283.

Vegas, A. (editor) (2011) *Inorganic 3D Structures. The Extended Zintl-Klemm Concept. Structure and Bonding, Vol.* **138**. Springer-Verlag, Berlin Heidelberg, 201 pp.

Vegas, A., Santamaría-Pérez, D., Marqués, M., Flórez, M., Baonzac, V.G. and Recio, J.M. (2006) Anions in metallic matrices model: application to the aluminium crystal chemistry. *Acta Crystallographica*, **B62**, 220–227.

Voronkov, A.A., Syzova, R.G., Ilyukhin, V.V. and Belov, N.V. (1973) Crystal chemistry of mixed anionic frameworks. I. Principles of their formation. *Kristallografiya*, **18**, 112–121.

Voronkov, A.A., Ilyukhin, V.V. and Belov, N.V. (1974) Principles governing the formation of mixed frameworks and their formula. *Doklady AN SSSR*, **219**, 600–603.

Voronkov, A.A., Ilyukhin, V.V. and Belov, N.V. (1975) Crystal chemistry of mixed frameworks. Principles of their formation. *Kristallografiya*, **20**, 556–566.

Voronkov, A.A., Rastsvetaeva, R.K., Ilyukhin, V.V. and Mukhtarova, N.N. (1983) Crystal chemistry of sulfates. Sulfates of trivalent elements in the light of the concept of mixed radicals. *Mineralogicheskii Zhurnal*, **5(2)**, 39–47.

Warren, B.E. and Modell, D.I. (1931) The structure of vesuvianite $Ca_{10}Al_4(Mg,Fe)_2Si_9O_{34}(OH)_4$. *Zeitschrift fuer Kristallographie*, **78**, 422–432.

Weakley, T.J.R., Ylvisaker, E.R., Yager, R.J., Stephens, J.E., Wiegel, R.D., Mengis, M., Wu, P., Photinos, P. and Abrahams, S.C. (2004) Phase transitions in $K_2Cr_2O_7$ and structural redetermination of phase II. *Acta Crystallographica*, **B60**, 705–715.

Weber, Th., Dshemuchadse, J., Kobas, M., Conrad, M., Harbrecht, B. and Steurer, W. (2009) Large, larger, largest – a family of cluster-based tantalum copper aluminides with giant unit cells. I. Structure solution and refinement. *Acta Crystallographica*, **B65**, 308–317.

Welch, M., Cooper, M.A., Hawthorne, F.C. and Criddle, A.J. (2000) Symesite, $Pb_{10}(SO_4)O_7Cl_4(H_2O)$, a new PbO-related sheet mineral: description and crystal structure. *American Mineralogist*, **85**, 1526–1533.

Wells, A.F. (1954) The geometrical basis of crystal chemistry. Part 1. *Acta Crystallographica*, **7**, 535–544.

Wells, A.F. (1970) *Models in Structural Inorganic Chemistry*. Clarendon Press, Oxford, UK, 186 pp.

Wondratchek, H., Merker, L. and Schubert, K. (1964) Beziehungen zwischen der Apatit-Struktur und der Struktur der Verbindungen vom Mn_5Si_3-(D88)Typ. *Zeitschrift für Kristallographie*, **120**, 393–395.

Yakovenchuk, V.N., Nikolaev, A.P., Selivanova, E.A., Pakhomovsky, Y.A., Korchak, J.A., Spiridonova, D.V., Zalkind, O.A. and Krivovichev, S.V. (2009) Ivanyukite-Na-T, ivanyukite-Na-C, ivanyukite-K, and ivanyukite-Cu: New microporous titanosilicates from the khibiny massif (Kola peninsula, Russia) and crystal structure of ivanyukite-Na-T. *American Mineralogist*, **94**, 1450–1458.

Yakubovich, O.V. and Dadachov, M.S. (1992) Synthesis and crystal structure of ammonium analog of leucophosphite, $NH_4\{Fe_2(PO_4)_2(OH)(H_2O)\}(H_2O)$. *Soviet Physics Crystallography*, **37**, 757–760.

Zoltai, T. (1960) Classification of silicates and other minerals with tetrahedral structures. *American Mineralogist*, **45**, 960–973.

Zvyagin, B.B. and Merlino, S. (2003) The pyroxene-spinel polysomatic system. *Zeitschrift für Kristallographie*, **218**, 210–220.

EMU Notes in Mineralogy, Vol. 19 (2017), Chapter 2, 79–138

Methods of crystallography: powder X-ray diffraction

ANGELA ALTOMARE[1,*], CORRADO CUOCCI[1], G. DIEGO GATTA[2], ANNA MOLITERNI[1] and ROSANNA RIZZI[1]

[1]*Istituto di Cristallografia, Sede di Bari, Via G. Amendola 122/o, Bari, Italy, e-mail: angela.altomare@ic.cnr.it*
[2]*Dipartimento di Scienze della Terra, Università degli Studi di Milano, Via Botticelli 23, Milan, Italy*

In the last twenty-five years, relevant theoretical, methodological and experimental advances have been made in the development and application of the X-ray powder diffraction (XRPD) method. In particular, attention has been devoted to the interpretation of XRPD data. The XRPD approach is used currently in mineralogical as well as in many other scientific fields (solid-state chemistry, pharmacology, materials science, *etc.*) to address a wide range of scientific purposes: qualitative analysis for the identification of the crystalline phases constituting a powder sample; quantitative analysis for estimating the weight fraction of each phase in a mixture; structure solution; microstructural analysis for the inspection of crystalline domain size effects and lattice defects; investigation of highly complex materials: compounds with incommensurate structures, nanoparticles, amorphous materials; studies at non-ambient conditions, *in situ*, time-resolved and *in operando* for the description of thermal or compressional behaviour, phase stability and structural evolution.

The aim of this chapter is to provide an overview of some basic principles and significant aspects of the XRPD method and examples of its applications to mineralogical problems.

1. Introduction

X-ray diffraction is the most appropriate technique for investigating crystalline substances. Crystalline materials, natural or synthetic, are not always available as single crystals of quality and/or dimensions suitable for study by single-crystal diffraction techniques. Often, crystalline materials are available as fine-grained powder. In this case, the X-ray powder diffraction (XRPD) technique is the best alternative to single-crystal diffraction (Dinnebier and Billinge, 2008). However, single-crystal and powder-diffraction techniques are complementary, with different domains of applicability.

The powder diffraction method was developed as early as 1916 by Debye and Scherrer. For nearly half a century, it was devoted to qualitative, semi-quantitative and macroscopic stress measurement studies only. In 1969, Hugo Rietveld proposed an interesting method to refine a structure model using the whole powder diffraction profile from neutron data (Rietveld, 1969). Since then, the interest in powder diffraction has increased considerably.

DOI: 10.1180/EMU-notes.19.3

In the last twenty five years, XRPD has been commonly and widely used for solving crystal structures. Experimental, methodological and theoretical advances have tackled successfully the difficulties of powder characterization, so that the number of crystal structures solved by XRPD is growing. Moreover, studies at non-ambient conditions (with temperature or pressure variable), *in situ*, time-resolved and *in operando*, unthinkable a few decades back, are executed currently for monitoring the structural evolution of crystalline samples. A decisive step forward came also with the advent of high-speed modern computers and the use of highly collimated and very intense synchrotron radiation, available nowadays.

XRPD is widely used in mineralogical as well as in many other scientific fields (solid-state chemistry, pharmaceutical, materials science, *etc.*) for the following main purposes: qualitative analysis, in order to identify the crystalline species, often called 'phases', constituting a powder (see Section 3); quantitative analysis, which estimates the weight fraction of each phase in a mixture (see Section 3); structure solution, to determine the crystal structure at atomic level (see Section 4); study of microstructural properties, such as crystalline domain size effects and lattice defects (for this subject the reader is referred to specialized papers, *e.g.* Popa, 2008; Scardi, 2008).

A powder is formed ideally by many microcrystallites (usually in the $1-10$ μm size range) and randomly oriented. As a consequence, the powder diffraction pattern consists of a one-dimensional profile formed by a series of peaks whose intensities reflect the atomic arrangement. The interpretation of the profile is not straightforward (see Section 2). In particular, in the case of structure solution, the task is arduous: the three-dimensional crystal structure must be recovered starting from the one-dimensional pattern. The difficulties are ascribed mainly to the following problems (see Section 2): diffraction peak overlap (overlapping, depending mainly on structure complexity and peak experimental resolution, increases with the θ angle); it is often difficult to determine background noise correctly; powder microcrystallites may align along preferred orientations, causing alteration in the registered peak intensities. These problems hinder the accurate interpretation of the pattern and, consequently, the success of the investigation. Note that several software packages have been developed and improved significantly over the last decade (by graphical and computing tools) in order to perform powder characterization in a short time.

In spite of the recent progress, structure solution by XRPD is often a challenge. For single-crystal diffraction studies, small and medium size structures (up to 300 non-hydrogen atoms in the asymmetric unit), as well as small proteins (assuming that high-quality data are available to ~1.1 Å in resolution), can be solved routinely. In contrast, in the case of powder data, even simple crystal structures (*i.e.* <30 non-hydrogen atoms in the asymmetric unit) can be unsolvable. Efforts are being focused on reducing this gap. At the same time, it must be mentioned that the single-crystal technique does not give information on the bulk material and neither is it used routinely for microstructural studies.

XRPD is currently used for studying highly complex materials: compounds with incommensurate structures, nanoparticles, amorphous materials, *etc.* For these materials, the 'total scattering' including Bragg and diffuse scattering (*i.e.* scattering from not fully ordered atoms) must be considered. The atomic pair distribution function (PDF) theory has been developed successfully and applied to these problems (*e.g.* Billinge, 2008 and references therein).

This chapter aims to provide: the basic practical principles of the powder diffraction technique (Section 2); the approaches for qualitative and quantitative analysis, with particular attention to the most recent developments (Section 3); the methods for carrying out the full pathway of the solution process, starting from the determination of crystal cell parameters up to the structure model refinement by the Rietveld method (Section 4); basic concepts of mineral physics, techniques and instrumentation in order to describe elastic behaviour and the main deformation mechanisms (at the atomic level) at non-ambient pressure or temperature (Section 5). Examples of application of the XRPD technique to mineralogical problems will be presented with reference to the computing programs used most widely in powder crystallography.

Some concepts, fundamental in crystallography, will be used in the following paragraphs. For a more in-depth treatment the reader is referred to Giacovazzo (2011).

2. Powder XRD technique

Diffraction theory asserts that a single crystal, stricken by an incoming X-ray beam, diffracts X-rays at an angle θ (the Bragg angle) depending on the distance between the crystallographic planes. According to the Ewald construction, in order to bring different reciprocal lattice points on to the Ewald surface and to detect the diffraction intensity the orientation of the crystal must be changed. Every time the Ewald sphere intersects a reciprocal space point, the Bragg law is verified and a diffraction 'spot' results.

2.1. The powder diffraction pattern

The random orientation of the crystallites constituting the powder does not require the rotation of the sample for diffraction to occur, as is needed in the single crystal case. If we consider the Ewald sphere, the vector $\boldsymbol{r}_{\mathbf{h}}^*$ of each randomly oriented crystallite forms a sphere of vectors radiating outward from the origin of the reciprocal lattice (the number of vectors is equal to the number of crystallites). Of all the $\boldsymbol{r}_{\mathbf{h}}^*$ vectors, only those which intersect the Ewald sphere (they form a circle) are able to diffract, as shown in Fig. 1a in a two-dimensional projection.

As a consequence, a cone of diffraction (the Debye cone) subtending its Bragg angle corresponds to each point in the reciprocal lattice. The intersection of these cones with a two-dimensional detector (*e.g.* film, image plate or CCD) placed normal to the beam produces a set of concentric circles called Debye rings (see Fig. 1b). The radius of each ring is proportional to 2θ; different rings have different intensities. The uniform distribution of the circles is correlated to the random distribution of crystallites (this condition is not fulfilled if preferred orientation is present; see Section 2.2)

In order that the difference from the single-crystal case may be appreciated, in Fig. 2 we show the overlay of the two diffraction images from a single crystal (the black spots) and from powder (the circles).

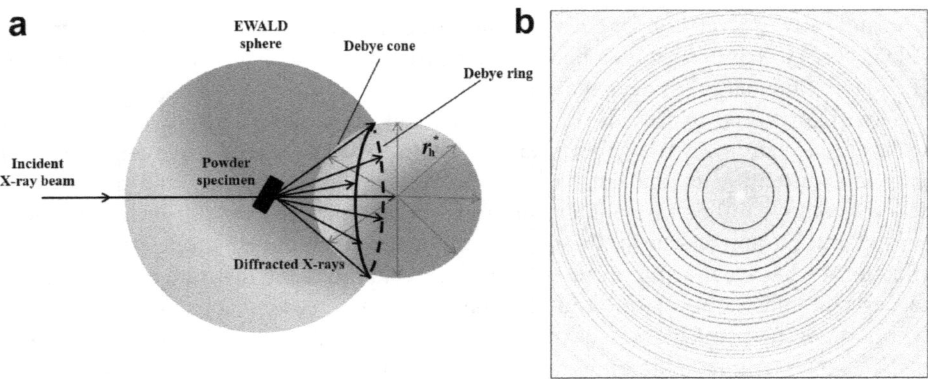

Figure 1. Debye cones and rings.

Whereas for single crystals the experimental diffraction information is three-dimensional (a spot for every $\boldsymbol{r}_h^*$), for powders the orientation of each $\boldsymbol{r}_h^*$ is lost and the vector space is reduced to one dimension of the modulus of $\boldsymbol{r}_h^*$. This means that the diffraction rings can be projected into the one-dimensional diffraction profile (intensity *vs*. 2θ) consisting of peaks the intensities of which decrease usually with increasing 2θ, superimposed on a usually non-linear background. The decreasing intensity in the diffraction profile is due to the X-ray scattering factor decay with sinθ/λ. The most relevant difference between the powder diffraction technique and the single-crystal one is, therefore, the collapse of information on to one dimension in a powder pattern. This might cause problems in the interpretation of the experimental data (see Section 2.2). In Fig. 3 a typical powder diffraction profile is shown.

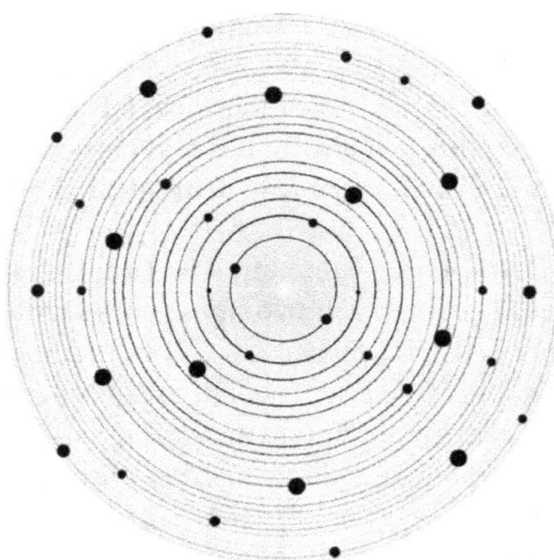

Figure 2. Single-crystal spots and powder rings of the same crystalline material.

Powder diffraction data are collected by scanning a 2θ range in a step mode and amount of time per step or in a continuous mode. They consist of several datapoints: an intensity count Y_i and a $2\theta_i$ angle value corresponding to each *i*-th datapoint. Powder diffraction experiments must be carried out by adopting suitable experimental conditions which best suit the problem to be investigated and are able to maximize the recovery of experimental information. A large variety of experimental setups is available depending on: type of radiation (*e.g.* X-rays or neutrons), source of X-ray radiation (*e.g.* conventional or synchrotron), characteristics of

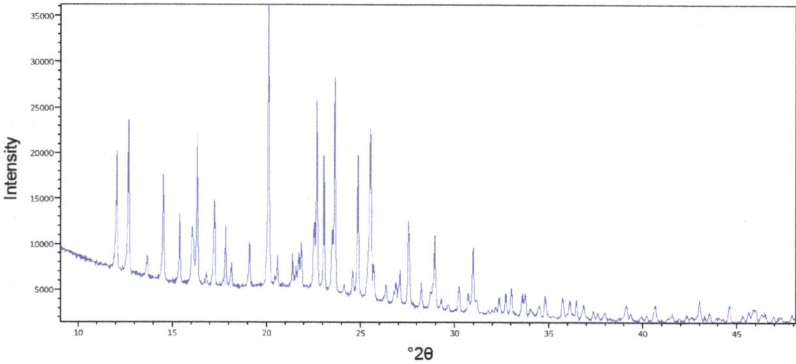

Figure 3. A typical X-ray powder diffraction profile.

the radiation (*e.g.* pseudo-monochromatic for angular dispersive mode or white beam for energy dispersive mode), wavelength, instrument geometry (*e.g.* reflection or transmission), data collection (*e.g.* 2θ range, step size, counting time), detectors (Cockcroft and Fitch, 2008).

The 'structure' of a typical powder diffraction pattern may be described by the following features of the Bragg peaks: (1) position; (2) intensity; (3) shape and width. Numerous factors play a central or secondary role in determining the peak components and they can be grouped as: (1) structural factors, which depend on the crystal structure; (2) sample factors, determined by shape and size of the sample, preferred orientation, grain size, microstructure, *etc.*; and (3) instrumental factors, such as the properties of radiation and of detector, type of focusing geometry, slit and/or monochromator geometry.

Let us describe in more detail the characteristic components of a Bragg peak:

(1) *Peak position.* This is determined by the size, shape and symmetry of the unit cell and then, once experimentally evaluated, it can be used to determine the unknown unit-cell parameters $\{a,b,c,\alpha,\beta,\gamma\}$ *via* an indexing approach (see Section 4.2). In reality, some instrumental variables (*e.g.* axial divergence of the incident beam, sample displacement from the goniometer axis, 2θ zero-shift error, *etc.*) and specimen features can distort the peak positions with respect to the specific 2θ angle values satisfying Bragg's law. In this case, it is convenient to correct the peak positions before their use, *e.g.* by an appropriate calibration of the instrument by using an internal standard. If these effects are not suitably taken into account, peak positions are determined approximately with serious consequences in the powder pattern interpretation.

(2) *Peak intensity.* In the definition of the intensities of the individual diffraction peaks, the sample and instrumental factors can be considered less critical than structural ones. In fact, the intensities depend on periodicity of the unit cell and on the unit-cell content, along with the arrangement of atoms within the cell.

In a powder diffraction experiment what is really measured is the so-called profile intensity (the quantity Y_i in Fig. 4), which represents the intensity at different $2\theta_i$ angle values of the pattern. Looking at the Bragg peak illustrated in Fig. 4, the value of the peak maximum ($Y_{\max}$) is intuitive and often indicated as peak intensity. It can be measured easily and used in many applications where relative intensities are required, *e.g.* for qualitative phase analysis (see Section 3). This way of measuring the intensity is, however, inadequate when quantitative estimates are required. In these cases, the intensity $I_{\mathbf{h}}$ corresponding to the reflection **h** is calculated as the area under the peak (shaded area in Fig. 4) and is known as 'integrated intensity'. It represents the intensity of a Bragg peak in powder diffraction.
Due to peak overlap (see next paragraph), the intensity value Y_i at $2\theta_i$ contains information about more than one independent observation: *i.e.* the contributions coming from the overlapped individual peaks at $2\theta_i$. To recover the individual contributions means it is necessary to decompose, *via* profile fitting, the full powder pattern into single peaks (see Section 4.4). As stated above, $I_{\mathbf{h}}$ is a function of the atomic structure and of instrumental parameters. The $I_{\mathbf{h}}$ value associated with diffraction line **h**, is given by the following general expression:

$$I_{\mathbf{h}} = \mathrm{K} m_{\mathbf{h}} L_{\mathbf{h}} P_{\mathbf{h}} A E_{\mathbf{h}} O_{\mathbf{h}} |F_{\mathbf{h}}|^2, \qquad (1)$$

where:

- K is a constant;
- $m_{\mathbf{h}}$ is the multiplicity factor, *i.e.* the number of symmetrically equivalent reflections;
- $L_{\mathbf{h}}$ is the Lorentz factor, defined by the geometry of diffraction;
- $P_{\mathbf{h}}$ is the polarization factor; it accounts for a partial polarization of the scattered electromagnetic wave;

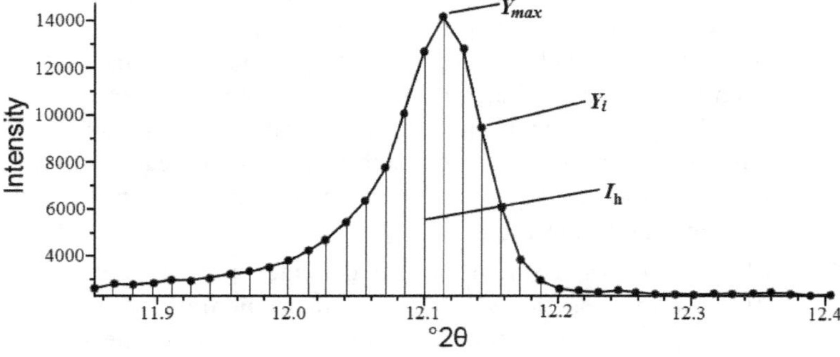

Figure 4. For a Bragg peak: Y_i is the profile intensity in the i-th datapoint, $Y_{\max}$ represents the value of the peak maximum and $I_{\mathbf{h}}$ is the integrated intensity.

- A is the absorption factor, which determines absorption of both the incident and diffracted beams;
- $E_{\mathbf{h}}$ is the extinction factor, which accounts for deviations from the kinematical diffraction model;
- $O_{\mathbf{h}}$ is the preferred orientation factor corresponding to an incompletely random distribution of the crystallites in the powder compound;
- $|F_{\mathbf{h}}|$ is the amplitude of the structure factor.

The expression of $I_{\mathbf{h}}$ as reported in equation 1, is valid for both of the experimental geometries used in powder diffraction data collection: (1) flat plate, also called reflection or Bragg-Brentano geometry, and (2) capillary or transmission or Debye-Sherrer geometry.

(3) *Peak shape and width.* Because the Ewald sphere has finite thickness (due to wavelength aberrations) and reciprocal lattice points are far from being point-like, the Bragg peaks always have non-zero widths as a function of 2θ: *i.e.* for a given $d_{\mathbf{h}}$, the peak is expanded around the $2\theta_{\mathbf{h}}$ position which fulfills the Bragg condition. This enlargement is taken into account in the shape functions in terms of the Full Width at Half Maximum (FWHM) of the peak; it varies with 2θ and its dependence on the Bragg angle is commonly represented by the Caglioti formula (Caglioti *et al.*, 1958):

$$\mathrm{FWHM} = \sqrt{U \tan^2\theta + V \tan\theta + W} \tag{2}$$

where U, V and W are refinable variables.

The peak shape function can be considered as the convolution of various individual functions related to the instrumental parameters (*e.g.* optical aberrations, de-focusing effects, characteristics of the primary beam, *etc.*) and to the features of the sample (*e.g.* crystallite size and strain, lattice defects). The analytical description of the peak shape, after the instrumental contribution has been subtracted, leads to microstructure information in terms of crystallite size and/or lattice strain. In particular, the instrument-subtracted peak-broadening is inversely proportional to the crystallite size (Scherrer, 1918) and directly proportional to the lattice strain.

The most commonly used peak-shape functions are: Gaussian, Lorentzian, Pseudo-Voigt and Pearson-VII. The first two are the simplest and their distributions differ mainly in the tails: the Lorentzian function is sharp near its maximum, but has long tails on each side near its base while the Gaussian distribution has no long tails at the base, but has a rounded maximum. For XRD in particular, simple Gaussian or Lorentzian distributions generally do not describe well the real observed peak shape which usually ranges from the Gaussian to the Lorentzian limits, *i.e.* the shape can be described better as the mixture of the two distributions. For this purpose, the Pseudo-Voigt, which is the weighted mean between a Lorentzian and a Gaussian, and the Pearson-VII, of which Lorentzian and Gaussian distributions are special cases, are used in general.

The peak-shape functions are symmetrical with respect to a vertical line intersecting the peak maximum while, mainly due to the axial divergence of the X-ray beam, the peaks in the powder pattern can be asymmetrical. The asymmetry occurs usually in the small-angle side of a Bragg peak, which is considerably broader than its large-angle side. In addition, peak asymmetry is usually prominently visible at small Bragg angles ($2\theta < 20-30^\circ$) while it disappears at $2\theta \approx 90^\circ$. Although the peak asymmetry cannot be eliminated completely, a proper configuration and alignment of the instrument can reduce it substantially. It must be stressed that an accurate peak description is highly desirable for any kind of investigation by powder pattern.

Finally, one more observable can be singled out in a powder diffraction pattern: the background. It describes the contribution arising from instrumental factors such as: scattering from the atmosphere, non-crystalline sample holder, detector noise, incomplete monochromatic nature of the incident beam, *etc*., and from sample factors: inelastic scattering, X-ray fluorescence, amorphous components in the sample, *etc*. Although the background may be used to extract information about the crystallinity of the sample, in the majority of powder diffraction applications it is an inconvenience which must be dealt with (see next paragraph). Carefully designed experimental setups can suitably modify the features of the Bragg peaks.

2.2. Problems in the XRPD technique

2.2.1. Peak overlap

Due to the intrinsic one-dimensionality of a powder diffraction pattern, peaks corresponding to different diffraction planes with similar inter-planar distances ($d_{\mathbf{h}}$) are commonly overlapped. This effect may be accidental or may be the unavoidable consequence of the symmetry.

In recent decades, considerable improvements in experimental powder diffraction techniques have resulted in the availability of exceptionally precise and high-resolution data, especially when synchrotron radiation sources are employed. To this, innovative methodologies have been added, thus making it possible to recover, from a powder diffraction pattern, a sufficient amount of information to perform studies on structure and properties of crystalline materials of great complexity. Despite that, accuracy in the estimation of the integrated intensities from the powder pattern can be very poor due mainly to the unavoidable peak overlapping which is a consequence of the collapse of the three dimensions of reciprocal space onto the one dimension of the powder diffraction pattern.

The peak overlap implies that the number of visible peaks in a powder pattern is less than the number of Bragg peaks and peaks in strong overlap cannot be considered as statistically independent (Pawley, 1981; Sivia and David, 1994). As a consequence, the determination of the number of independent peaks, N_{ind}, even if difficult to assess, can be useful in providing information about the quality of the data and the possibility of successfully solving a crystal structure from powder data.

Some approaches have been developed to deal with the problem: Altomare and co-workers, proposed a geometrical algorithm to transform the correlation among peak intensities of N reflections into $N_{ind} \leqslant N$ statistically independent observations (Altomare *et al.*, 1995). David (1999) approached the problem by deriving the number of statistically independent Bragg peaks from statistical measurement of overlap.

Peak overlap not only reduces the number of independent observations but also means that the process of extraction of the integrated intensity associated with each reflection is not straightforward.

2.2.2. Background

In a powder diffraction pattern the presence of background is unavoidable. As a consequence, it must be taken into account, and every attempt must be made to minimize it during the experiment and to increase the peak/background ratio before any further analysis. A typical background is enhanced at small Bragg angles and gradually decreases toward large 2θ values. For a non-crystalline sample holder and/or significant amounts of amorphous material in the compound under investigation, no Bragg peaks occur but only diffuse scattering which contributes to the background with a broad maximum.

The background can be handled manually (by selecting points interpolated by the noise curve) and/or automatically (by using smooth analytical functions whose coefficients are refinable variables). If it is reasonably constant over the 2θ range under investigation, the simplest way to proceed is to subtract an average noise value to have the net peak intensities. Unfortunately, the background is generally nonlinear and, to describe it, the application of *ad hoc* analytical functions is needed. The most commonly used are: polynomial function, Chebyshev polynomial (type I or type II), Fourier polynomial, diffuse background function specific for scattering from amorphous phases or from a non-crystalline sample holder.

In a powder diffraction pattern, the correct description of the background and the job of differentiating peaks from background is generally difficult. For example, errors in the model of the peak shape cause errors in the background and, consequently, errors in the estimation of the integrated intensities associated with each reflection. In the case of X-ray data, the high-angle region of the pattern appears to be particularly critical because in this part of the pattern the decay of the scattering factor with $\sin\theta/\lambda$ and the peak broadening make the errors in the background definition critical. In this region, small errors in the background modelling can produce non-negligible changes in the intensity estimation, thus making it difficult to separate the peak contribution from the noise.

2.2.3. Preferred orientation

The conventional theory of powder diffraction assumes that, in a polycrystalline compound, the amount of infinite constituent crystallites are oriented completely randomly, this means that all reflections have an equal probability of fulfilling the diffrasction condition. In practice, the number of crystallites may be very large, but the crystallite orientations may be not completely random. This is because the shape of the crystallites, their morphology, also plays an important role in achieving randomness of orientation.

If the crystallites have an isotropic shape, random distribution of their orientations is guaranteed but, when the shapes are anisotropic, *e.g.* for materials that crystallize as plates or needles, the result is a non-random orientation of crystallites due to their natural tendency to align themselves along a particular crystallographic direction (the so called 'preferred orientation axis').

The non-random crystallite orientation is called preferred orientation (PO) or texture effect, and it is one of the most serious causes of distortion of the scattered intensities.

In fact, when PO is present, uniform distribution of the intensity around each Debye ring, as represented in Fig. 1, is not achieved; because of the non-random orientation, the circles appear as a distribution of enlarged spots. From the point of view of the diffraction pattern, we observe that the intensities of Bragg peaks which correspond to reciprocal lattice points with vectors parallel (platey sample) or perpendicular (needle sample) to the preferred orientation axis are greater or smaller than correct ones.

Preferred orientation effects are addressed by introducing the preferred orientation factor $O_{\mathbf{h}}$ in equation 1. The most commonly used correction is represented by the March-Dollase model (March, 1932; Dollase, 1986):

$$O_{\mathbf{h}}(\alpha) = \left(r^2 \cos^2 \alpha + \frac{\sin^2 \alpha}{r}\right)^{-\frac{3}{2}} \tag{3}$$

where α is the angle between the PO direction and the reciprocal lattice vector direction of the Bragg peak being corrected, and r is a refinable parameter representing the degree of PO.

Unfortunately, if strong PO effects are present, equation 3 can be inadequate and the best way to proceed is to prepare the powder sample carefully. This goal may not be achieved easily due to the difficulty of forcing the crystallites to have a random orientation.

In recent years, methods have been developed to exploit the texture effect in order to estimate better the relative intensities of overlapping reflections, thus giving rise to the 'texture method'. In a textured sample, the reflections are concentrated in some regions of the reciprocal space: *i.e.* not all crystallite orientations are equally represented, the powder pattern intensities will be dependent on the orientation of the sample in the X-ray beam. One solution is to measure the diffraction pattern with the sample oriented in several different directions, so that additional information on the intensities of reflections that overlap in 2θ but not in orientation space is obtained (Baerlocher *et al.*, 2004).

3. Qualitative and quantitative analysis

Phase identification of a polycrystalline mixture (qualitative analysis) and determination of the weight fraction for each identified phase (quantitative analysis) by XRPD are two methods applied widely for characterizing an unknown crystalline compound. Their application is interdisciplinary and involves many fields: inorganic, metallorganic and organic chemistry, pharmaceutics, mineralogy, archaeometry, forensic science, *etc*.

The aim of this section is to provide the basics of qualitative and quantitative analysis: an outline of some traditional methods; a description of the most recent approaches and related software; and examples of applications.

3.1. Qualitative analysis

A powder diffraction pattern can be reduced to a 'stick' pattern [*i.e.* a set of (d, I) values] which is a 'finger print' of a crystalline compound, being the interplanar d spacings (derived from the diffraction angles) and the peak intensities (I) characteristic of a crystalline phase. d spacings are fixed by the lattice parameters (see Section 4.2), peak intensities by the unit-cell content. This is the basic idea that suggested the use of the experimental (d, I) values for phase identification, by comparing them with the diffraction patterns of known reference single-phase crystalline compounds. The use of a 'stick' pattern has the advantage of limiting the amount of reference data to be involved in qualitative analysis and, at the same time, the disadvantage of losing some diffraction pattern information.

Scientists became aware of the potential of the XRPD technique for qualitative analysis in 1919 when Hull presented XRPD as a new method of chemical analysis (Hull, 1919) and showed that "substances with different crystalline structure will give entirely different patterns of lines", while "substances of similar chemical nature, on the other hand, will in general have similar crystal structure, and give similar patterns, so that it is often possible to identify a photograph at a glance as belonging to a certain type of element or compound". Some years later, Winchell (1927) defined a powder diffraction pattern as an 'autograph' or 'finger print' of a crystalline compound and announced the availability of a collection of standard diffraction patterns for 170 minerals. In 1936, Hanawalt and Rinn published a paper describing a scheme for classifying and using X-ray powder diffraction patterns for the identification of crystalline materials (Hanawalt and Rinn, 1936). Two years later, in a pioneering and milestone paper, Hanawalt *et al.* (1938) provided a reference library of standard diffraction patterns of known compounds by publishing a list of 1000 patterns, described in terms of a two-column table (d and I values), classified *via* a system based on the d values associated with the three strongest intensity lines. This work was the precursor of the modern digitized Powder Diffraction File (PDF) (Smith and Jenkins, 1996; Faber and Fawcett, 2002; Kabekkodu *et al.*, 2002). PDF is a commercial database containing hundreds of thousands of reduced diffraction patterns [measured or calculated (d, I) values]. Each diffraction pattern, together with additional available and useful information concerning the compound (*e.g.* chemical name, crystal system, cell parameters, space group, *etc.*) and/or the experimental conditions constitute a database 'entry', historically called a 'card'. PDF database cards contain a large number and a wide variety of compounds (*e.g.* minerals, inorganic, organic and metallorganic materials, alloys, pharmaceuticals, *etc.*). The PDF database is maintained by the International Center of Diffraction Data (ICDD) (http://www.icdd.com) and is growing continuously and updated. Other databases are available nowadays, some of them free of charge, *e.g.* the Crystallographic Open

Database (COD) (Gražulis *et al.*, 2009, 2012), a collection of crystal structures of inorganic, metal-organic, organic compounds and minerals, freely downloadable from the Web and continuously growing thanks to the efforts of an international team of researchers (the current number of COD entries is >379.000). For a complete list of databases see http://www.iucr.org/resources/data.

Despite the age and apparent simplicity of the idea inspiring qualitative analysis, it is still far from trivial, due to the typical powder diffraction problems (see Section 2.2) which cause unavoidable errors in peak positions and in intensities. These problems grow with increasing complexity of the mixtures analysed (*e.g.* in case of a sample containing many crystal phases). The success of qualitative analysis depends heavily on the accuracy of determination of the (d, I) values of both the reference and unknown compound [care has to be taken in the diffraction experiment, see, *e.g.* Whitfield and Mitchell (2008) for some useful suggestions].

For a more in-depth overview of qualitative analysis the reader is referred to Jenkins and Snyder (1996), Langford and Louër (1996), Whitfield and Mitchell (2008) and Percharsky and Zavalij (2009).

3.2. The main traditional approaches to qualitative analysis

Qualitative analysis can be described in terms of a three-step procedure: (1) search, according to a guiding criterion (see below); (2) match, usually done with respect to patterns belonging to a reference library (*e.g.* the PDF database); (3) identification by an evaluation of the reliability of the match. This last step can take into account a suitable mathematical tool (*i.e.* a figure of merit) and/or prior knowledge about sample chemistry (or any other available information) and/or experience. This three-step procedure can be performed manually (Hanawalt, 1986 and references therein), as done originally, by consulting published search manuals, or automatically, *via* computer software.

The main assumptions in qualitative analysis are:

(1) The diffraction pattern of a mixture is the sum of patterns of each component crystalline phase.
(2) The intensities of each component crystalline phase are proportional to the amount of the phase in the mixture. Absorption, extinction and preferred orientation effects may be sources of error.
(3) All the crystalline phases of the mixture are present in the investigated reference database.
(4) Tolerances of experimental line positions and intensities are taken into account.

The pioneering idea of Hanawalt and Rinn of tabulating and recovering diffraction patterns is at the basis of many search/match approaches.

A traditional search can be performed using one of the following methods:

(a) *Hanawalt method*, based on the three most intense lines. This method is 'intensity-driven', consequently when non-negligible errors

in intensity occur (*e.g.* in the case of a strong preferred orientation and/ or absorption effects) the results can be erroneous.
(b) *Fink method.* This is a useful alternative to the Hanawalt method and is based on the lines with the largest *d* values (the first eight lines in terms of increasing 2θ values), consequently it is a '*d* spacing-driven' approach.
(c) *Alphabetic method.* This takes advantage of chemical information and enables a systematic search of all the database cards with a given substance name.
(d) *Boolean method.* This exploits a lot of information (*d* spacings, chemistry, *etc.*) and has become fruitful thanks to the development of a computer readable version of the PDF database. Computer programs, *via* the Boolean method, enable the database to be searched actively using restraints to guide the search through a restricted number of database cards, *e.g.* by imposing restraints on 'minerals', the search of the PDF database will involve only the subset of cards concerning minerals, making phase identification much easier and thus improving the final result.

The first three search methods were used originally for manual searching. This kind of search can be tedious, time consuming and, in case of complex mixtures, unlikely to be successful. To overcome these problems, starting in the mid-1960s, search/match computer programs were developed (Frevel, 1965; Nichols, 1966; Johnson and Vand, 1967; Jenkins and Snyder, 1996; Langford and Louër, 1996). Jenkins (2000) observed that "a fourth generation of software is currently evolving where all interactions with the user are done through a 'point and click' video interface with a mouse".

Analysis of the 'Search/match Round Robin 2002' outcomes, led Le Meins *et al.* (2003) to emphasize that "the present range of software is not an efficient substitute for good staff experience and training".

3.3. Advances in qualitative analysis

A new approach for qualitative analysis has been developed recently (Barr *et al.*, 2004a,b,c,d,e, 2009a,b; Gilmore *et al.*, 2004); it opens the door to new opportunities for qualitative analysis and is particularly useful in the case of high-throughput crystallography and for managing and analysing huge quantities of diffraction patterns. The new and innovative procedure has been implemented in a family of fourth generation commercial software: *SNAP-1D* (Barr *et al.*, 2004e), *PolySNAP* (Barr *et al.*, 2004c) and *PolySNAP3* (Barr *et al.*, 2009b). Among the main features of this new approach are:

(1) it has the advantage of exploiting the full pattern information instead of the 'stick' pattern and, at the same time, the disadvantage of requiring more computing time and the availability of a database of single-phase full diffraction profiles. The number of reference diffraction patterns

available for this approach is much smaller than, *e.g.* that of the PDF database;

(2) it uses full-pattern-matching tools based on the following statistics: non-parametric Spearman rank, Pearson parametric linear correlation coefficient, Kolmogorov-Smirnov test, suitably combined in order to provide an overall measure of similarity;

(3) it can manage a large number of powder diffraction patterns and classify them. Clustering methods, including dendograms, principal component analysis (PCA) and metric multidimensional scaling (MMDS) analysis, are used for automatically grouping similar diffraction patterns into clusters, characterizing each cluster and recognizing any unusual sample that can be dissimilar (*e.g.* because it can contain an unknown polymorph);

(4) it can apply silhouettes and fuzzy cluster approaches to validate cluster membership;

(5) it is particularly useful in the case of quality control (when the same results are usually expected for all of the samples analysed). It has been tested successfully on pharmaceuticals and inorganic compounds;

(6) it can be applied to data obtained by Raman, IR, solid-state NMR spectroscopy and Differential Scanning Calorimetry (DSC) also combined with the XRPD technique.

High-throughput analysis is one of the new features of other commercial packages, *e.g. X'pert Highscore Plus* (PANalytical, www.panalytical.com), *Win*^{Pow} (Stoe, www.stoe.com), and *PDXL* (Rigaku, www.rigaku.com).

Among the freely available computer programs devoted to qualitative analysis by XRPD, we recall: *PULWIN* (Brückner, 2000), *QUALX2.0* (Altomare *et al.*, 2015), the updated version of *QUALX* (Altomare *et al.*, 2008a) and *MacDiff* (http://www.geol-pal.uni-frankfurt.de/Staff/Homepages/Petschick/MacDiff/MacDiffInfoE.html). For additional information on the available search-match software see the current web site http://www.ccp14.ac.uk/solution/search-match.htm and Cranswick (2008a).

QUALX2.0 represents the fourth generation of the search-match computer software. Due to our detailed knowledge of this software, we will present here its main features to explain how this program works. *QUALX2.0* is able to interrogating both the PDF-2 and POW_COD databases. This last is a freely available database, generated by the structure information contained in the CIF files of COD and exported in SQLite3 format. *QUALX2.0* searches for the phase(s) (belonging to PDF-2 or POW_COD) best matching the experimental pattern. The search/match approach uses the (d, I) values and is characterized by a high degree of automatism. A user-friendly graphic interface enables the default choices of the program to be changed and to select the reference database. In a default run, *QUALX2.0* carries out the following steps:

(a) background estimation *via* a polynomial curve;

(b) subtraction of background contribution from the experimental pattern;

(c) peak location *via* the Savitzky and Golay approach (1964);

(d) search for the phase(s) of the reference database.

The software also enables: (a) Kα_2 stripping, (b) smoothing of the experimental pattern; (c) changing the background describing curve; (d) modifying peak-search results; (e) correcting data by zero-shift error; (f) display of the database card; (g) setting graphical restraints (on chemical composition, chemical name, type of compound, symmetry, unit cell, crystal properties) to be used in the database investigation; and (h) saving the output file in HTML format.

In addition, similarly to other search/match programs, *QUALX2.0* is able to carry out a semi-quantitative analysis (see Section 3.5).

The daily usage of fourth generation software *QUALX2.0* is shown on an X-ray diffraction pattern that belongs to a set of test mixtures prepared for the "Search/match Round Robin 2002" on qualitative phase analysis organized by Le Meins, Cranswick and Le Bail. The geological sample consists of four major phases. Seventy three experimental peaks have been selected in the 10–76° 2θ range. Peak positions have been corrected by zero-shift error using the reflection-pair method (Dong *et al.*, 1999). The related *(d, I)* values were used actively for phase analysis. The four selected crystalline phases are (see Fig. 5): quartz (silicon oxide), fluorapatite (calcium phosphate fluoride chloride), siderite (iron carbonate) and gormanite (iron aluminium phosphate hydroxide hydrate), corresponding to those reported in the literature. The Round Robin results are described in the paper by Le Meins *et al.* (2003).

3.4. Quantitative analysis

Once the component phases of a mixture have been identified, the next step for characterizing the sample is to estimate their abundance by quantitative analysis.

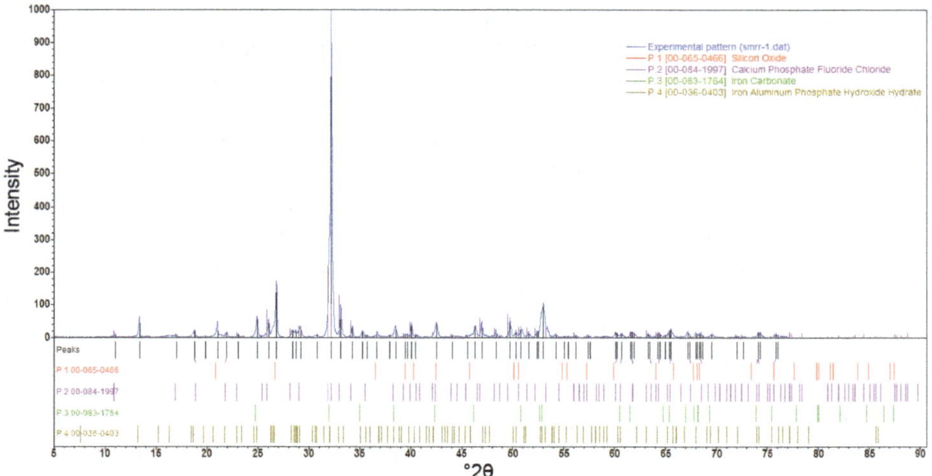

Figure 5. Qualitative analysis by *QUALX2.0* for a geological sample (SMRR-1) (laboratory data, Bragg-Brentano geometry – CuKα radiation, experimental 2θ range 5–77°, 2θ step = 0.02°).

The use of XRPD for quantitative phase analysis (QPA) was foreseen by Hull in his milestone paper of 1919. Among the first examples of QPA by XRPD were: Navias (1925) determined the amount of mullite in fired clays; and Clark and Reynolds (1936) applied an internal standard method (see Section 3.5) aimed at analysing mine dust by a film technique. The potential of XRPD and its application for QPA was enhanced by the development of the Geiger-counter diffractometer in 1945 (Hubbard and Snyder, 1988). A great contribution to the theoretical basis of quantitative analysis was given by Brindley (1945) who studied the effects of grain or particle size on the relative diffracted intensities, and Alexander and Klug (1948, reprinted in 1989) who analysed the X-ray absorption aspects in quantitative analysis in the case of a flat sample. Many approaches have been developed subsequently, among them the 'spiking', 'dilution', 'internal standard' and 'Reference Intensity Ratio' (RIR); the methods used most commonly are the internal standard and the RIR (see Section 3.5).

All these approaches belong to the so-called 'single-peak' methods, relying on the intensity of a particular peak (or a group of peaks) for each component phase of interest. They require the use of a reference phase (internal or external standard).

An important approach, alternative to the 'single-peak' one, has been developed: a whole-pattern method based on the comparison of the experimental data with the calculated pattern, obtained as a summation of the component-phase contributions; these last can be calculated from crystallochemical information on the pure phase (provided that unit-cell parameters, space group, atomic coordinates and thermal factors are known), or, alternatively, measured from a pure phase compound. Among the methods based on this approach, a relevant role is played by the Rietveld method (see Section 4.6). The Rietveld method, used originally for refining single-phase crystal structures using neutron data, has been adopted for quantitative analysis (Werner *et al.*, 1979; Hill and Howard, 1987) leading to a rebirth of QPA. Hill and Howard (1987) demonstrated the existence of a simple relationship between the scale factor (associated with each phase in the multicomponent Rietveld analysis) and the phase abundance in the mixture. They successfully applied this approach to neutron powder diffraction data collected from binary mixtures of rutile/corundum, rutile/quartz and silicon/quartz. QPA using the Rietveld method has been used widely, with renewed and growing interest in the case of XRD data. Thanks to the development of stable, efficient and sophisticated software, and to the availability of powerful synchrotron radiation sources and advanced conventional X-ray diffractometers, the Rietveld method can be considered nowadays to be the preferred, most accurate and flexible approach for QPA.

Data quality is a fundamental requirement for obtaining reliable quantitative phase results, whatever the method adopted. The accuracy of QPA can be influenced significantly by instrumental and sample-related effects such as preferred orientation (see Section 2.2), extinction (Sabine, 1985; Cline and Snyder, 1983), microabsorption effects (Cline and Snyder, 1983; Brindley, 1945), the presence of amorphous components (Kern *et al.*, 2012). Care has to be taken in sample preparation to minimize preferred-orientation effects and microabsorption; this last effect can be limited by grinding the specimen to a particle size of <10 μm (Jenkins and Snyder, 1996).

QPA has had a growing and wide application not only in the scientific, but also in the industrial field [Cranswick and Kruger (2002), Walenta and Füllmann (2003), Cranswick (2008b), Madsen and Scarlett (2008)] such as (a) cement production (*via* an on-line X-ray diffraction and suitably modified Rietveld method able to study the complex nature of cements); (b) mine waste rock; (c) control of drug quality.

For a detailed outline of the QPA topic, the reader is referred to a rich literature: Zevin and Kimmel (1995), Whitfield and Mitchell (2008), Madsen and Scarlett (2008), Madsen *et al.* (2012, 2013).

3.5. The main traditional approaches to quantitative analysis

The two main approaches for QPA ('single-peak' and whole-pattern) will be outlined.

Among the 'single-peak' approaches, we recall:

(1) *Internal standard method.* This consists of mixing the sample homogenously with a known amount of reference material (internal standard (s)), the weight fraction of which is w_s. The procedure aims to eliminate the dependence on absorption factor by considering the ratio $I_{\mathbf{h},i}/I_{\mathbf{k},s}$ between an **h** line intensity for the phase i ($I_{\mathbf{h},i}$) and a reference **k** line intensity for the standard s ($I_{\mathbf{k},s}$). The weight fraction of the unknown phase i (w_i) is related to $I_{\mathbf{h},i}/I_{\mathbf{k},s}$, according to:

$$I_{\mathbf{h},i}/I_{\mathbf{k},s} = Kw_i/w_s, \qquad (4)$$

K is a factor determined by an internal standard calibration curve for each component. This is done by collecting many experimental profiles, corresponding to two-phase mixtures obtained by using different internal standard weight fraction w_s values, and plotting $I_{\mathbf{h},i}/I_{\mathbf{k},s}$ *vs.* w_i/w_s. K is the slope of the internal standard calibration curve. Equation 4 is the basic equation of the internal standard method. Deviations from the linearity of the calibration curve can be due to non-negligible microabsorption effects.

The procedure requires care in the selection of the internal standard. The method enables quantification of a single crystalline phase, by deriving w_i *via* equation 4. It can also be applied in the case of amorphous material.

(2) *Reference Intensity Ratio (RIR) method.* This can be considered a special case of the internal standard procedure. It is the ratio between the intensity of the strongest peak of phase i and the intensity of the strongest peak of a reference standard in a 50:50 wt.% mixture. Usually the standard is corundum and the reference peak corresponds to the (113) line [according to equation 4, $I_i/I_c = K = \text{RIR}$]. Many papers have been devoted to the RIR method and its practical use (see, *e.g.* de Wolff and Visser 1964, 1988; Hubbard and Snyder, 1988; Davis *et al.*, 1990; Snyder, 1992). de Wolff and Visser first suggested the use of corundum as internal standard for several reasons: it is virtually unaffected by

preferred orientation effects, it is characterized by relatively strong lines over a wide range of *d* values, it is chemically stable, almost inert and commercially available in an appropriate crystallite size. de Wolff and Visser were aware of the roughness of the proposed method which provides a semi-quantitative analysis.

The first table of RIR values for quantitative analysis was provided by Davis and Smith (1988) with the intention of maintaining and publishing periodic updates (Davis *et al.*, 1989). The ICDD has introduced, when available, RIR values in the PDF cards to meet the need of users. Instead of collecting data by using a standard material and applying the conventional RIR method, literature values of RIR can be exploited for a semi-quantitative analysis *via* the normalized RIR method. The presence of amorphous and/or unidentified crystalline phases invalidates the semi-quantitative analysis. Using this approach, qualitative analysis software is able to carry out a semi-quantitative analysis.

An example of semi-quantitative analysis is described here. *QUALX2.0* has been applied to an XRD pattern concerning a three-phase crystalline mixture (code name CPD-1A) that belongs to a set of test mixtures prepared for a Round Robin on QPA sponsored by the International Union of Crystallography (IUCr) Commission on Powder Diffraction (CPD). The outcomes of this QPA Round Robin are described in papers by Madsen *et al.* (2001) and Scarlett *et al.* (2002). The selected test sample is a mixture 'ideal' for semi-quantitative studies, with no specific problems; there are no strong preferred orientation effects, no amorphous content, and only a small number of component phases, with none of them overlapping severely.

Twenty nine experimental peaks have been selected in the range 25−139° and the related (*d, I*) values are used actively for phase analysis. In order to perform a semi-quantitative analysis, only cards containing information on RIR values have been chosen. The three selected PDF cards are: 00-070-2049 for calcium fluoride (PDF card RIR value = 3.84), 00-89-0511 for zinc oxide (PDF card RIR value = 5.37) and 00-083-2080 for aluminium oxide (PDF card RIR value = 1.00); the three accepted crystalline phases (see Fig. 6) correspond to those reported in the literature. The contribution of each is scaled suitably on the experimental pattern and, by taking into account the corresponding PDF card RIR values, the following weight fractions (w_{QA}) are provided by *QUALX2.0*: 94.50% for fluorite (calcium fluoride), 3.80% for zincite (zinc oxide) and 1.70% for corundum (aluminium oxide). The true weight fractions (w_T) are 94.81%, 4.04% and 1.15%, respectively. By comparing w_{QA} and w_T values it can be observed easily that the semi-quantitative analysis by *QUALX2.0*, in the case of a simple mixture, can be considered satisfactory.

Because many Round Robin participants have applied the RIR method for QPA, Madsen *et al.* (2001) provided a table showing the RIR values reported by participants (different RIR values have been used) to emphasize one of the limits of the RIR method: the QPA results are dependent on the chosen RIR. The drawbacks of a semi-quantitative approach increase when more complex compounds sre being examined.

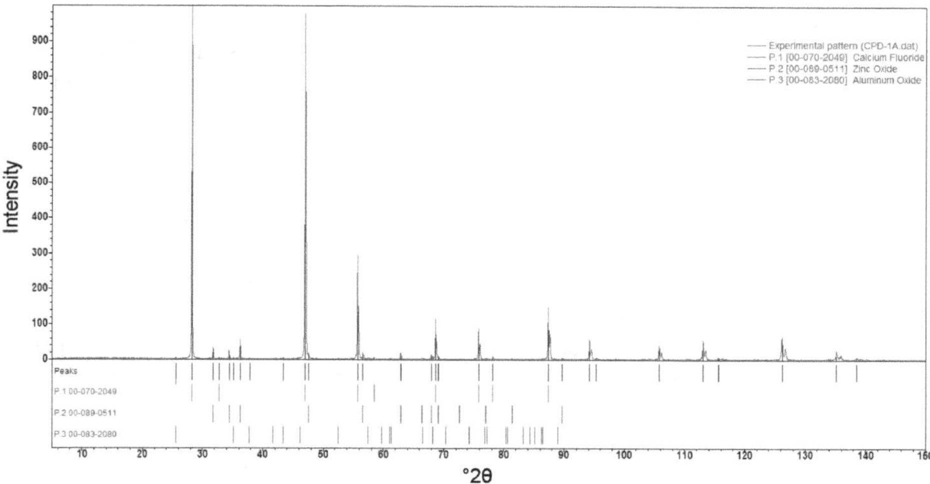

Figure 6. Semi-quantitative analysis by *QUALX2.0* for a three-phase mixture (CPD-1A) (laboratory data collected using a Philips X'Pert X-ray diffractometer, CuKα radiation, experimental 2θ range 5–150°, 2θ step = 0.02°).

Among the methods belonging to the whole-pattern based QPA approach, the Rietveld method is the most widely applied. It has the advantage over the 'single-peak' approach of being able to determine the phase weight fractions without requiring an experimental calibration curve or the use of an internal or external standard.

The total calculated profile y_i(calc) corresponding to the i-th step is given by:

$$y_i(\text{calc}) = \sum_{j=1}^{p} Y_{ij}(\text{calc}) + y_i(\text{back}) \tag{5}$$

where Y_{ij}(calc) is the calculated pattern for the component phase j, given by

$$Y_{ij}\ (\text{calc}) = S_j \sum_{\mathbf{h}} m_{\mathbf{h}} LP_{\mathbf{h}} |F_{\mathbf{h}}|^2 P_{i,\mathbf{h}} O_{\mathbf{h}} A \tag{6}$$

s_j is the scale factor associated with the j-th phase and p is the total number of identified phases in the mixture (see Section 4.4 for the meaning of the rest of terms in equations 5 and 6). As for structure solution, refinement is done by minimizing the sum of the weighted squared difference between observed and calculated profiles (see equation 21). For QPA, during the refinement step, the crystal structure taken from the literature is usually kept fixed (structure parameters are not refined) unlike the rest of the parameters. This refinement strategy is valid for 'simple' crystal structures (such as quartz, calcite, dolomite, hematite, *etc.*) but not for more complex structures such as, *e.g.* zeolites or clay minerals (Gualtieri, 2000), for which, in some cases, site occupancies have to be varied in order to obtain accurate scale factors (Bish and Post, 1993).

Hill and Howard (1987) proposed an important and, at the same time, simple relationship that enables us to derive the weight fraction (w_j) of the j-th phase in a mixture directly from the refined (by the Rietveld method) scale factor s_j, as follows:

$$w_j = s_j(Z_j M_j V_j) / \sum_{k=1}^{p} s_k(Z_k M_k V_k) \tag{7}$$

where Z_j, M_j and V_j are the number of molecules per unit cell, the molecular weight and the unit-cell volume of phase j.

A relationship similar to equation 7 and expressed in terms of the calculated density, ρ_j, of phase j has been proposed by Bish and Howard (1988).

Equation 7 does not take into account the absorptive power of each component phase. According to Brindley theory, equation 7 becomes (Taylor and Matulis, 1991):

$$w_j = s_j(Z_j M_j V_j) / \left[\sum_{k=1}^{p} s_k(Z_k M_k V_k / \tau_k) \tau_j \right] \tag{8}$$

where τ_j is the particle absorption factor for phase j.

Among the advantages of the Rietveld method for QPA (Bish and Post, 1993) are:

(1) It uses all reflections in the pattern instead of just the strongest ones, consequently it minimizes the uncertainty in the calculated weight fractions and the effects of preferred orientation.
(2) It can correct the experimental data for preferred orientation effects.
(3) In case of failure in phase identification and omission of a phase, the analysis will yield differences between observed and calculated profiles that can suggest the possible presence of unsuspected additional phases.

The Rietveld method has the disadvantage of requiring knowledge of the crystal structure for each crystalline phase in the mixture; consequently conventional Rietveld method applications in the case of poorly determined, disordered or unknown structures are not possible. Some variants of the traditional QPA Rietveld approach have been developed in order to overcome these limitations, proving useful in the case of minerals and the presence of disorder structures (*e.g.* clay minerals). Among these approaches:

(1) the method described by Taylor and Rui (1992) is able to include in the Rietveld refinement a phase with an imperfectly known or unknown structure, by applying empirical amplitudes. The approach has been implemented in *SIROQUANT* (Taylor, 1991), a commercial program for quantitative analysis of minerals.
(2) The method proposed by Scarlett and Madsen (2006) is a generalization of the work by Taylor and Rui; it can carry out QPA in case of (i) a structure model that deviates from the published one (*e.g.* clays); (ii) partial structures for which only the unit-cell

parameters and space group are known; and (iii) completely unknown crystal structures. The approach has been introduced in *TOPAS* (Bruker AXS, 2003).

For a more complete list of the available QPA computer programs see Cranswick (2008a) and the web site programming.ccp14.ac.uk/solution/rietveld-software.

Among the freely available QPA Rietveld-based programs are:

(1) *GSAS-II* (Toby and Von Dreele, 2013), the latest version of *GSAS* (Larson and Von Dreele, 2004), a widely used package which, in addition to QPA, can perform data reduction, structure solution and structure refinement in the case of both single-crystal and powder-diffraction data. Readers can refer to the file www.ccp14.ac.uk/solution/gsas/files/expgui_quant_gualtieri.pdf, which is a useful guided training exercise on quantitative analysis using *GSAS* and *EXPGUI* programs (Toby, 2001), the latter implements a graphical user interface for *GSAS*.

(2) *Quanto* (Altomare *et al.*, 2001). Among its main features are: minimal user intervention, a high level of automation and a user-friendly graphical interface.

3.6. Latest advances in methodology and software for quantitative analysis

A new whole-pattern-based QPA approach has been proposed recently (Barr *et al.* 2004c,e, 2009b; Gilmore *et al.*, 2004). It is an alternative to the Rietveld method and needs a library of single-phase standard full patterns.

If S is the sample pattern (consisting of a mixture of up to N component phases), $\{S_i\}$, $i = 1,...m$, the set of the experimental points, x_{ij} the measured point i of the pattern corresponding to the phase j and $p_1,..., p_N$ the fractions of the sample pattern, the following system of linear equations can be built:

$$\begin{aligned} x_{11}\,p_1 + x_{12}\,p_2 + \ldots\ldots + x_{1N}\,p_N &= S_1 \\ x_{21}\,p_1 + x_{22}\,p_2 + \ldots\ldots + x_{2N}\,p_N &= S_2 \\ & \\ x_{m1}\,p_1 + x_{m2}\,p_2 + \ldots\ldots + x_{mN}\,p_N &= S_m \end{aligned} \tag{9}$$

It can be written in matrix form:

$$\boldsymbol{x}{\cdot}\boldsymbol{p} = \boldsymbol{S} \tag{10}$$

System 9 is over dimensioned ($N << m$); it is not always possible to invert the $m \times N$ matrix; the Single Value Decomposition (SVD) approach (Press *et al.*, 1992) ensures stability for the matrix inversion and provides the solution $\boldsymbol{p}$. The method was applied successfully to mixtures prepared for the Round Robin on QPA sponsored by the IUCr CPD. This approach has been implemented in the commercial programs *SNAP-1D* and *PolySNAP*. The method has been improved recently: instead of SVD, linear regression has been adopted. The revised approach has been introduced in the commercial software *PolySNAP3*. The new procedure tends to give more reliable standard uncertainties for the results, even if they are still overestimated. Whichever procedure is applied (SVD or linear regression), the authors are aware that the new approach cannot compete with the Rietveld method, which can give more accurate results, even if it reveals itself as useful when uncertainties of ~10% are tolerated, as can often be the case with high-throughput experiments. Very recently, a new QPA method has been proposed (Toraya, 2016), able to derive the weight fraction of each component crystalline phase from their observed integrated intensities and chemical composition. The method, called "direct derivation" (DD) method, does not need crystallographic databases and reference intensity ratios. The accuracy of the DD method has been demonstrated to be almost comparable to that of the Rietveld method.

4. Structure solution

4.1. The steps of structure solution

One of the most challenging objectives of the powder diffraction technique is structure solution. In the last 25 years, interest in powder solution has increased remarkably and surprising success has been achieved in this area. The reasons for this progress can be found in the evolution of diffractometric instrumentation, proliferation of new and efficient theories and methods and, most of all, a significant demand for knowledge of the structural properties of materials used in diverse scientific fields: chemistry, physics, biology, materials science and geology.

Structure solution by powders is carried out by a complex pathway which requires first the determination of cell parameters and space group. Then the structure model is attained by using reciprocal space methods or, alternatively, by direct space methods (Caliandro *et al.*, 2008). All these steps are non-trivial; some are insoluble, as mentioned in Section 2.2.

In this section, we highlight two main problems: (1) the 'phasing problem', regarding structure solution in general, including the single-crystal case. Structure factors are physical quantities in terms of complex numbers from which the electron density map, and consequently the structure model, is derived. Unfortunately, from a diffraction experiment we can derive the structure factor moduli but not their phases. These last are recovered by statistical and probabilistic methods which use the experimentally available moduli, in particular by Direct methods or Patterson methods (Giacovazzo, 2011), the outcome of which depends mainly on structure complexity and experimental data quantity and quality; and (2) the 'overlapping problem'. This point was discussed in Section 2.2. Here it is important to remember that the overlapping problem hinders the accurate estimation of the structure factor moduli values extracted from the experimental pattern. As a consequence, the phasing methods and the attainment of the structure model may fail.

In the following sections we give the main theoretical concepts of each step of the solution process (they are valid in general, not only in the mineralogical field) and, as an example of application, the corresponding results obtained in solving the nepheline structure (idealized formula $Na_3K[Al_4Si_4O_{16}]$) of a sample from the Somma-Vesuvius volcanic complex (southern Italy) (Balassone *et al.*, 2014). The solution process of nepheline (from now on we will adopt the code name NEPHE) has been executed fully by the computer program *EXPO2013* (Altomare *et al.*, 2013b) which provided the results reported below. *EXPO2013* (free for academic and non-profit research institutions), supported by a very user-friendly graphical interface, is able to carry out all the steps of the solution, from determination of cell parameters up to the Rietveld refinement (Rietveld, 1969), in a largely automatic manner.

See the web site http://www.ccp14.ac.uk/mirror/mirror.htm for the complete list of the available free programs devoted to the different steps of structure solution (from single-crystal and powder data).

4.2. Indexing

Structure solution requires that the crystal cell is known; therefore the primary step in the solution process is a correct indexation. The indexation process starts from the experimental powder diffraction pattern and aims to recover the cell parameters ($a, b, c, \alpha, \beta, \gamma$) by assigning the appropriate triple (hkl or $\mathbf{h}$) of Miller indices of the crystal-lattice plane to each one of the experimental diffraction peaks. In other words, it consists of reconstruction of the three-dimensional elementary cell by starting from the list of the $d_\mathbf{h}$ values extracted from the one-dimensional pattern collected. The $d_\mathbf{h}$ interplanar spacing is related to the diffraction angle by the well known Bragg law:

$$d_\mathbf{h} = \lambda / 2\sin\theta_\mathbf{h}, \tag{11}$$

where $\theta_\mathbf{h}$ is the Bragg angle and λ is the measurement wavelength. The passage from the experimental $d_\mathbf{h}$ values to the crystal cell is articulated as follows (see Altomare et *al.*, 2008b and references therein). By assuming:

$$Q(hkl) = \frac{1}{d_\mathbf{h}^2} \tag{12}$$

the expression of $Q(hkl)$ in terms of reciprocal-cell parameters and (hkl) indices depends on the crystal system (Table 1).

By considering the most general case (triclinic system):

$$Q(hkl) = h^2 A_{11} + k^2 A_{22} + l^2 A_{33} + hk\, A_{12} + hl\, A_{13} + kl\, A_{23} \tag{13}$$

where

$$A_{11} = \boldsymbol{a}^{*2},\ A_{22} = \boldsymbol{b}^{*2},\ A_{33} = \boldsymbol{c}^{*2},\ A_{12} = 2\boldsymbol{a}^*\boldsymbol{b}^*\cos\gamma^*,$$
$$A_{13} = 2\boldsymbol{a}^*\boldsymbol{c}^*\cos\beta^*,\ A_{23} = 2\boldsymbol{b}^*\mathbf{c}^*\cos\alpha^* \tag{14}$$

$\boldsymbol{a}^*$, $\boldsymbol{b}^*$ and $\boldsymbol{c}^*$ are three vectors of the reciprocal lattice related to the direct lattice values ($\boldsymbol{a}$, $\boldsymbol{b}$ and $\boldsymbol{c}$) by:

$$\boldsymbol{a} = (\boldsymbol{b}^* \times \boldsymbol{c}^*)/V^*,\ \boldsymbol{b} = (\boldsymbol{c}^* \times \boldsymbol{a}^*)/V^*,\ \boldsymbol{c} = (\boldsymbol{a}^* \times \boldsymbol{b}^*)/V^* \tag{15}$$

where

$$V^* = \boldsymbol{a}^* \times \boldsymbol{b}^* \times \boldsymbol{c}^* = 1/V \tag{16}$$

is the reciprocal unit-cell volume.

Table 1. The expression of $Q(hkl)$ depending on the crystal system.

Crystal system	$Q(hkl)$
Cubic	$(h^2+k^2+l^2)\ A_{11}$
Tetragonal	$(h^2+k^2)\ A_{11}+l^2\ A_{133}$
Hexagonal	$(h^2+hk+k^2)\ A_{11}+l^2\ A_{33}$
Orthorhombic	$h^2\ A_{11}+k^2\ A_{22}+l^2\ A_{33}$
Monoclinic	$h^2\ A_{11}+k^2\ A_{22}+l^2\ A_{33}+hl\ A_{13}$
Triclinic	$h^2\ A_{11}+k^2\ A_{22}+l^2\ A_{33}+hk\ A_{12}\ +hl\ A_{13}+kl\ A_{23}$

Indexation is based on the determination of six A_{ij} values (less in case of symmetry greater than triclinic) and three crystallographic indices (*hkl*) different for each peak, in such a way that equation 13 is verified, within a suitable tolerance, for each experimental $Q(hkl)$ value.

Traditional methods are based on equation 13 and they are adopted by the most widely used indexing computer programs: *DICVOL* (Boultif and Louër, 2004), *TREOR* (Werner *et al.*, 1985) and *ITO* (Visser, 1969).

More recently, alternative methods to the traditional approaches using direct space have been proposed. Their rationale is to find the lattice parameters (a, b, c, α, β, γ) by trial and error, providing the best agreement between the experimental and the calculated patterns in terms of a suitable figure of merit R_p. The search (performed in direct space) for the best cell, corresponding to the global minimum of the hypersurface $R_p[a, b, c, \alpha, \beta, \gamma]$, is driven by the Genetic Algorithm or Monte Carlo approach (Altomare *et al.*, 2008b).

Several computing programs (Cranswick, 2008a), based on different strategies, are available. They are able to: (1) search and locate peaks in the measured pattern; (2) create the list of the experimental $d_{\mathbf{h}}$ values. In particular, a number of peaks reduced with respect to the total number of peaks in the pattern is considered because it is important to use reliable peak positions (see below). In practice, only the first 20−30 observed lines are significant because of lower sensitivity to small changes in cell constants; (3) solve equation 13 by using different algorithms and strategies; and (4) provide the cell most compatible with the experimental data.

The indexation step usually ends with more than one plausible cell. A figure of merit (FOM) is associated with each cell provided in order to evaluate its physical plausibility and agreement with the pattern. The classical FOM is M_{20} (de Wolff, 1968), but alternative and/or more recent FOMs have been proposed (Ishida and Watanabe, 1967, 1971; Smith and Snyder, 1979; Bergmann, 2007; Le Bail, 2008; Altomare *et al.*, 2009a). A careful check of the largest FOM cell (in terms of agreement between the reflection $2\theta_{\mathbf{h}}$ values corresponding to the cell and experimental peak positions) is recommended. In fact, failure to identify the cell correctly may occur depending on the presence of two main sources of error: (1) low precision and accuracy in detecting experimental peak positions. This circumstance is frequently due to peak overlap, poor peak resolution, 2θ zero shift, errors in measurement, low peak/background ratio; and (2) the possible presence of impurity lines in the experimental pattern, *e.g.* arising from more than one chemical phase present in the compound under examination.

Among the computing programs available for indexing, we mention *N-TREOR09* (Altomare *et al.*, 2009a) which has been integrated in *EXPO2013*. Based on the traditional method, it searches for the correct cell by carrying out the following steps: (1) selection of a subset of low-resolution observed *Q*(*hkl*) values (*i.e.* basis lines) among the set of experimental $d_{\mathbf{h}}$ values; (2) assignment of trial Miller indices to the basis lines; and (3) solution of equation 13.

The first (up to 25) $d_{\mathbf{h}}$ values are used by *N-TREOR09* for finding a plausible cell. The search for the correct cell is carried out from cubic to lower-symmetry crystal systems. The number of the basis lines is a function of the symmetry of the crystal system investigated. The results of indexation of NEPHE follow. The experimental powder diffraction pattern was collected at room temperature by using an automated Rigaku RINT2500 laboratory diffractometer (50 kV, 200 mA) equipped with the silicon strip Rigaku D/teX Ultra detector. An asymmetric Johansson Ge crystal was used to select the monochromatic CuKα_1 radiation (λ = 1.54056 Å). The angular range 8–120° (2θ) was scanned with a step size of 0.02° (2θ) and counting time of 2.4 s/step. The measurements were executed in transmission mode, by introducing the sample in a special glass capillary with a diameter of 0.3 mm and mounted on the axis of the diffractometer. In order to reduce the effect of possible preferred orientation, the capillary was rotated during measurements to improve the randomization of the orientations of the individual crystallites. The experimental profile collected is shown in Fig. 7, where the red bars on the top of the peaks correspond to the positions of peaks automatically detected by *EXPO2013*.

Sixty two lines of the diffraction pattern, with intensity values greater than a default threshold, were selected automatically by *EXPO2013* for indexing (the first 25 low-angle well defined peaks were used actively for solving equation 13). Three cells (two hexagonal and one orthorhombic) were provided. For them, not more than one observed line is unindexed. One of the two hexagonal cells corresponded to a reliable value of M_{20} (M_{20} = 107), much larger than the M_{20} values of the other two cells. We remember a basic rule about M_{20}: if the number of unindexed peaks among the first twenty is not larger than 2 and if M_{20}>10 then the indexing process is physically reliable and the cell found is correct (de Wolff, 1968; Werner, 2002). The following cell parameters a = 9.9887(2), $b = a$, c = 8.3717(3) Å, $\alpha = \beta$ = 90°, γ = 120°, corresponding to a cell volume V = 723.37(3) Å^3 were therefore recognized as the correct ones.

Let us consider, in addition to the correct hexagonal cell, the second plausible orthorhombic cell [a = 8.6508(2) Å, b = 19.9766(11) Å, c = 8.3719(4) Å, $\alpha = \beta = \gamma$ = 90°, V =1446.77(2) Å^3, M_{20} = 24] in the list of the three solutions found by *EXPO2013*. Reflections corresponding to the two cells are generated: we assumed the space group

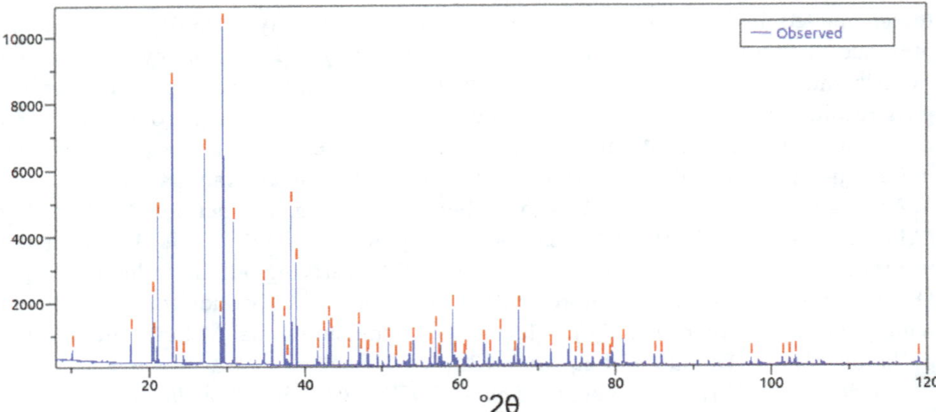

Figure. 7. The experimental powder diffraction pattern of the nepheline structure of a sample from the Somma-Vesuvius volcanic complex. The upper red bars correspond to the positions of peaks automatically detected by *EXPO2013*.

with the largest Laue symmetry and no extinction conditions (*i.e. P6/mmm* and *Pmmm*, in the case of the hexagonal and orthorhombic cells, respectively). In Fig. 8 (for a reduced 2θ interval), at the bottom, the 2θ positions of the reflections generated are shown (red and grey vertical bars for hexagonal and orthorhombic cells, respectively). The orthorhombic cell, with a larger volume and lower symmetry than the hexagonal cell, corresponds to a large number of reflections lying in 2θ intervals where no pronounced diffraction peaks are present (note that an impurity peak occurs at 24.54°2θ). The different agreement between reflection positions and peaks confirms the gap between the two M_{20} values.

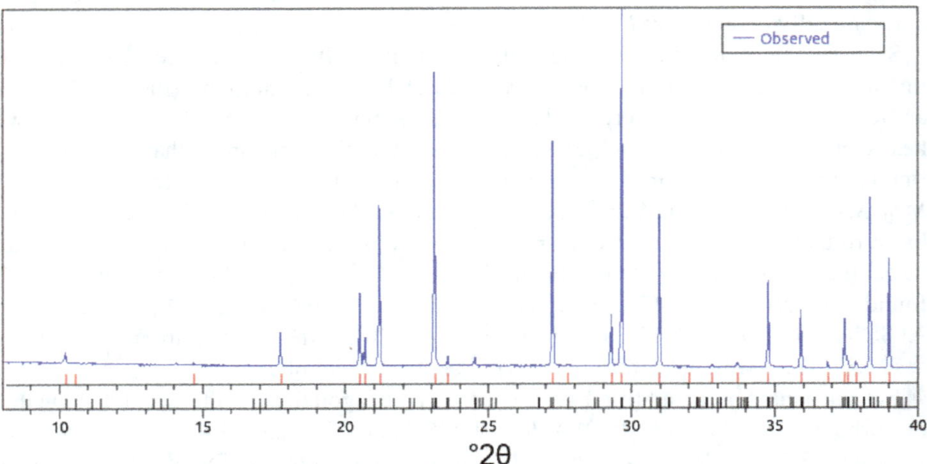

Figure 8. NEPHE structure: generated reflections in the case of the correct hexagonal cell (red vertical bars) and the second ranked orthorhombic cell (grey vertical bars).

4.3. Space-group determination

Once the unit cell and the crystal system have been identified, the next step is the determination of the space group using statistical analysis of the diffraction intensities.

The combination of information about the Laue group [see the *International Tables for Crystallography*, Vol. A (2002)] and the systematically absent reflections (Altomare *et al.*, 2008b) leads to the determination of the extinction symbol (ES). The list of the extinction symbols is given in the *International Tables for Crystallography* and it depends on the crystal system (14 extinction groups for the monoclinic, 111 for orthorhombic, 31 for tetragonal, 12 for trigonal-hexagonal and 18 for the cubic system). The ES does not define the space group unambiguously because more than one space group can correspond to an ES: *e.g.* in the monoclinic crystal system, three space groups ($P2$, Pm, $P2/m$) correspond to the P 1 - 1 extinction symbol; but only the $P2_1/n$ space group corresponds to the $P\ 1\ 2_1/n\ 1$ extinction symbol. In the case of space-group ambiguity we can: (1) carry out the solution process for each possible space group and finally select the correct solution; (2) choose the space group which best fits the structural properties of the material to be investigated. Therefore, the strategy of space-group determination considers first the ES identification which is achieved by checking the systematically absent reflections (they correspond to null experimental diffraction intensities) in the experimental diffraction pattern. This last operation is not trivial (either manually or by automatic calculation) and not always free of errors because it is based on the decomposition of the diffraction profile from which integrated intensities of each individual reflection are extracted. The overlapping problem compromises the intensity estimation process and in addition the background can hinder the correct identification of the systematically absent reflections in particular in the large 2θ regions of the pattern, where the diffraction signal is weak. As a consequence, the ES is not easily identified.

For this reason, theories have been developed (Markvardsen *et al.*, 2001; Altomare *et al.*, 2004, 2005, 2007) to evaluate a probability value for each extinction symbol compatible with the crystal system provided by the indexation step. The ES corresponding to the greatest probability should be the correct one but exceptions to this rule may occur because of the low accuracy of the diffraction integrated intensity estimates. A careful check of the result is always advisable.

As an example of application, for NEPHE, in Table 2, we give the results of the automatic space-group determination procedure by *EXPO2013*. The 12 extinction symbols (hexagonal crystal system) are ranked according to the calculated probability value (the ideal probability is 1). The choice of $P6_3$ - - as the correct ES is supported by: (a) the quite large value of the estimated probability which is also much larger than the second ranked value; (b) careful inspection of the experimental pattern. Three space groups ($P6_3$, $P6_3/m$ and $P6_322$) correspond to $P6_3$ - -. The solution process is then continued by assuming $P6_3$ as space group of the structure under study (the other possible space groups will be considered later if $P6_3$ fails to provide the correct solution).

Table 2. NEPHE structure. The 12 extinction symbols ranked according to the calculated probability value.

Extinction symbol	Probability
$P6_3$ - -	0.899
$P6_1$ - -	0.382
$P6_2$ - -	0.339
$P3_1$ - -	0.339
P - - -	0.100
P - - *c*	0.001
P - *c* -	0.000
P - *c c*	0.000
R (obv) - -	0.000
R (rev) - -	0.000
R (obv) - *c*	0.000
R (rev) - *c*	0.000

4.4. Extraction of the diffraction integrated intensities

An important process in the solution pathway is the extraction, from the experimental pattern, of the integrated intensity $I_{\mathbf{h}}$ of each **h** reflection from which the structure factor modulus $|F_{\mathbf{h}}|$ is derived ($I_{\mathbf{h}} \propto F_{\mathbf{h}}|^2$) (see Section 2.1). This operation requires that the observed experimental profile y_i(obs) at the *i*-th step is described mathematically by means of the calculated profile y_i(calc):

$$y_i(\text{calc}) = s\sum_{\mathbf{h}} y_{i,\mathbf{h}}(\text{calc}) + y_i(\text{back}) = s\sum_{\mathbf{h}} m_{\mathbf{h}} LP_{\mathbf{h}} |F_{\mathbf{h}}|^2 P_{i,\mathbf{h}} O_{\mathbf{h}} A + y_i(\text{back}) \quad (17)$$

where s is a scale factor; the summation is over the reflections contributing to the profile intensity; $y_{i,\mathbf{h}}$(calc) represents the contribution of the **h** reflection; $m_{\mathbf{h}}$ is the reflection multiplicity; $LP_{\mathbf{h}}$ is the Lorentz-polarization factor; $P_{i,\mathbf{h}}$ and $O_{\mathbf{h}}$ are the peak shape and preferred orientation function corresponding to the peak associated with the **h** reflection, respectively. Several kinds of reflection profile analytical functions can be used for describing $P_{i,\mathbf{h}}$; A is the absorption factor depending on the instrument geometry; y_i(back) is the background term which can be described by a polynomial function. The quantity $m_{\mathbf{h}} LP_{\mathbf{h}} |F_{\mathbf{h}}|^2$ gives $I_{\mathbf{h}}$.

Two main methods are used to evaluate the $I_{\mathbf{h}}$ values:

- the *Pawley method* (Pawley, 1981).

  This is based on a non-linear least-squares procedure that refines scale factor, lattice parameters ($2\theta_{\mathbf{h}}$ depends on them), peak shape and asymmetry parameters and the integrated intensities. Sometimes this method can be faulty because the correlation of the integrated intensities, due to peak overlap, commonly leads to ill-conditioning least-squares which provide negative intensities. The method has been improved, *e.g.* by the introduction of a Bayesian probability approach, capable of providing positive intensities (Sivia and David, 1994).

- the *Le Bail method* (Le Bail *et al.*, 1988).

This is based on the iterative use of the following equation:

$$I_{\mathbf{h}} = \sum_{i}\{y_i(\text{obs}) - y_i(\text{back})\}y_{i,\mathbf{h}}(\text{calc})/\sum_{\mathbf{k}} y_{i,\mathbf{k}}(\text{calc}) \qquad (18)$$

where the summation at the numerator is over the peak range corresponding to the **h** reflection and the summation at the denominator is over the reflections contributing to the i-th step observation. The formula starts with arbitrary integrated intensity values in $y_{i,\mathbf{h}}$(calc), then the algorithm is applied iteratively. Moreover, the profile parameters are optimized by least squares. The method is highly convergent, provides positive intensities but tends to equiportion the peak intensity of a group of strongly overlapping reflections.

Both the Pawley and Le Bail methods are based on the minimization of the difference between the observed and calculated profiles. The R_p profile reliability parameter is used for evaluating the efficiency of the powder decomposition process:

$$R_p = \frac{\sum_i |y_i(\text{obs}) - y_i(\text{calc})|}{\sum_i y_i(\text{obs})} \qquad (19)$$

The summation is extended over the number of experimental profile counts.

Irrespective of whether the Pawley or Le Bail method is used, the accuracy of the $I_{\mathbf{h}}$ values is commonly low, especially for the overlapping reflections, depending on the structure complexity and experimental resolution. This occurrence is so critical that it can reduce the success probability of solving crystal structures when the so called ‘traditional’ approaches such as Direct methods are used (see Section 4.5).

4.5. Structure solution

The solution process can be carried out by following one of the methods described below.

4.5.1. Reciprocal space (RS) methods

Reciprocal space methods are also called ‘traditional’ because, as with the single-crystal *ab initio* structure-solution process, only minimal information is required (in addition to the cell parameters and space group, only the chemical formula and the experimental pattern are required). They work in the reciprocal space (the space of the reflections) and are based on two main steps: (1) the decomposition of the experimental profile for extracting the integrated intensities (see Section 4.4); (2) the solution of the 'phase problem': the phase $\varphi_{\mathbf{h}}$ of the structure factor $F_{\mathbf{h}}$ $[F_{\mathbf{h}} = |F_{\mathbf{h}}|\exp(i\varphi_{\mathbf{h}})]$ is estimated probabilistically using Direct methods (DM) (Giacovazzo, 1998, 2013), the most widely used of which are the Patterson methods (Rius and Frontera, 2007; Rius *et al.*, 2007; Rius, 2011, 2014) or maximum entropy methods (Gilmore, 1996). The inverse Fourier transform of $F_{\mathbf{h}}$ (known in moduli and phases) provides the electron density map $\rho(\boldsymbol{r})$

$$\rho(\boldsymbol{r}) \cong \frac{1}{V}\sum_{\mathbf{h}}|F_{\mathbf{h}}|\exp(i\varphi_{\mathbf{h}})\exp(-2\pi i\mathbf{h}\cdot\boldsymbol{r}) \tag{20}$$

The structure model is obtained by the $\rho(\boldsymbol{r})$ interpretation.

Direct methods are fast and reliable but their efficiency is limited by the quality of the $|F_{\mathbf{h}}|$ values extracted the probabilistic analysis of which is essential for phasing. The phasing process usually ends with several phasing trials which are ranked according to a combined figure of merit (*e.g.* CFOM (Cascarano *et al.*, 1987, 1992)). The ideal CFOM value is equal to 1.0 and should correspond to the correct phasing, but inadequacy in terms of $|F_{\mathbf{h}}|$ may lead to incorrect phases and/or meaningless CFOM values. Consequently, the selection of the best phasing set by means of CFOM may fail. New procedures have been developed aimed at strengthening the efficacy of the phasing process and at discriminating incorrect solutions from promising trials (Altomare *et al.*, 2003, 2011, 2013b).

The outcome of DM depends not only on the quality of the intensities extracted but also on the experimental resolution. In fact, the limited experimental resolution distorts the Fourier synthesis even at atomic resolution: when the resolution is >1 Å, the electron density map is an imperfect representation of the true $\rho(\boldsymbol{r})$. This is particularly true in the case of organic compounds for which, because of the low atomic scattering factors of the light atoms constituting the molecule, the experimental diffraction intensity at high resolution is very weak and unusable. Resolution bias errors combine with uncertainty on moduli and/or phases of structure factors and often provide an approximate map which corresponds to a structure model which is far from correct: some peaks are missed, false positions occur and some other positions are misplaced from the true ones. A structure model of such quality requires optimization before final refinement by the Rietveld method (the low convergent rate of the Rietveld method does not guarantee the attainment of the correct solution if the starting model to be refined is very rough) (see Section 4.6). Therefore, DM usually end with procedures aimed at improving the model: recovering the missed atoms and locating atom peaks in better positions.

Among the methods developed for model optimization are: (a) the Weighted Least-Squares Fourier recycling (WLSQ-FR) method (Altomare *et al.*, 2006). This consists of cyclic combination of least squares (for improving the atom positions) and $2|F_{\mathbf{h}}|_{\mathrm{o}}-|F_{\mathbf{h}}|_{\mathrm{c}}$ Fourier map calculation (for recovering the missing atoms), where $|F_{\mathbf{h}}|_{\mathrm{o}}$ is the structure factor modulus observed and $|F_{\mathbf{h}}|_{\mathrm{c}}$ is that calculated from the current model. Least squares use $|F_{\mathbf{h}}|_{\mathrm{o}}$ values modulated by a suitable weighting scheme which takes into account the low reliability of the experimental structure factor moduli of overlapping reflections. They refine the atomic fractional coordinates and the isotropic thermal factors. The WLSQ-FR method is effective in particular for the optimization of inorganic compounds; (b) the Resolution Bias Modification algorithm (RBM) (Altomare *et al.*, 2008c,d, 2009b; 2010). Developed in reciprocal and direct spaces, and based on the assumption that each peak in the electron density map consists of a main peak and ripples, it is a procedure capable of correcting the electron density map by typical resolution-bias errors: peak broadening, peak shift, intensity distortion. In particular, organic molecules benefit from the RBM application; (c) the COVMAP approach (Altomare *et al.*, 2012). Inspired by the theory that calculates the covariance and the correlation between two points of the electron density map, it recovers the missing atoms by modifying $\rho(\boldsymbol{r})$ according to basic crystal chemistry rules, essentially the expected bond distances. The method has been combined with WLSQ-FR and RBM to strengthen its capacity for structure model improvement. In this way, the correct solution is obtained even when the DM model is very approximate. This surprising result has led workers to attempt the solution process by starting from a fully random structure model and to propose a new approach, the RAMM method (Altomare *et al.*, 2013a) which skips the Direct methods step and substitutes the DM model with a random model which attempts to provide the correct solution using the COVMAP-based approach.

For the NEPHE structure, in Fig. 9 we give the structure model obtained by *EXPO2013* just after the phasing process by DM (Fig. 9a) and after WLSQ-FR optimization (Fig. 9b) which leads to the correct solution.

4.5.2. Direct space (DS) methods

Direct space methods have been developed more recently than RS methods in order to overcome the limitation on RS methods of their strong dependence on the quality and resolution of the structure factor moduli extracted from the experimental profile. In fact, DS methods avoid the extraction process, but they need prior information about

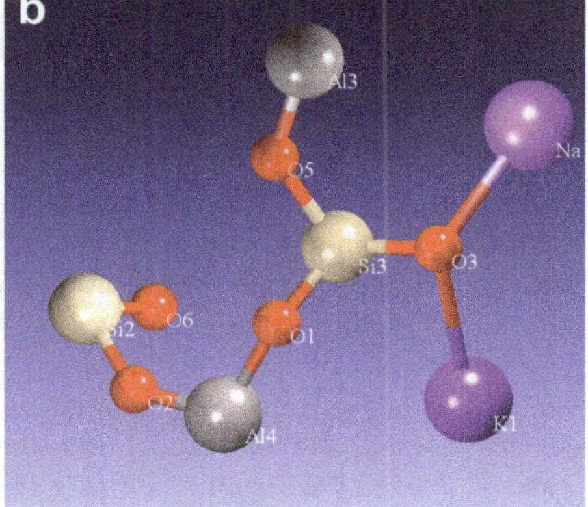

Figure 9. NEPHE compound. (a) The crystal structure obtained just after the phasing process by Direct methods; (b) the correct solution automatically provided by *EXPO2013*.

the expected molecular geometry (for this reason they are referred to as non-*ab initio*, because they require additional information, not always available).

Three main steps are involved: (1) the generation of several trial models all compatible with the expected molecular geometry. A starting model is built by molecular editor software or, alternatively, imported from the literature (an already solved structure with similar molecular geometry); the model is described in terms of bond distances and angles and torsion angles; then numerous models are generated by varying the location of the model in the crystal cell (three degrees of freedom for each fragment), the orientation of the model in the space (three degrees of freedom for each fragment) and the internal conformation expressed in terms of torsion angles; (2) in parallel with each random model a cost function (R_{wp}) is calculated giving the weighted agreement between the observed profile and that calculated by the current model; (3) the global minimum is sought in the N-dimensional hypersurface of R_{wp} (N is the number of structural parameters to be varied). This corresponds to the best structure model.

The DS methods adopt one of the following global optimization algorithms: Grid Search (Chernyshev and Schenk, 1998), Monte Carlo (Harris *et al.*, 1994; Andreev *et al.*, 1997; Tremayne *et al.*, 1997), Simulated Annealing (Kirkpatrick, 1983; Andreev and Bruce, 1998; Engel *et al.*, 1999; Coelho, 2003), Big-Bang Big-Crunch (Altomare *et al.*, 2013c), Genetic algorithm (Goldberg, 1989; Kariuki *et al.*, 1997; Shankland *et al.*, 1997).

Among the available computing programs based on DS methods, *DASH* (David, *et al.*, 2006) and *FOX* (Favre-Nicolin and Cerny, 2004) are used widely. *EXPO2013* is also able to solve structures by DS methods, in particular, by Simulated Annealing.

With respect to RS methods, another advantage of DS methods is that they do not use the integrated intensities extracted from the powder profile and they do not require high-resolution data. Usually, a resolution of 2.5 Å is sufficient for most small molecules/inorganic structures, less if rigid bodies are used. The time required by DS methods for attaining the correct solution depends heavily on the number of degrees of freedom varied in the global optimization process.

We report a successful example of structure solution using Simulated Annealing in *EXPO2013*: $Sb_2(PO_4)_3$ (Jouanneaux *et al.*, 1991) (cell parameters: a = 11.936(1) Å, b = 8.7354(8) Å, c = 8.3185(8) Å, β = 91.12(2)°, space group: $P2_1/n$, radiation type: synchrotron, λ = 1.45072 Å). The starting model was built using the graphical tools of *EXPO2013* combining two octahedra of SbO_6 and three tetrahedra of PO_4 (Fig. 10a). The Dynamical Occupancy Correction (Favre-Nicolin and Cerny, 2002) algorithm was applied in order to take into account the fact that oxygen atoms share different coordination polyhedra. Polyhedral representation of the final model is displayed in Fig. 10b (c-axis projection, a horizontal). The value of the cost function (R_{wp}) corresponding to the final model was 13.87% and it was attained in ~1 h. The structure was also solved successfully starting from isolated atoms of Sb, P and O positioned randomly in the unit cell. In this last case, R_{wp} was 17.62% and the solution was obtained in ~8 h.

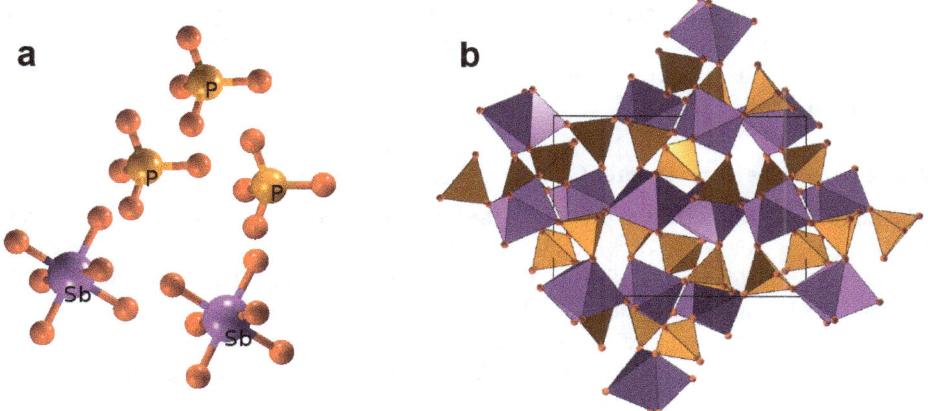

Figure 10. $Sb_2(PO_4)_3$ compound. (a) The starting model combining two octahedra of SbO_6 and three tetrahedra of PO_4; (b) final model obtained by Simulated Annealing in *EXPO2013* (*c*-axis projection, *a* horizontal).

4.5.3. Hybrid methods

Hybrid methods combine direct and reciprocal-space methods (Johnston *et al.*, 2002; Brenner *et al.*, 1997; Altomare *et al.*, 2008e). Among them, we note: (a) 'Charge Flipping'. Developed first for single-crystal solutions (Oszlányi and Sütó, 2004) and then transferred to powders (Wu *et al.*, 2006; Baerlocher *et al.*, 2007), charge flipping starts from an electron-density map calculated by the observed structure factor moduli and random phases. Then the map is modified in direct space (flipping the sign of all the densities below a very small positive threshold value) and inverted successively. New phases are thus obtained and a new map is created by using the current phases and the observed structure factor moduli and so on, switching back and forth between real and reciprocal space. In the adaptation to powder data, the method also tries to improve the estimates of the observed moduli by repartitioning the overall intensity of groups of overlapping reflections according to the information coming from the inverted map. Moreover, a histogram-matching/repartitioning-based approach (Baerlocher *et al.*, 2007) has been suitably integrated; (b) 'Focus' (Grosse-Kunstleve *et al.*, 1997). This approach is powerful for solving zeolites. High symmetry, a large unit cell and a high degree of reflection overlapping make the solution of these structures a non-trivial matter. The method combines an automatic Fourier recycling algorithm with a topology (framework) search specific to zeolites. It starts by an electron-density map calculated by random phases. Then the zeolite-specific crystal chemical information is used in the interpretation of the map; (c) *POLPO* (Altomare *et al.*, 2000; Giacovazzo *et al.*, 2002). It can be used for solving polyhedrally coordinated structures. The method starts from the following conditions: the positions of the heavy atoms located by DM are usually well positioned contrary to what is found for light atoms. For that reason the expected polyhedral coordination information (the type of coordination, tetrahedral or octahedral, the polyhedral coordination distance and tolerance values on polyhedral distances and angles) of each of the heavy atom is introduced. Several structure models are generated by Monte Carlo algorithms, all compatible with the starting conditions and from them the model corresponding to the best R_p value is selected as the most meaningful.

4.6. The Rietveld refinement

The final step of the solution process is the structure model refinement by the Rietveld method (Rietveld, 1969) of which we give here only the fundamentals. For details of this subject the reader is referred to specialized papers, *e.g.* Bish and Post (1989), Young (1993) and McCusker *et al.* (1999).

The Rietveld method is a whole-pattern-fitting technique based on cycles of non-linear least-squares aimed at refining both structure and profile parameters. The quantity to be minimized in the least squares is:

$$\sum_i w_i[y_i(\text{obs}) - y_i(\text{calc})]^2 \tag{21}$$

where the summation is over all data points, y_i(obs) and y_i(calc) are the observed and calculated (according to equation 17) profile intensities at the i-th step, respectively; $w_i = 1/y_i$(obs) is the weight associated with the i-th observed count.

The model parameters which can be refined are: atom fractional coordinates, isotropic and anisotropic thermal factors, site occupancies; and profile parameters: 2θ zero, asymmetry, background but also scale factor, preferred orientation, absorption, specimen displacement, specimen transparency, crystallite size and microstrain.

Due to the non linearity of the relationships between the intensities and the parameters adjustable by least-squares, the structure model to be refined must be close to the correct one, otherwise the refinement process can diverge or lead to a false minimum. This condition, generally true (also for single-crystal refinement), is particularly critical in case of powder for which the loss of information due to the one-dimensional collapse and the correlation between the several refined parameters cause false minima to be prevalent. Moreover the good outcome of the Rietveld method depends on both the quality of the collected data and the description of the observed profile by analytical functions. For these reasons care must be taken in carrying out the refinement and, in particular, in defining a strategy of refinement (McCusker *et al.*, 1999).

The best way of monitoring the progress of Rietveld refinement is to check the difference between the observed and calculated pattern by the profile numerical agreement factor:

$$R_{\text{wp}} = \{\sum_i w_i[y_i(\text{obs}) - y_i(\text{calc})]^2 / \sum_i w_i[y_i(\text{obs})]^2\}^{1/2} \tag{22}$$

Other statistical R values are also used.

The criteria for judging the goodness of a Rietveld refinement result are: (a) the fit of the calculated pattern to the observed data, supported not only by a small R_{wp} value, but also by seeing the difference plot; (b) the chemical sense of the refined structure model (reliable bond distances, angles and population parameters).

One way to compensate for the loss of information mentioned previously is to incorporate additional observations in the Rietveld refinement process. They can be included as constraints (hard constraints) which are imposed rigorously and/or restraints (soft constraints) which are imposed approximately. An example of constraint is the symmetry constraint placed on atoms in special positions; examples of restraints are bond distances and angles. Restraints are useful for stabilizing the refinement, avoiding false minima, reaching convergence quickly and increasing the number of parameters to be refined. For this reason they are usually applied in the case of large structures.

Structures with up to 200 structural parameters can be refined successfully by the Rietveld method if the experimental data are good. Note that the method has been applied appropriately in refining protein structures also (we give here only the pioneering and the most recent paper on this subject: Von Dreele, 1999; Margiolaki *et al.*, 2013; Valmas *et al.*, 2015). Knowledge of the unit-cell parameters, space group and refined crystal structure is basic in mineralogy for different purposes; one of these is the study of minerals at non-ambient pressure/temperature, as described below.

5. Powder diffraction at extreme conditions

Over the last 60 years, *in situ* X-ray and neutron powder diffraction experiments have enabled the study of matter in response to 'intensive' thermodynamic parameters, in particular, pressure and temperature. The range of pressure or temperature at which diffraction experiments can be conducted currently is vast, ranging from a few degrees Kelvin to kiloKelvin or from a few bars to Mbars, reflecting the extraordinary developments in high-temperature and high-pressure techniques over recent decades. The research in high-pressure/high-temperature mineralogy has been aimed mainly at describing the physical and chemical properties of materials: thermal expansion, compressibility, thermal and compressional anisotropy, *P*/*T*-phase stability, *P*/*T*-induced phase transitions, *P*/*T*-induced structure evolution (*e.g.* *T*-induced order-disorder processes, atomic migration, bond libration; *P*-induced distortion mechanisms, hybridization, magnetic states and superconductivity, amorphization). The investigation of mineral behaviour under extreme conditions is the basis for predicting the properties of multi-phase systems of geological interest. Seismology, for example, is one of the disciplines based on mineral physics. Seismic techniques provide the highest-resolution measurements of Earth structures, and seismic velocities reflect the density and the elasticity of mineral assemblages in response to the pressure and temperature applied. Modern models of the Earth (*e.g.* Preliminary Reference Earth Model 'PREM', Dziewonski and Anderson, 1981) are based on the interpretation of seismic data in the light of the results of *P*/*T*-laboratory experiments on mineral stability and seismic velocity measurements. On the other hand, experiments in high-*P*/*T* mineralogy often have implications beyond the Earth sciences, *e.g.* in materials sciences or metallurgy. Several studies on the optimization of synthesis protocols, thermo-elastic properties and *P/T* stability of oxides, ceramic compounds, alloys, microporous or mesoporous materials and hard or super-hard materials have been conducted in the context of mineralogical investigations. In this chapter, we focus our attention on the basic concepts of mineral physics used to describe the *P/T*-induced phenomena, on the principal devices able to generate non-ambient conditions and on experimental findings based on X-ray and neutron powder diffraction in the range of pressure and temperature conditions of planetary geophysics.

5.1. High-temperature elastic behaviour of crystalline materials

In response to applied temperature, a crystal increases its energy mainly through oscillation of atoms about their equilibrium position. This effect is usually described in terms of 'lattice vibrations' or 'phonons'. Melting begins when the amplitude of the vibrations exceeds a critical value, with a breakage of atomic bonds. On the other hand, the amplitude decreases at low temperature, reaching a minimum when $T = 0$ K (zero-point oscillations). A simple model based on the classical harmonic oscillators used to describe the motion of atoms is sufficient to explain some of the *T*-induced phenomena in a crystal (*e.g.* order-disorder processes, electrical conductivity, melting). However, in classic harmonic oscillators, the equilibrium bond lengths are, in theory, unaffected by temperature. To describe all the phenomena in response to high temperature, we need to consider also the anharmonic vibrational contribution: the thermal expansion of crystals is, for example, one of the effects of the anharmonic vibrational term, along with the difference between the isothermic and adiabatic elastic constants of a solid and their dependence on temperature (and pressure).

Powder diffraction experiments are used currently to describe the thermo-elastic behaviour of crystalline materials on the basis of the variation of the unit-cell parameters in response to the applied temperature. Fractional changes of the unit-cell parameters represent the strains. Strains form a second-rank tensor (ε_{ij}, Nye, 1985), which can be calculated directly from the sets of cell parameters measured *e.g.* at different temperatures or pressures. The diagonal components of ε_{ij} are the 'normal strains', whereas the other components represent the 'shear strains'. There is no unique definition of strain: Eulerian and Lagrangian strains, for example, are specified with respect to the final and the initial states, respectively, thus providing a different description of the deformation. In addition, finite and infinitesimal strain tensors may be used, according to Schlenker *et al.* (1978). However, at the limit of small strains, the four aforementioned tensors are practically indistinguishable. A further variable is the Cartesian axial set used to describe the strain ellipsoid orientation: strain tensor on a set of Cartesian axes with X // a^* and Z // c is the conventional orientation specified by the *Institute of Radio Engineers* (Schlenker *et al.*, 1978); however, further orientations have also been adopted (*i.e.* Y // b and Z // c^*, X // a^* and Y // b) for reasons of convenience in specific studies.

Under elastic regime, and for a modest T change (ΔT), the components of the strain tensor are proportional to the applied temperature, with:

$$\varepsilon_{ij} = \alpha_{ij}\Delta T \tag{23}$$

where α_{ij} are the thermal expansion coefficients (expressed in K^{-1} or in $°C^{-1}$), and are constants in a given material; ε_{ij} and α_{ij} are both symmetrical tensors. For any material there are three mutually perpendicular directions which remain perpendicular after expansion. These are the three principal axes of strain, along the directions of which the following equations are valid:

$$\varepsilon_1 = \alpha_1\Delta T,\ \varepsilon_2 = \alpha_2\Delta T,\ \varepsilon_3 = \alpha_3\Delta T \tag{24}$$

with α_1, α_2 and α_3 defined as the principal expansion coefficients. In their infinitesimal form, the principal expansion coefficients are defined as:

$$\alpha_i = (1/l_i)(\partial l_i/\partial T) = \partial \ln l_i/\partial T \text{ (} l \text{ is the length, } i = 1, 2, 3) \tag{25}$$

Thus, if a sphere of material is heated, in general it becomes an ellipsoid, with principal axes of lengths proportional to $(1 + \alpha_1\Delta T)$, $(1 + \alpha_2\Delta T)$, and $(1 + \alpha_3\Delta T)$, respectively.

The property of thermal expansion can also be represented by a surface of the form:

$$\alpha_1x_1^2 + \alpha_2x_2^2 + \alpha_3x_3^2 = 1 \tag{26}$$

(x_1, x_2, x_3 are three mutually perpendicular axes).

The volume thermal expansion coefficient is defined as:

$$\alpha_V = (\alpha_1 + \alpha_2 + \alpha_3) = (1/V)(\partial V/\partial T) = \partial \ln V/\partial T \text{ (} V \text{ is the volume)} \tag{27}$$

Strain and thermal ellipsoids of a crystal follow Neumann's principle ("The symmetry elements of any physical property of a crystal must include the symmetry elements of the point group of the crystal", in Nye, 1985). In cubic crystals, the three principal axes of thermal expansion are identical, and so the 'ellipsoid' is actually a sphere. In hexagonal and tetragonal crystals, the ellipsoid is uniaxial, and the axis is aligned with the unique crystallographic axis. In orthorhombic crystals, the three principal axes of the ellipsoid are parallel to the three mutually orthogonal crystallographic axes. In monoclinic crystals, one of the axes of the ellipsoid is parallel to the 2-fold axis (conventionally [010]), whereas the other two axes of the ellipsoid are dispersed on the plane perpendicular to the 2-fold axis (conventionally (010)). In triclinic crystals, the orientation of the ellipsoid is completely independent of the orientation of the crystallographic axes. In monoclinic or triclinic crystals the strain ellipsoid may rotate with changing T (or P). A simple protocol to obtain magnitude and orientation of the principal axes of the strain ellipsoid over a finite range in T (or P), on the basis of the unit-cell parameters variation, is given in Hazen *et al.* (2000), based on the findings of Ohashi and Burnham (1973), Schlenker *et al.* (1975, 1978), Ohashi (1982) and Boisen and Gibbs (1990). A program to calculate strain tensors from unit-cell parameters, *Win_Strain*, was written by R. Angel and is available from http://www.rossangel.com.

Nowadays, *in situ* powder diffraction experiments allow data to be collected over a significantly large T-range, allowing the opportunity to model even the variation in the axial or volume thermal expansion coefficients with temperature. Experimental findings suggest that the evolution of α with T in minerals may be described adequately by the following polynomial function (Pawley *et al.*, 1996; Holland and Powell, 1998):

$$\alpha(T) = s_0 + s_1 T^{-1/2} \tag{28}$$

where s_0 and s_1 are refinable parameters. Equations 27 and 28 lead to:

$$V(T) = V_0 \exp[s_0(T - T_0) + 2s_1(T^{1/2} - T_0^{1/2})] \approx V_0[1 + s_0(T - T_0) + 2s_1(T^{1/2} - T_0{}^{1/2})] \tag{29}$$

(V_0 and T_0 represent the initial state).

The same formalism can be used to describe the axial thermo-elastic behaviour, with:

$$l(T) = l_0[1 + s_0(T - T_0) + 2s_1(T^{1/2} - T_0^{1/2})] \tag{30}$$

(l_0 represents the initial state).

The thermo-elastic anisotropy of a crystalline material can be described, at a first approximation, by $s_0(a)$:$s_0(b)$:$s_0(c)$, with *a, b, c* being the three crystallographic axes.

As there is no physical restriction on the formalism to describe the evolution of α with *T*, in the literature such an evolution is also modelled (Fei, 1995) with:

$$\alpha(T) = s_0 + s_1 T + s_2 T^{-2} \tag{31}$$

A few recent examples of experiments aimed at describing the thermo-elastic behaviour of minerals, or their synthetic counterparts, based on the aforementioned protocols are found in Rinaldi *et al.* (2005; Co-Mg-olivine, *in situ* neutron powder diffraction up to 1273 K), Gatta *et al.* (2007; orthopyroxene, *in situ* neutron powder diffraction up to 1473 K), Gatta *et al.* (2012; Cs-ABW zeolite, *in situ* synchrotron powder diffraction up to 1270 K). A list of the thermal expansion coefficients of the most important minerals can be found in Fei (1995) and Holland and Powell (1998).

5.2. High-pressure elastic behaviour of crystalline materials

Under a linear elastic regime, the strains in a crystal are linearly dependent on the applied stress (Hooke's Law). The stress is describable by a second-rank tensor (σ_{ij}), and the stress-strain relationship is the following tensor equation (Nye, 1985):

$$\varepsilon_{ij} = s_{ijkl}\sigma_{kl} \tag{32}$$

where s_{ijkl} is the elastic compliance fourth-rank tensor. As an alternative to the previous equation, the stress-strain relationship can be expressed as:

$$\sigma_{ij} = c_{ijkl}\,\varepsilon_{kl} \tag{33}$$

here c_{ijkl} is the elastic stiffness fourth-rank tensor. Compliance (usually expressed in GPa^{-1}) and stiffness (usually in GPa) are reciprocal in nature, but individual components of the two tensors are not. Under hydrostatic conditions, the off-diagonal components of the stress tensor are zero (*i.e.* there are no shear stresses) and the diagonal terms of the tensor are all equal to the applied pressure (*P*), so that:

$$\sigma_{kl} = -P \text{ (for } k = l) \tag{34}$$

and $\sigma_{kl} = 0$ for $k \neq l$.

Thus, the hydrostatic conditions lead to:

$$\varepsilon_{ij} = -Ps_{ijkk} \tag{35}$$

The volume compressibility (β_V) is defined as the proportional decrease in volume of a crystal when subjected to unit hydrostatic pressure, so that:

$$\beta_V = (1/V)(\partial V/\partial P) = \partial \ln V/\partial P \quad (36)$$

The isothermal bulk modulus at ambient pressure (P_0) and temperature (T_0) is defined as:

$$K_{P_0,T_0} = \beta_V^{-1} = V_0(\partial P/\partial V)_{P_0,T_0} \quad (37)$$

The fractional volume change of the crystal under stress is given by the sum of the diagonal terms of the strain tensor; at hydrostatic conditions:

$$\Delta V/V = \varepsilon_{ii} = -Ps_{iikk} \quad (38)$$

The aforementioned equations lead to the following relationship between the isothermal elastic compliances (with the individual written out in matrix notation) of the crystal and its isothermal bulk modulus, often termed the Reuss bulk modulus:

$$K_{P_0,T_0} = (s_{11} + s_{22} + s_{33} + 2s_{12} + 2s_{13} + 2s_{23})^{-1} = \beta_V^{-1} \quad (39)$$

This relationship is valid for all crystal systems.

The linear compressibility (β_x) of a crystal is the relative decrease in length of a line (l) when the crystal is subjected to unit hydrostatic pressure, so that:

$$\beta_x = (1/l)(\partial l/\partial P) = \partial \ln l/\partial P \quad (40)$$

If the isothermal elastic compliances are known, the linear compressibility along any direction in a crystal is defined by its direction cosines (u_i), so that:

$$\beta_x = s_{ijkk}u_iu_j \quad (41)$$

For the seven crystal systems, the expressions of the linear compressibilities of the axes and the individual elastic compliances can be obtained (see Nye, 1985). *In situ* powder diffraction experiments allow variations in the unit-cell parameter to be measured in response to the applied pressure. The orientation and magnitude of the strain ellipsoid can be calculated as previously described for the T-induced strain.

However, these expressions have been derived for the linear elastic regime, which means implicitly that only small stresses and small resulting strains can be considered. At high pressures, linearity breaks down, and the total strain is no longer proportional to the applied total pressure. The expressions given above do, of course, remain valid for small pressure changes δP beyond a given pressure P, for which

$$\varepsilon_{ij} = -\delta P s_{ijkk} \quad (42)$$

and the s_{ijkk} are a function of pressure. Thus, while linear elasticity can describe successfully the elastic response of a material at any pressure to a small increment in pressure, it cannot address the total response of the material to a large pressure change. For that, the concept of an equation of state is required.

5.3. *P-V* and *P-T-V* equations of state

The compressional behaviour of crystalline materials at constant temperature is usually described using the formalism of isothermal equations of state (EoS). The simplest EoS is that proposed by Murnaghan (1937), based on the assumption that the isothermal bulk modulus varies linearly with pressure:

$$K_P = K_{P_0} + (\partial K/\partial P)P = K_{P_0} + K'P \qquad (43)$$

This simple EoS reproduces successfully both P-V data and the correct values of the room-pressure bulk modulus for limited compressional regimes in which the volume decrease is <10%. The Murnaghan EoS is used commonly by the mineralogical thermodynamic databases (*e.g.* Holland and Powell, 1998) because it is readily inverted:

$$V_P = V_{P_0}\,[1 + (K'P/K_{P_0})]^{-1/K} \qquad (44)$$

or

$$P = (K_{P_0}/K')[(V_{P_0}/V_P)^{K'} - 1] \qquad (45)$$

where V_{P_0} and V_P represent the unit-cell volume at ambient and high-P conditions, respectively.

The Birch-Murnaghan EoS (Birch, 1947) is another isothermal equation used commonly in mineralogy. This EoS is based on the assumption that the high-pressure strain energy in a solid can be expressed as a Taylor series in the Eulerian finite strain, defined as:

$$f_E = [(V_{P_0}/V_P)^{2/3} - 1]/2 \qquad (46)$$

Expansion in the Eulerian strain polynomial has the following form:

$$P(f_E) = 3K_{P_0} f_E\,(1 + 2f_E)^{5/2}\,\{1 + 3/2(K' - 4)\,f_E + 3/2[K_{P_0}K'' + (K' - 4)(K' - 3) + 35/9]\,f_E^2 + ...\} \qquad (47)$$

(where $K' = \partial K_{P_0}/\partial P$; $K'' = \partial^2 K_{P_0}/\partial P^2$).

Other isothermal EoS have been proposed, described by Anderson (1995), Angel (2000), Stacey and Davis (2004). A review with the description of the EoS formalisms, fitting protocols, prediction of the uncertainties and practical considerations was given by Angel (2000).

The evolution of V *vs.* T through the thermal expansion coefficient (α, equation 27) and V *vs.* P through the compressibility coefficient (β_V, equations 36 and 37) lead to:

$$V_{P,T} \approx V_{P_0,T_0}\,(1 + \alpha_P \Delta T)(1 - \beta_T \Delta P) \qquad (48)$$

Because:

$$\alpha_P = \alpha_{P_0} + (\partial\alpha/\partial P)_{T_0}\Delta P \qquad (49)$$

and

$$\beta_T = \beta_{T_0} - (\partial\beta/\partial T)_{P_0}\Delta T \qquad (50)$$

we obtain:

$$V_{P,T} \approx V_{P_0,T_0}\,[1 + \alpha_{P_0}\Delta T - \beta_{T_0}\Delta P + (\partial\alpha/\partial P)_{T_0}\Delta P\Delta T - (\partial\beta/\partial T)_{P_0}\Delta P\Delta T] \qquad (51)$$

This is valid only for limited pressure intervals as it relies on the assumption of linear elasticity. It is also possible to use the same formalism of the aforementioned isothermal EoS (equations 45 and 47) considering the evolution of K with T (Fei, 1995):

$$K_{P_0,T} = K_{P_0,T_0} + (\partial K/\partial T)\Delta T \tag{52}$$

However, determination of the parameters of the EoS in this approach requires *in situ* experiments at combined high *P-T*. Often, experiments can be performed only at high *P* (at room *T*) or at high *T* (at room *P*), and so only isothermal compressibility and isobaric thermal expansion coefficients are available. In this case, an approximate EoS can be used:

$$V_{P,T} \approx V_{P_0,T_0}[1 - \beta_{T_0}\Delta P + \alpha_{P_0}\Delta T] \tag{53}$$

A few recent examples of experiments aimed at describing the *P-T-V* EoS of minerals, or their synthetic counterparts, based on the aforementioned protocols, can be found in Comodi *et al.* (2002: muscovite and paragonite, *in situ* synchrotron powder diffraction at combined *P/T*, up to 1270 K and 5 GPa) and Gatta *et al.* (2011a: epidote, *in situ* synchrotron powder diffraction up to 1200 K and up to 10 GPa).

5.4. Structure evolution at high pressure and temperature

Changes in unit-cell parameters in response to the applied temperature or pressure reflect small atomic displacements, atomic ordering, magnetization, *etc.* Structure refinements allow a description of the main *T*- or *P*-induced mechanisms at the atomic scale.

Disorder of two (or more than two) chemical species across different crystallographic sites at high temperature leads to an entropic stabilization of mineral phases if compared to low-temperature ordered structures. Structure refinements observe the long-range order in a crystalline material. The ordering can be described on the basis of the direct measurement of scattering amplitudes at crystallographic sites or can be deduced on the basis of bond-lengths. *T*-induced ordering was described as 'convergent' or 'non convergent' (*e.g.* Thompson, 1969). A convergent ordering occurs if, in response to *T*, distinct crystallographic sites become symmetrically equivalent when their average occupancy becomes identical. In this case, the order-disorder process leads to a symmetry change. In non-convergent ordering, the *T*-induced disordering does not transform distinct crystallographic sites to be symmetrically equivalent, even when the occupancies are identical on each. In this second case, no symmetry change occurs. The knowledge of cation partitioning in response to *T, P* and time can be used to obtain geothermometric, geobarometric or geospeedometric information. A thorough description of the *T*-induced order-disorder processes in minerals, with a thermodynamic modelling of the phenomenon, was provided by Redfern (2000). An excellent example of the issues is provided by the series of the neutron powder diffraction experiments devoted to the *T*-induced octahedral ordering in olivines (*i.e.* Mg/Mn and Fe/Mn by Redfern *et al.*, 1997; Mg/Fe by Redfern *et al.*, 2000; Mg/Ni by Henderson *et al.*, 2001; Mg/Co by Rinaldi *et al.*, 2005).

The inter-atomic bond distances, deduced on the basis of a structure refinement, are affected by the thermal vibrations of the atoms. This is due to the fact that the coordinates of a given crystallographic site represent the centroid of the electron density distribution, as a result of the combined effects of atomic structure and thermal displacement (Busing and Levy, 1964). The distance between the centroids of two bonded atoms does not provide a reliable picture of the bonding evolution with T, with an unrealistic shortening with increasing T. Busing and Levy (1964) provided the first protocol designed to solve this problem, showing that a more realistic measure of interatomic distance between two atomic sites is their mean separation, which will always be greater than the interatomic distance as determined by refinement. The calculation of the inter-atomic mean separation values following Busing and Levy (1964) is based on a model of the correlation of thermal motions between the two bonded atoms, and different corrections were proposed (*i.e.* lower bound corrections, upper bound corrections, riding motion corrections, non-correlated motion corrections). A further correlated motion is the so-called 'rigid-body motion', for atoms that are strongly bonded in a coordination polyhedron, for example in Si- or Al-tetrahedra, or Al- or Mg-octahedra. In this protocol, the atoms are assumed to vibrate as a group (*i.e.* as if connected by rigid rods), oscillating with correlated translational and librational motion (*e.g.* Schomaker and Trueblood, 1968; Bartelmehs *et al.*, 1995). A thorough description of the correction of bond distances based on the rigid-body motion was given by Downs (2000), with some examples applied to minerals. The aforementioned corrections were developed using the anisotropic displacement parameters of atoms. With a few exceptions, structure refinements based on powder diffraction data do not usually provide anisotropic displacement parameters. Downs *et al.* (1992) derived a simple equation ('simple rigid bond correction', SRB) for bond-length correction between a cation and an anion connected by a strong bond, based on the assumption that the cation undergoes only translational motion:

$$R_{SRB}^2 = R_{obs}^2 + (3/8\pi^2)\,(B_{eq}(Y) - B_{eq}(X)) \tag{54}$$

where X is the cation and Y is the anion, B_{eq} is the equivalent displacement parameter (in Å^2) and R_{SRB} and R_{obs} are the corrected and observed bond lengths (in Å), respectively. This model can be applied easily to structure data with isotropic displacement parameters.

The effect of temperature can lead also to drastic structure transformation through a phase transition to a different crystalline polymorph, stable at the new T conditions, or to the amorphous state. But the effect of temperature can lead also to a chemical transformation of the materials: *e.g.* hydrous materials tend to dehydrate in response to the applied T. Some mineralogical examples, in this respect, can be found in zeolites (which tend to lose H_2O molecules at $T < 650$ K, *e.g.* Cruciani, 2006) or in micas (which show dehydroxylation at $T > 800$ K, *e.g.* Gemmi *et al.*, 2008).

In response to applied pressure, crystalline materials react through three main P-induced mechanisms to accommodate the volume compression: inter-polyhedral tilting, polyhedral distortion and polyhedral compression. Tilting of the polyhedra occurs around the oxygens shared between adjacent polyhedra, which act as 'hinges'. This mechanism does not imply any distortion of the polyhedra. The second mechanism involves distortion of the polyhedra but maintenance of the average intra-polyhedral M–O bond length. The third mechanism is represented by the P-induced compression of the single M–O bond lengths. These three mechanisms act simultaneously at any given pressure, but with a different magnitude. Polyhedral tilting, if allowed by the structural topology, is usually dominant, with polyhedra distortions and compression of the intra-polyhedral M–O bond distances being minor mechanisms. This hierarchy can be explained by the fact that the first mechanism is energetically less costly than the other two, although in less-strongly bonded polyhedra, such as some perovskites, the compression of the polyhedra is an essential component in determining the thermodynamic properties (*e.g.* Angel *et al.*, 2005). Several parameters have been proposed in order to describe analytically the distortion of polyhedra, as a deviation from regularity. The most used are: the 'quadratic elongation' ($<\lambda>$) and the 'bond angle variance' (σ^2) (Robinson *et al.*, 1971), or the 'polyhedral volume distortion' according to Balic-Zunic and Vickovic (1996) and Makovicky and Balic-Zunic (1998).

5.5. Devices for *in situ* experiments under extreme conditions

In situ high- and low-*T* X-ray or neutron powder diffraction experiments can be performed using a series of devices able to provide a homogeneous temperature distribution in the polycrystalline sample. Comprehensive reviews have been done by Hazen and Finger (1982), Chung *et al.* (1993), Peterson and Yang (2000) and Miletich *et al.* (2005), with diagrams and technical details of the most common devices used in diffraction experiments. In general, the high-*T* devices can be distinguished as (1) radiative and conductive resistance heaters, (2) gas-stream heaters, (3) open-flame heaters and (4) mirror furnaces. Conductive resistance heaters currently available for conventional X-ray powder diffractometers allow even 1850 K to be reached; gas-flow devices provide temperatures up to ~1450 K, and are often used at large-scale facilities; open-flame gas torches are able to generate temperatures up to 2700 K, but temperature stability suffers from gas-flow fluctuations coupled with large thermal gradients within the flame; mirror furnaces can generate temperatures up to 2200 K, though with significant thermal gradients. For experiments designed to describe the thermo-elastic behaviour of materials, conductive resistance heaters and gas-stream heaters are therefore the most appropriate. Temperature control is usually achieved using thermocouples, in contact with (or close to) the polycrystalline sample, or using an internal standard (*i.e.* a material with a well known coefficient of thermal expansion).

A further issue concerns low-*T* experiments. Diffraction experiments at low *T* are usually performed in organic chemistry or bio-crystallography, in order to stabilize and preserve molecular crystalline materials, and to reduce thermal motion of the atoms and thus boost diffracted beam intensities. In mineralogy, low-*T* experiments are performed also in order to reduce the atomic displacements around equilibrium positions, and in particular for hydrous materials. Low-*T* experiments at $T \geqslant 80$ K are usually conducted with liquid nitrogen. Low-*T* devices operating with liquid nitrogen (open-flow systems) are available currently as commercial products. For experiments at temperature down to a few degrees Kelvin, liquid helium cryostats are usually used. Open-flow cryostats are more common, these enable temperatures of ~20 K to be attained. Closed-cycle cryostats (*i.e.* cryostats with a two-stage cooling system with liquid nitrogen as the outer thermal boundary) enable temperatures of 3–4 K to be reached. Costs and technical requirements mean that helium cryostats are usually available only at large-scale facilities, and are found only rarely in conventional X-ray diffraction laboratories.

Devices able to generate high-pressure conditions for *in situ* X-ray or neutron diffraction experiments belong to two main categories: devices for large-volume samples (LVS) (> 1 mm^3) and devices for small-volume samples (SVS) (< 0.01 mm^3). LVS devices are represented mainly by piston-cylinder presses, opposed-anvil presses, and multi-anvil systems. Such devices are described by Besson (1997), Fei and Wang (2000), Keppler and Frost (2005) and Miletich *et al.* (2005). All of them consist of (1) a hydraulic press with a pressure (and temperature) control system, and (2) a pressure module within which a sample assembly is compressed under quasi-hydrostatic pressure. The main differences among the aforementioned LVS devices are in the different pressure modules. All the LVS devices are able to generate high-*T* conditions, by resistive heating. Thermocouples are used to measure temperature. Internal standards, with well known coefficients of compressibility and thermal expansion, allow *P* and *T* to be measured. One of the most useful opposed-anvil devices is the Paris-Edinburgh press, which is used for synchrotron and neutron powder diffraction experiments (*e.g.* Besson *et al.*, 1992; Loveday, 2004). Paris-Edinburgh presses are currently used for experiments up to 12–15 GPa and 2000 K. The multi-anvil devices are the only LVS devices able to reach 30 GPa and 2000 K. Experiments with LVS devices can be performed currently only at large-scale facilities because of the need for intense radiation sources to penetrate the components of the device surrounding the sample in order to obtain diffraction.

The most common SVS device designed to perform *in situ* high-pressure experiments is the diamond-anvil cell (DAC). The description of the technical aspects of the different types of DACs can be found in Hazen and Finger (1982) and Miletich *et al.* (2000, 2005). In general, in a DAC, two diamonds are used as windows to compress a high-pressure chamber (usually confined in a metal gasket) which contains the polycrystalline sample under investigation, the pressure-transmitting medium, and a standard to measure the *P*. Diamonds are 'transparent' windows for electromagnetic radiations, and so not only can *in situ* diffraction experiments be performed with a DAC but Raman or IR experiments can also be conducted, for example. The use of a liquid *P*-transmitting media in DACs allows us to perform *in situ* experiments under truly hydrostatic regimes (*e.g.* Angel *et al.*, 2007; Klotz *et al.*, 2009). The measurement of pressure in a DAC is usually performed by a laser-induced fluorescence technique (*i.e.* *P*-dependent spectral shifts of bands in the fluorescence spectra of various sensor materials; the precision is better than 0.05 GPa) or using an internal standard with a well known isothermal EoS (precision: better than 0.01 GPa) (Miletich *et al.*, 2000). The nature of the *P* medium governs not only the *P* range open to investigation, but also the kind of materials under investigation; *P* media can interact with the materials under investigation in response to the applied pressure. An excellent example in this respect is that of zeolites compressed in 'penetrating' media (Gatta, 2008, 2010; Gatta and Lee, 2014). Modified DACs are currently used for H*P*/H*T*-experiments. The high *T* is provided by heatable modules integrated in the DAC body. Combined H*P*/H*T*-measurements have been performed successfully up to ~15 GPa and ~1200 K. In this case, independent *P/T* calibrants are needed. Extreme H*P*/H*T* experiments can be performed by laser heating. YAG (λ = 1.06 μm), YLF (λ = 1.05 μm) or CO_2 (λ = 10.6 μm) lasers (beam diameter: 1–15 μm, laser power: 50–100 W) are used to generate very high *T*. The temperature is usually estimated by the Planck radiation function (the incandescent light is usually collected in the spectral range 500–800 nm). With this experimental setup, experiments up to 200 GPa and 4000 K have been performed successfully. Further technical details pertaining to heating DACs, with a series of case studies, can be found in Dubrovinsky and Dubrovinskaia (2004) and Boehler (2005a,b). Overall, DACs are the only devices able to reach pressures of the order of hundreds of GPa. Diffraction experiments with DAC in the *P* range of petrological interest (*i.e.* 0–10 GPa) can be conducted in conventional laboratories.

5.6. X-ray *vs.* neutron powder diffraction under extreme conditions

X-rays and neutrons are different but complementary probes for condensed matter research. In mineralogy, the use of neutron diffraction is modest compared to X-ray usage. A thorough analysis of the different properties of neutrons and X-rays can be found in Bacon (1975), Parise (2006) and Rinaldi *et al.* (2009). Briefly:

(1) The neutron has zero net charge and so does not interact with the charge of the electron. The very weak interaction with matter allows deep penetration into a sample, or through the sample encapsulation such as a high-pressure device.

(2) Neutrons have a magnetic moment, a spin, which interacts with the unpaired electron spins.

(3) Neutrons show a wave-like nature (with wavelengths in the range of interatomic distances), and so they can be diffracted following the Bragg's law.

(4) The mass of the neutron is relatively large (*i.e.* it is similar to that of atomic nuclei). Neutrons can be considered as powerful probes of the nuclear and magnetic structure of condensed matter.

In mineralogy, neutron powder diffraction experiments have been used mainly for structure refinements by the Rietveld method (see Section 4.6) in order to

(1) locate light elements (in particular hydrogen);

(2) describe order-disorder processes even when a site population is represented by elements of neighbouring atomic number (as the non-linear dependence of neutron scattering for different atomic species allows discrimination between quasi-iso-electronic species, *e.g.* in minerals: Si/Al, Mg/Al, Fe/Mn);

(3) provide a better description of atomic displacement around the equilibrium position (as neutrons are scattered by the nucleus and so the scattering power does not fall off with angle, making possible measurements at high values of $\sin\theta/\lambda$);

(4) investigate magnetic materials.

A further issue concerns the difference between energy-dispersive X-ray diffraction mode (data collection as a function of λ at fixed θ) and angle-dispersive X-ray diffraction mode (data collection as a function of θ at fixed λ). Energy-dispersive diffraction experiments are faster, but have lower resolution of the Bragg peaks and have improper handling of intensities if compared to angle-dispersive diffraction data (Fei and Wang, 2000; McMahon, 2004, Miletich *et al.*, 2005). In this light, the majority of experiments at non-ambient conditions at synchrotron sources are currently performed in angle-dispersive mode. For neutron diffraction experiments, constant wavelength or time-of-flight diffraction modes are used currently for *in situ* low-*T* or high-*T*/*P* experiments and provide comparable data quality. The Paris-Edinburgh press is the most used device for neutron high-*P* or (high-*P*/high-*T*) experiments.

5.7. A case study: high-pressure and high-temperature behaviour of the zeolite $CsAlSiO_4$

Gatta *et al.* (2008) reported the synthesis protocol and the structure refinement of $CsAlSiO_4$ by X-ray single-crystal diffraction. $CsAlSiO_4$ is a zeolite (ABW framework type). Its structure is orthorhombic, with $Pc2_1n$ space group and unit-cell constants: a = 9.414(1), b = 5.435(1), c = 8.875(1) Å. The framework consists of tetrahedral sheets parallel to (001), in which 6-membered rings (6mRs) of corner-sharing tetrahedra lie (Fig. 11). Apical oxygens of three neighbouring tetrahedra in a ring point upwards, whereas apical oxygens of the other three tetrahedra point down (Fig. 11), with distorted eight-membered ring channels (8mRs) running along [010], where the extra-framework sites lie. The Si/Al-distribution in the tetrahedral framework of $CsAlSiO_4$ was found to be fully ordered.

Gatta *et al.* (2012) investigated the elastic behaviour and the *P*/*T*-induced structure evolution of $CsAlSiO_4$ by *in situ* high-temperature and high-pressure synchrotron powder diffraction. H*P* synchrotron X-ray powder diffraction data were collected at up to 9 GPa, using a membrane-type diamond anvil cell, at the ID09 beamline at the European Synchrotron Radiation Facility (ESRF, Grenoble, France). A mixture of methanol:ethanol (4:1) was used as hydrostatic pressure-transmitting medium, and pressure was estimated by the ruby fluorescence method (Miletich *et al.*, 2000). A Le Bail full-profile fit (see Section 4.4) was performed in order to measure the cell parameters at high pressure. All attempts to refine the crystal structure of $CsAlSiO_4$ at high pressure by the Rietveld method were unsuccessful. *In situ* H*T* synchrotron X-ray powder diffraction data were collected at the MCX beamline at ELETTRA (Trieste, Italy). A powder sample of $CsAlSiO_4$ was contained in a quartz-glass capillary. The use of a gas blower (Oxford Danfysik DGB-0002) allowed the sample temperature to be controlled at up to 1000°C. The diffraction data were analysed by the Rietveld method. Further details of the experiments at high pressure and temperature, protocols for data treatment and structure refinements were given by Gatta *et al.* (2012).

The evolution of the cell parameters of $CsAlSiO_4$ with pressure is shown in Fig. 12. There is no clear evidence of a phase transition within the *P* range investigated. A subtle change of the elastic behaviour appears to occur at ~4 GPa along [100] and [010], though without any supporting evidence from the unit-cell volume (Fig. 12). *P*-*V* data were fitted with a third-order Birch-Murnaghan EoS (equation 47), giving: V_0 =

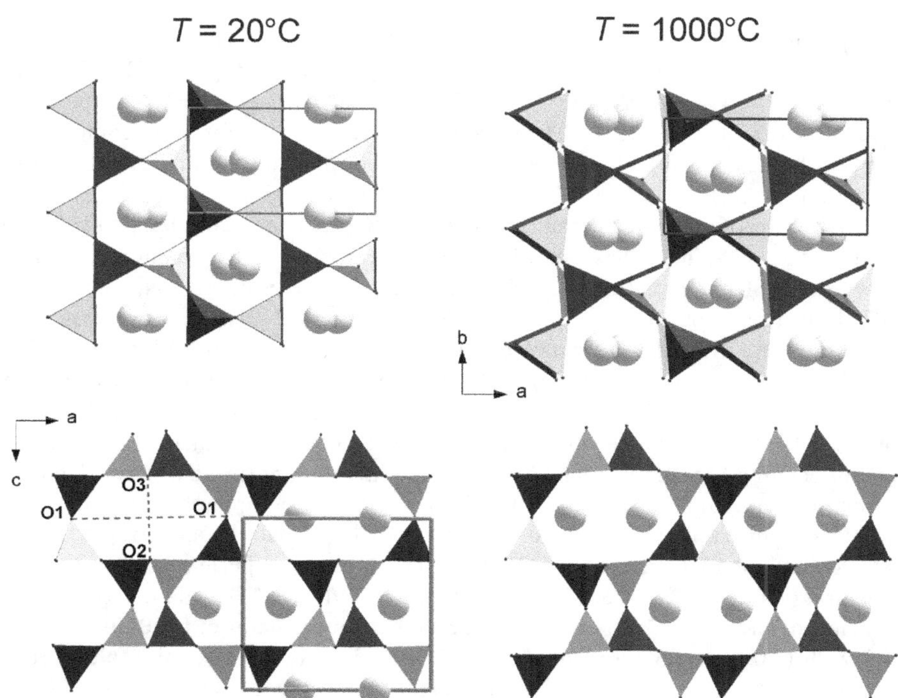

Figure 11. The crystal structure of $CsAlSiO_4$ viewed down [001] and down [010] based on the structure refinements by Gatta *et al.* (2012) at 20°C (*left side*) and 1000°C (*right side*). Si tetrahedra in light grey; Al tetrahedra in dark grey. The 'diameters' of the 8-membered rings, O1–O1 and O2–O3, are given as dashed lines.

457.9(4) $Å^3$, K_{P_0} = 42(1) GPa and K' = 3.9(3). The evolution of the cell parameters with pressure shows a remarkably anisotropic compressional behaviour (Fig. 12). The

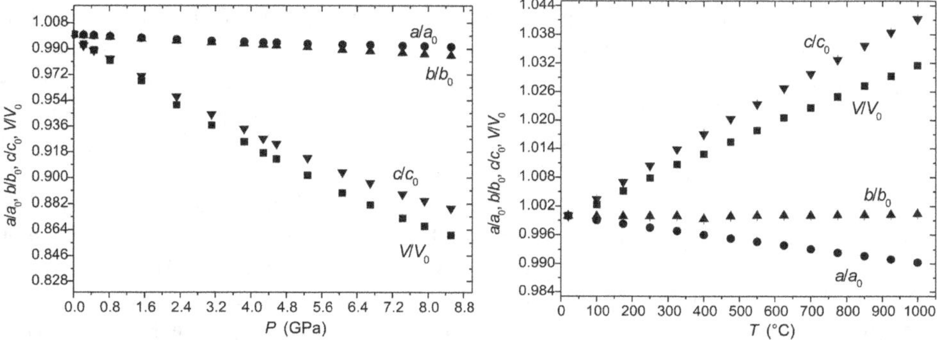

Figure 12. Evolution of the unit-cell parameters of $CsAlSiO_4$ with *P* and *T*, based on the data reported by Gatta *et al.* (2012).

elastic parameters calculated with a 'linearized' BM-EoS are: $K_{P_0}(a) = 244(11)$ GPa for the a axis ($K(a)' = 4$) (EoS valid between 0.0001 and 4 GPa); $K_{P_0}(b) = 181(3)$ GPa for the b axis ($K(b)' = 4$), and $K_{P_0}(c) = 14.5(5)$ GPa and $K(c)' = 2.6(1)$ for the c axis [$K_{P_0}(a)$:$K_{P_0}(b)$:$K_{P_0}(c)$ ~16:12:1]. The structure was found to be almost uncompressible along [100]. The data collected in decompression show that any P-induced deformation mechanism in $CsAlSiO_4$ up to 8.5 GPa is completely reversible.

The evolution of the lattice parameters of $CsAlSiO_4$ with temperature up to 1000°C is shown in Fig. 12. The behaviour of the cell parameters T is monotonic, with no evidence of a phase transition. $CsAlSiO_4$ shows a negative thermal expansion along [100], almost no expansion along [010] and a pronounced positive expansion along [001] (Fig. 12). Fitting the V-T data (equation 29), Gatta *et al.* (2012) obtained $V(T)/V_0 = 1 + 3.63(1) \times 10^{-5} \times \Delta T$. Extending the same formalism to the unit-cell axes (equation 30): $a(T)/a_0 = 1 - 9.97(1) \times 10^{-6} \times \Delta T$; $b(T)/b_0 = 1 + 0.36(1) \times 10^{-6} \times \Delta T$; $c(T)/c_0 = 1 + 47.46(6) \times 10^{-6} \times \Delta T$ (the second-order terms were not significant). The diffraction data collected at room T after the HT experiment confirm that any T-induced deformation mechanism in $CsAlSiO_4$ up to 1000°C is completely reversible. The Rietveld structure refinements performed at high T explained the anisotropic thermo-elastic behaviour of $CsAlSiO_4$. At high T, the structure experiences a pronounced tetrahedral tilting, which leads (1) to a contraction of the O1–O1 'diameter' and to an expansion of the O2–O3 'diameter' of the 8mR channel parallel to [010] (Fig. 11), along with (2) a 'corrugation' of the tetrahedral double-chain running along [010] (Fig. 11). The first mechanism ("1") controls the unit-cell contraction along [100] and the expansion along [001], whereas the second mechanism ("2") governs the modest expansion along [010]. Inter-tetrahedral rotation on (001) leads to a pronounced di-trigonalization of the 6mR (Fig. 11). The coordination polyhedron of Cs is significantly irregular at room T (*i.e.* the nominal coordination of Cs in $CsAlSiO_4$ is at least 14-fold). The HT structure refinements show a slight change in the bonding configuration of the Cs site in response to the applied temperature. The Cs-polyhedral volume increases with T, with a significant expansion between 20 and 325°C (+1.9%), followed by a saturation at higher T (+1.3% between 325 and 1000°C).

The experiments conducted at high pressure and temperature by Gatta *et al.* (2012) confirm that $CsAlSiO_4$ maintains its crystallinity at least up to 9 GPa (at room-T) and up to 1000°C (at room-P) under elastic regime. These experimental findings are somewhat surprising if we consider the microporous structure of this material. Combining the P-T-V data of $CsAlSiO_4$, Gatta *et al.* (2012) obtained the following EoS (equation 53): $V_{P,T} \approx V_{P_0,T_0}[1 - \beta\Delta P + \alpha\Delta T] = V_{P_0,T_0}[1 - 0.0242(2) \times \Delta P + 3.63(1) \times 10^{-5} \times \Delta T]$ (with $\beta = 1/K_{T_0}$ in GPa^{-1} and α in °C^{-1}).

A pronounced thermal and compressional anisotropy was observed more recently by Gatta *et al.* (2014) for $TlAlSiO_4$, isotypic with $CsAlSiO_4$. *In situ* synchrotron powder diffraction experiments conducted at up to 8 GPa (at room T) and up to 1000°C (at room P) led to the following thermal and compressional schemes (describing the structure in the space group $Pc2_1n$): a negative thermal expansion along [100] (*i.e.* $\alpha_0(a) = -8.5(1) \times 10^{-6}$ °C^{-1}), almost no expansion along [010] (*i.e.* $\alpha_0(b) = 0.9(1) \times 10^{-6}$ °C^{-1})

and a positive expansion along [001] (*i.e.* $\alpha_0(c) = 52.4(1) \times 10^{-6}$ °C^{-1}); a compressional scheme with: $K_{P_0}(a)$ = 68(1) GPa, $K_{P_0}(b)$ = 112(2) GPa and $K_{P_0}(c)$ = 21.96(7) GPa, $K_{P_0}(a)$:$K_{P_0}(b)$:$K_{P_0}(c)$ = 3.09:5.09:1. Both $TlAlSiO_4$ and $CsAlSiO_4$ show two directions which are significantly less compressible than a third. Such an anisotropic scheme is governed basically by the ABW topology and, in particular, by the geometrical configuration of the 8mR channels (which are parallel to [010] in $Pc2_1n$). In particular, the ABW open-framework leads to an elastic behaviour similar to that observed in 'layered materials' (*i.e.* the layers parallel to (001) being connected through bridging oxygen atoms, Fig. 11). The typical response of 'layered materials' to hydrostatic compression (*e.g.* phyllosilicates, Gatta *et al.*, 2009, 2010, 2011b) is with two co-planar directions drastically less compressible than that perpendicular to the layers.

6. Conclusions

In this chapter, we have provided an overview of the potentialities of the X-ray powder diffraction technique as a powerful tool for studying poly-crystalline materials. Qualitative and quantitative phase analysis, structure solution, description of the thermo-elastic behaviour and *P/T* phase stability of minerals are necessary for predicting the properties of systems of geological interest. On the other hand, experiments on minerals often have implications beyond the Earth sciences as a significant number of inorganic crystalline materials have natural analogues. The new diffraction sources, the efficiency of modern detectors, and the increasing power of software, described here, are opening new and fascinating opportunities for the powder diffraction technique.

Acknowledgments

The authors thank Prof. R.J. Angel for his critical reading of the part pertaining to the powder diffraction at extreme conditions. They also thank Dr Caterina Chiarella for her technical contribution.

References

Alexander, L. and Klug, H.P. (1948) Basic aspects of X-ray absorption in quantitative diffraction analysis of powder mixtures. *Analytical Chemistry*, **20**, 886–889.

Alexander, L. and Klug, H.P. (1989) Basic aspects of X-ray absorption in quantitative diffraction analysis of powder mixtures. *Powder Diffraction*, **4**, 66–69.

Altomare, A., Cascarano, G., Giacovazzo, C., Guagliardi, A. and Moliterni, A.G.G. (1995) On the number of systematically independent observations in a powder diffraction pattern. *Journal of Applied Crystallography*, **28**, 738–744.

Altomare, A., Giacovazzo, C., Guagliardi, A., Moliterni, A.G.G. and Rizzi, R. (2000) Completion of crystal structures from powder data: the use of the coordination polyhedra. *Journal of Applied Crystallography*, **33**, 1305–1310.

Altomare, A., Burla, M.C., Giacovazzo, C., Guagliardi, A., Moliterni, A.G.G., Polidori, G. and Rizzi, R. (2001) *Quanto*: a Rietveld program for quantitative phase analysis of polycrystalline mixtures. *Journal of Applied Crystallography*, **34**, 392–397.

Altomare, A., Caliandro, R., Cuocci, C., Giacovazzo, C., Moliterni, A.G.G. and Rizzi, R. (2003) A systematic procedure for the decomposition of a powder diffraction pattern. *Journal of Applied Crystallography*, **36**, 906–913.

Altomare, A, Caliandro, R, Camalli, M., Cuocci, C., da Silva, I., Giacovazzo, C., Moliterni, A.G.G. and Spagna, R. (2004) Space-group determination from powder diffraction data: a probabilistic approach. *Journal of Applied Crystallography*, **37**, 957–966.

Altomare, A, Camalli, M, Cuocci, C., da Silva, I., Giacovazzo, C., Moliterni, A.G.G. and Rizzi, R. (2005) Space group determination: improvements in *EXPO2004*. *Journal of Applied Crystallography*, **38**, 760–767.

Altomare, A., Cuocci C., Giacovazzo C., Moliterni A.G.G. and Rizzi R. (2006) Powder diffraction: the new automatic least-squares Fourier recycling procedure in EXPO2005. *Journal of Applied Crystallography*, **39**, 558–562.

Altomare, A., Camalli, M., Cuocci, C., Giacovazzo, C., Moliterni, A.G.G. and Rizzi, R. (2007) Advances in space-group determination from powder diffraction data. *Journal of Applied Crystallography*, **40**, 743–748.

Altomare, A., Cuocci, C., Giacovazzo, C., Moliterni, A. and Rizzi, R. (2008a) *QUALX*: a computer program for qualitative analysis using powder diffraction data. *Journal of Applied Crystallography*, **41**, 815–817.

Altomare, A., Giacovazzo, C. and Moliterni, A. (2008b) Indexing and space group determination. Pp. 206–226 in: *Powder Diffraction Theory and Practice* (R.E. Dinnebier and S.J.L. Billinge, editors). Royal Society of Chemistry Publishing, Cambridge, UK.

Altomare, A., Cuocci, C., Giacovazzo, C., Kamel, G.S., Moliterni, A. and Rizzi, R. (2008c) Minimally resolution biased electron-density maps. *Acta Crystallographica*, A**64**, 326–336.

Altomare, A., Cuocci, C., Giacovazzo, C., Moliterni, A. and Rizzi, R. (2008d) Correcting resolution bias in electron density maps of organic molecules derived by direct methods from powder data. *Journal of Applied Crystallography*, **41**, 592–599.

Altomare, A., Caliandro, R., Cuocci, C. Giacovazzo, C., Moliterni, A.G.G., Rizzi, R. and Platteau, C. (2008e) Direct methods and simulated annealing: a hybrid approach for powder diffraction data. *Journal of Applied Crystallography*, **41**, 56–61.

Altomare, A., Campi, G., Cuocci, C., Eriksson, L., Giacovazzo, C., Moliterni, A., Rizzi, R. and Werner, P.-E. (2009a) Advances in powder diffraction pattern indexing: *N-TREOR09*. *Journal of Applied Crystallography*, **42**, 768–775.

Altomare, A., Cuocci, C., Giacovazzo, C., Moliterni, A. and Rizzi, R. (2009b) Correcting electron-density resolution bias in reciprocal space. *Acta Crystallographica*, A**65**, 183–189.

Altomare, A., Cuocci, C., Giacovazzo, C., Moliterni, A. and Rizzi, R. (2010) The dual-space resolution bias correction algorithm: applications to powder data. *Journal of Applied Crystallography*, 43: 798–804.

Altomare, A., Cuocci, C., Giacovazzo, C., Moliterni, A. and Rizzi, R. (2011) Advances in the EXPO2009 systematic decomposition procedure: an atom-matching-based figure of merit. *Journal of Applied Crystallography*, **44**, 448–453.

Altomare, A., Cuocci, C., Giacovazzo, C., Moliterni, A. and Rizzi, R. (2012) COVMAP: a new algorithm for structure model optimization in the EXPO package. *Journal of Applied Crystallography*, **45**, 789–797.

Altomare, A., Cuocci, C., Giacovazzo, C., Moliterni, A. and Rizzi, R. (2013a) RAMM: a random-model-based method for solving *ab initio* crystal structure using *EXPO* package. *Journal of Applied Crystallography*, **46**, 476–482.

Altomare, A., Cuocci, C., Giacovazzo, C., Moliterni, A., Rizzi, R., Corriero, N. and Falcicchio, A. (2013b) EXPO2013: a kit of tools for phasing crystal structures from powder data. *Journal of Applied Crystallography*, **46**, 1231–1235.

Altomare, A., Corriero, N., Cuocci, C., Moliterni, A. and Rizzi, R. (2013c) The hybrid big bang-big crunch method for solving crystal structure from powder diffraction data. *Journal of Applied Crystallography*, **46**, 779–787.

Altomare, A., Corriero, N., Cuocci, C., Falcicchio, A., Moliterni, A. and Rizzi, R. (2015) *QUALX2.0*: a qualitative phase analysis software using the freely available database POW_COD. *Journal of Applied Crystallography*, **48**, 598–603.

Anderson, O.L. (1995) *Equations of State of Solids for Geophysics and Ceramic Science*. Oxford University Press, Oxford, UK, 432 pp.

Andreev, Y.G. and Bruce, P.G. (1998) Solving crystal structures of molecular solids without single crystals: a simulated annealing approach. *Journal of the Chemical Society, Dalton Transactions*, **24**, 4071–4080.

Andreev, Y.G., Lightfoot, P. and Bruce, P.G. (1997) A general Monte Carlo approach to structure solution from powder diffraction data: Application to poly(ethylene oxide)$_3$:LiN(SO_3CF_3)$_2$. *Journal of Applied Crystallography*, **30**, 294–305.

Angel, R.J. (2000) Equations of state. Pp. 35–59 in: *High-Temperature and High Pressure Crystal Chemistry* (R.M. Hazen and R.T. Downs, editors). Reviews in Mineralogy and Geochemistry, **41**. Mineralogical Society of America and Geochemical Society, Chantilly, Virginia, USA.

Angel, R.J., Zhao, J. and Ross, N.L. (2005) General rules for predicting phase transitions in perovskites due to octahedral tilting. *Physical Review Letters*, **95**, 025503.

Angel, R.J., Bujak, M., Zhao, J., Gatta, G.D. and Jacobsen, S.D. (2007) Effective hydrostatic limits of pressure media for high-pressure crystallographic studies. *Journal of Applied Crystallography*, **40**, 26–32.

Bacon, G.E. (1975) *Neutron Diffraction*, 3rd edition. Oxford-Clarendon Press, Oxford, UK. 636 pp.

Baerlocher, C., McCusker, L.B., Prokic, S. and Wessels, T. (2004) Exploiting texture to estimate the relative intensities of overlapping reflections. *Zeitschrift für Kristallographie*, **219**, 803–812.

Baerlocher, C., McCusker, L.B. and Palatinus, L. (2007) Charge flipping combined with histogram matching to solve complex crystal structures from powder diffraction data. *Zeitschrift für Kristallographie*, **222**, 47–53.

Balassone, G., Kahlenberg, V., Altomare, A., Mormone, A., Rizzi, R., Saviano, M. and Mondillo, N. (2014) Nephelines from the Somma-Vesuvius volcanic complex (Southern Italy): crystal-chemical, structural and genetic investigations. *Mineralogy and Petrology*, **108**, 71–90.

Balic-Zunic, T. and Vickovic, I. (1996) IVTON (Version 2) – Program for the calculation of geometrical aspects of crystal structures and some crystal chemical applications. *Journal of Applied Crystallography*, **29**, 305–306.

Barr, G., Dong, W. and Gilmore, C.J. (2004a) High-throughput powder diffraction. II. Applications of clustering methods and multivariate data analysis. *Journal of Applied Crystallography*, **37**, 243–252.

Barr, G., Dong, W., Gilmore, C. and Faber, J. (2004b) High-throughput powder diffraction. III. The application of full-profile pattern matching and multivariate statistical analysis to round-robin-type data sets. *Journal of Applied Crystallography*, **37**, 635–642.

Barr, G., Dong, W. and Gilmore, C.J. (2004c) *PolySNAP*, a computer program for analyzing high-throughput powder diffraction data. *Journal of Applied Crystallography*, **37**, 658–664.

Barr, G., Dong, W. and Gilmore, C.J. (2004d) High-throughput powder diffraction. IV. Cluster validation using silhouettes and fuzzy clustering. *Journal of Applied Crystallography*, **37**, 874–882.

Barr, G., Gilmore, C.J. and Paisley, J. (2004e) *SNAP-1D*, a computer program for qualitative and quantitative powder diffraction pattern analysis using the full pattern profile. *Journal of Applied Crystallography*, **37**, 665–668.

Barr, G., Cunnigham, G., Dong, W., Gilmore, C.J. and Kojima, T. (2009a) High-throughput powder diffraction V: the use of Raman spectroscopy with and without X-ray powder diffraction data. *Journal of Applied Crystallography*, **42**, 706–714.

Barr, G., Dong, W. and Gilmore, C.J. (2009b) *PolySNAP3*, a computer program for analyzing and visualizing high-throughput data from diffraction and spectroscopic sources. *Journal of Applied Crystallography*, **42**, 965–974.

Bartelmehs, K.L., Downs, R.T., Gibbs, G.V., Boisen, M.B., Jr. and Birch, J.B. (1995) Tetrahedral rigid-body motion in silicates. *American Mineralogist*, **80**, 680–690.

Bergmann J. (2007) EFLECH/INDEX – Another try of whole pattern indexing. *Zeitschrift für Kristallographie, Supplement*, **26**, 197–202.

Besson, J.M. (1997) Pressure generation. Pp. 1–45 in: *High-pressure Techniques in Chemistry and Physics. A Practical Approach* (W.B. Holzapfel and N.S. Isaacs, editors). Oxford University Press, Oxford, UK.

Besson, J.M., Nelmes, R.J., Hamel, G., Loveday, J.S., Weill, G. and Hull, S. (1992) Neutron powder diffraction above 10 GPa. *Physica* B, **180**, 907–910.

Billinge, S. (2008) Local structure from total scattering and atomic pair distribution function (PDF) analysis. Pp. 464–493 in: *Powder Diffraction Theory and Practice* (R.E. Dinnebier and S.J.L. Billinge, editors). Royal Society of Chemistry Publishing, Cambridge, UK.

Birch, F. (1947) Finite elastic strain of cubic crystals. *Physical Review*, **71**, 809–824.

Bish, D.L. and Howard, S.A. (1988) Quantitative phase analysis using the Rietveld method. *Journal of Applied Crystallography*, **21**, 86–91.

Bish, D.L. and Post, J.E. (editors) (1989) *Modern Powder Diffraction*. Reviews in Mineralogy, **20**. Mineralogical Society of America, Washington, DC.

Bish, D.L. and Post, J.E. (1993) Quantitative mineralogical analysis using the Rietveld full-pattern fitting method. *American Mineralogist*, **78**, 932–940.

Boehler, R. (2005a) Diamonds as optical windows to extreme conditions. Pp. 217–224 in: *Mineral Behaviour at Extreme Conditions* (R. Miletich, editor). EMU Notes in Mineralogy, **7**. Eötvös University Press, Budapest.

Boehler, R. (2005b) Laser heating at megabar pressures: Melting temperatures of iron and other transition metals. Pp. 273–280 in: *Mineral Behaviour at Extreme Conditions* (R. Miletich, editor). EMU Notes in Mineralogy, **7**. Eötvös University Press, Budapest.

Boisen, M.B., Jr. and Gibbs, G.V. (editors) (1990) *Mathematical Crystallography: An Introduction to the Mathematical Foundations of Crystallography*. Reviews in Mineralogy, **15**, Mineralogical Society of America, Chantilly, Virginia, USA.

Boultif, A. and Louër, D. (2004) Powder pattern indexing with the dichotomy method. *Journal of Applied Crystallography*, **37**, 724–731.

Brenner, S., McCusker, L.B. and Baerlocher, C. (1997) Using a structure envelope to facilitate structure solution from powder diffraction data. *Journal of Applied Crystallography*, **30**, 1167–1172.

Brindley, G.W. (1945) The effect of grain or particle size on X-ray reflections from mixed powders and alloys, considered in relation to the quantitative determination of crystalline substances by X-ray methods. *Philosophical Magazine*, **36**, 347–369.

Brückner, S. (2000) PULWIN: A program for analyzing powder X-ray diffraction patterns. *Powder Diffraction*, **15**, 218–219.

Bruker AXS (2003) *Topas V2.1: General Profile and Structure Analysis Software for Powder Diffraction Data*. Ed. Bruker AXS, Karlsruhe, Germany.

Busing, W.R. and Levy, H.A. (1964) The effect of thermal motion on the estimation of bond lengths from diffraction measurements. *Acta Crystallographica*, **17**, 142–146.

Caglioti, G., Paoletti, A. and Ricci, F.P. (1958) Choice of collimator for a crystal spectrometer for neutron diffraction. *Nuclear Instruments*, **3**, 223–228.

Caliandro, R., Giacovazzo, C. and Rizzi, R. (2008) Crystal structure determination. Pp. 227–265 in: *Powder Diffraction Theory and Practice* (R.E. Dinnebier and S.J.L. Billinge, editors). Royal Society of Chemistry Publishing, Cambridge, UK.

Cascarano, G., Giacovazzo, C. and Viterbo, D. (1987) Figures of merit in direct methods: a new point of view. *Acta Crystallographica*, **A43**, 22–29.

Cascarano, G., Giacovazzo, C. and Guagliardi, A. (1992) Improved figures of merit for direct methods. *Acta Crystallographica*, A**48**, 859–865.

Chernyshev, V.V. and Schenk, H. (1998) A grid search procedure of positioning a known molecule in an unknown crystal structure with the use of powder diffraction data. *Zeitschrift für Kristallographie*, **213**, 1–3.

Chung, D.D.L., De Haven, P.W., Arnold, H. and Gosh, D. (1993) *X-ray Diffraction at Elevated Temperatures*. VCH Publishers, New York, 268 pp.

Clark, G.L. and Reynolds, D.H. (1936) Quantitative analysis of mine dusts: An X-ray diffraction method. *Industrial & Engineering Chemistry Analytical Edition*, **8**, 36–40.

Cline, J.P. and Snyder, R.L. (1983) The dramatic effect of crystalline size on X-ray intensities. *Advances in X-ray Analysis*, **26**, 111–118.

Cockcroft, J.K.C and Fitch, A.N. (2008) Experimental setups. Pp. 20–57 in: *Powder Diffraction Theory and Practice* (R.E. Dinnebier and S.J.L. Billinge, editors). Royal Society of Chemistry Publishing, Cambridge, UK.

Coelho, A.A. (2003) *TOPAS Version 3.1*. Bruker AXS GmbH, Karlsruhe, Germany.

Comodi, P., Gatta, G.D., Zanazzi, P.F., Levy, D. and Crichton W. (2002) Thermal equations of state of dioctahedral micas on the join muscovite–paragonite. *Physics and Chemistry of Minerals*, **29**, 538–544.

Cranswick, L.M.D. (2008a) Computer software for powder diffraction. Pp. 494–570 in: *Powder Diffraction Theory and Practice* (R.E. Dinnebier and S.J.L. Billinge, editors). Royal Society of Chemistry Publishing, Cambridge, UK.

Cranswick, L.M.D. (2008b) An overview of powder diffraction. Pp. 1–72 in: *Principles and Applications of Powder Diffraction* (A. Clearfield, J.H. Reibenspies and N. Bhuvanesh, editors). Wiley-Blackwell, Singapore.

Cranswick, L. and Kruger, G. (editors) (2002) Powder diffraction in mining and minerals. *International Union of Crystallography Commission on Powder Diffraction Newsletter*, No. 27, http://www.iucr.org/resources/commissions/powder-diffraction/newsletter.

Cruciani, G. (2006) Zeolites upon heating: Factors governing their thermal stability and structural changes. *Journal of Physics and Chemistry of Solids*, **67**, 1973–1994.

David, W.I.F. (1999) On the number of independent reflections in a powder diffraction pattern. *Journal of Applied Crystallography*, **32**, 654–663.

David, W.I.F., Shankland, K., van de Streek, J., Pidcock, E., Motherwell, W.D.S. and Cole J. C. (2006) DASH: a program for crystal structure determination from powder diffraction data. *Journal of Applied Crystallography*, **39**, 910–915.

Davis, B.L. and Smith, D.K. (1988) Tables of experimental reference intensity ratios. *Powder Diffraction*, **3**, 205–208.

Davis, B.L., Smith, D.K. and Holomany, M.A. (1989) Tables of experimental reference intensity ratios. Table No. 2, December, 1989. *Powder Diffraction*, **4**, 201–205.

Davis, B.L., Kath, R. and Spilde, M. (1990) The reference intensity ratio: Its measurement and significance. *Powder Diffraction*, **5**, 76–78.

de Wolff, P.M. (1968) A simplified criterion for the reliability of a powder pattern indexing. *Journal of Applied Crystallography*, **2**, 108–113.

de Wolff, P.M. and Visser, J.M. (1964) Absolute intensities – Outline of recommended practice. Techniques for obtaining absolute intensities with internal standard. *Powder Diffraction*, **3**, 202–204.

de Wolff, P.M. and Visser, J.M. (1988) Absolute intensities – Outline of recommended practice. Techniques for obtaining absolute intensities with internal standard. *Report* 641.109, Technisch Physische Dienst TNO ENT-H, The Netherlands.

Dinnebier, R.E. and Billinge, S.J.L. (editors) (2008) *Powder Diffraction Theory and Practice*. Royal Society of Chemistry Publishing, Cambridge, UK.

Dollase, W.A. (1986) Correction of intensities for preferred orientation in powder diffractometry: application of the March model. *Journal of Applied Crystallography*, **19**, 267–272.

Dong, C., Wu, F. and Chen, H. (1999) Correction of zero shift in powder diffraction patterns using the reflection-pair method. *Journal of Applied Crystallography*, **32**, 850–853.

Downs, R.T. (2000) Analysis of harmonic displacement factors. Pp. 61–117 in: *High-Temperature and High-Pressure Crystal Chemistry* (R.M. Hazen and R.T. Downs, editors). Reviews in Mineralogy and Geochemistry, **41**. Mineralogical Society of America and Geochemical Society, Chantilly, Virginia, USA.

Downs, R.T., Gibbs, G.V., Bartelmehs, K.L. and Boisen, M.B. Jr. (1992) Variations of bond lengths and volumes of silicate tetrahedra with temperature. *American Mineralogist*, **77**, 751–757.

Dubrovinsky, L. and Dubrovinskaia, N. (2004) High-pressure crystallography at elevated temperatures. Pp. 393–410 in: *High-Pressure Crystallography* (A. Katrusiak and P. McMillan, editors). NATO Science Series II, **140**. Kluwer, Dordrecht, The Netherlands.

Dziewonski, A.M. and Anderson, D.L. (1981) Preliminary reference Earth model. *Physics of Earth and Planetary Interiors*, **25**, 297–356.

Engel, G.E., Wilke, S., König O., Harris, K.D.M. and Leusen, F.J.J. (1999) PowderSolve – a complete package for crystal structure solution from powder diffraction patterns. *Journal of Applied Crystallography*, **32**, 1169–1179.

Faber, J. and Fawcett, T. (2002) The Powder Diffraction File: present and future. *Acta Crystallographica*, **B58**, 325–332.

Favre-Nicolin, V. and Cerny, R. (2002) FOX, 'free objects for crystallography': a modular approach to ab initio structure determination from powder diffraction. *Journal of Applied Crystallography*, **35**, 734–743.

Favre-Nicolin, V. and Cerny, R. (2004) A better FOX: using flexible modelling and maximum likelihood to improve direct-space *ab initio* structure determination from powder diffraction. *Zeitschrift für Kristallographie*, **219**, 847–856.

Fei, Y. (1995) Thermal expansion. Pp. 29–44 in: *Mineral Physics and Crystallography, A Handbook of Physical Constants* (T.J. Ahrens, editor). AGU Reference Shelf, **2**. American Geophysical Union, Washington, D.C.

Fei, Y. and Wang, Y. (2000) High-temperature and high-pressure powder diffraction. Pp. 521–557 in: *High-Temperature and High-Pressure Crystal Chemistry* (R.M. Hazen and R.T. Downs, editors). Reviews in Mineralogy and Geochemistry, **41**. Mineralogical Society of America and Geochemical Society, Chantilly, Virginia, USA.

Frevel, L.K. (1965) Computational aids for identifying crystalline phases by powder diffraction. *Analytical Chemistry*, **37**, 471–482.

Gatta, G.D. (2008) Does porous mean soft? On the elastic behaviour and structural evolution of zeolites under pressure. *Zeitschrift für Kristallographie*, **223**, 160–170.

Gatta, G.D. (2010) Extreme deformation mechanisms in open-framework silicates at high-pressure: Evidence of anomalous inter-tetrahedral angles. *Microporous and Mesoporous Materials*, **128**, 78–84.

Gatta, G.D. and Lee, Y. (2014) Zeolites at high pressure: A review. *Mineralogical Magagazine*, **78**, 267–291.

Gatta, G.D., Rinaldi, R., Knight, K.S., Molin, G. and Artioli G. (2007) High temperature structural and thermoelastic behaviour of mantle orthopyroxene: an in situ neutron powder diffraction study. *Physics and Chemistry of Minerals*, **34**, 185–200.

Gatta, G.D., Rotiroti, N., Zanazzi, P.F., Rieder, M., Drabek, M., Weiss, Z., and Klaska, R. (2008) Synthesis and crystal structure of the feldspathoid $CsAlSiO_4$: an open-framework silicate and potential nuclear waste disposal phase. *American Mineralogist*, **93**, 988–995.

Gatta G.D., Rotiroti N., Pavese A., Lotti P., and Curetti N. (2009) Structural evolution of a $3T$ phengite mica up to 10 GPa: an in-situ single-crystal X-ray diffraction study. *Zeitschrift für Kristallographie*, **224**, 302–310.

Gatta G.D., Rotiroti N., Pavese A., Lotti P. and Curetti N. (2010) Structural evolution of a $2M_1$ phengite mica up to 11 GPa: an in-situ single-crystal X-ray diffraction study. *Physics and Chemistry of Minerals*, **37**, 581–591.

Gatta, G.D., Merlini, M., Lee, Y. and Poli, S. (2011a) Behavior of epidote at high pressure and high temperature: a powder diffraction study up to 10 GPa and 1,200 K. *Physics and Chemistry of Minerals*, **38**, 419–428.

Gatta G.D., Merlini M., Rotiroti N., Curetti N., and Pavese A. (2011b) On the crystal chemistry and elastic behavior of a phlogopite $3T$. *Physics and Chemistry of Minerals*, **38**, 655–664.

Gatta, G.D., Merlini, M., Lotti, P., Lausi, A. and Rieder, M. (2012) Phase stability and thermo-elastic behavior of $CsAlSiO_4$ (ABW): A potential nuclear waste disposal material. *Microporous and Mesoporous Materials*, **163**, 147–152.

Gatta G.D., Lotti P., Merlini M., Caputo D., Aprea P., Lausi A. and Colella C. (2014) Thermo-elastic behavior and P/T phase stability of $TlAlSiO_4$ (ABW). *Microporous and Mesoporous Materials*, **197**, 262–267.

Gemmi, M., Merlini, M., Pavese, A. and Curetti, N. (2008) Thermal expansion and dehydroxylation of phengite micas. *Physics and Chemistry of Minerals*, **35**, 367–379.

Giacovazzo C. (1998) *Direct Phasing in Crystallography*. Oxford University Press, Oxford, UK.

Giacovazzo, C. (editor) (2011) *Fundamentals of Crystallography*, 3rd edition. Oxford University Press, Oxford, UK.

Giacovazzo, C. (2013) *Phasing in Crystallography: A Modern Perspective*. IUCr/Oxford University Press, Oxford, UK.

Giacovazzo, C., Altomare, A., Cuocci, C., Moliterni, A.G.G. and Rizzi, R. (2002) Completion of crystal structures from powder data: a new method for locating atoms with polyhedral coordination. *Journal of Applied Crystallography*, **35**, 422–429.

Gilmore, C. (1996) Maximum entropy and Bayesian statistics in crystallography: a review of practical applications. *Acta Crystallographica*, **A52**, 561–589.

Gilmore, C.J., Barr, G. and Paisley, J. (2004) High-throughput powder diffraction. I. A new approach to qualitative and quantitative powder diffraction pattern analysis using full pattern profiles. *Journal of Applied Crystallography*, **37**, 231–242.

Goldberg, D.E. (1989) *Genetic Algorithms. In Search, Optimization, and Machine Learning*. Addison-Wesley, New York.

Gražulis, S., Chateigner, D., Downs, R.T., Yokochi, A.F.T., Quirós, M., Lutterotti, L., Manakova, E., Butkus, J., Moeck, P. and Le Bail, A. (2009) Crystallography Open Database – an open-access collection of crystal structures. *Journal of Applied Crystallography*, **42**, 726–729.

Gražulis, S., Daskevic, A., Merkys, A., Chateigner, D., Lutterotti, L., Quirós, M., Serebryanaya, R. N., Moeck, P., Downs, R. T. and Le Bail, A. (2012) Crystallography Open Database (COD): an open-access collection of crystal structures and platform for world-wide collaboration. *Nucleic Acids Research*, **40** (Database Issue), D420–D427.

Grosse-Kunstleve, R.W., McCusker, L.B. and Baerlocher, C. (1997) Powder diffraction data and crystal chemical information combined in an automated structure determination procedures for zeolites. *Journal of Applied Crystallography*, **30**, 985–995.

Gualtieri, A.F. (2000) Accuracy of XRPD QPA using the combined Rietveld-RIR method. *Journal of Applied Crystallography*, **33**, 267–278.

Hanawalt, J.D. (1986) Manual search/match methods for powder diffraction in 1986. *Powder Diffraction*, **1**, 7–13.

Hanawalt, J.D. and Rinn, H.W. (1936) Identification of crystalline materials – Classification and use of X-ray diffraction patterns. *Industrial & Engineering Chemistry Analytical Edition*, **8**, 244–247.

Hanawalt, J.D., Rinn, H.W. and Frevel, L.K. (1938) Chemical analysis by X-ray diffraction – classification and use of X-ray diffraction patterns. *Industrial & Engineering Chemistry Analytical Edition*, **10**, 457–512.

Harris, K.D.M, Tremayne, M., Lightfoot, P. and Bruce, P.G. (1994) Crystal structure determination from powder diffraction data by Monte Carlo methods. *Journal of the American Chemical Society*, **116**, 3543–3547.

Hazen, R.M. and Finger, L.W. (1982) *Comparative Crystal Chemistry*. Wiley, New York, 231 pp.

Hazen, R.M., Downs, R.T. and Finger, L.W. (2000) Principles of comparative crystal chemistry. Pp. 1–33 in: *High-Temperature and High-Pressure Crystal Chemistry* (R.M. Hazen and R.T. Downs, editors). Reviews in Mineralogy and Geochemistry, **41**. Mineralogical Society of America and Geochemical Society, Chantilly, Virginia, USA.

Henderson, C.M.B., Redfern, S.A.T., Smith, R.I., Knight, K.S. and Charnock, J.M. (2001) Composition and temperature dependence of cation ordering in Ni-Mg olivine solid solutions: a time-of-flight neutron powder diffraction and EXAFS study. *American Mineralogist*, **86**, 1170–1187.

Hill, R.J. and Howard, C.J. (1987) Quantitative phase analysis from neutron powder diffraction data using the Rietveld method. *Journal of Applied Crystallography*, **20**, 467–474.

Holland, T.J.B. and Powell, R. (1998) An internally consistent thermodynamic data set for phases of petrological interest. *Journal of Metamorphic Geology*, **16**, 309–343.

Hubbard, C.R. and Snyder, R.L. (1988) RIR – Measurement and Use in Quantitative XRD. *Powder Diffraction*, **3**, 74–77.

Hull, A.W. (1919) A new method of chemical analysis. *Journal of the American Chemical Society*, **41**, 1168–1175.

International Tables for Crystallography (2002) *Volume A: Space-group symmetry*, 5th edition (Th. Hahn, editor). Kluwer Academic Publishers, Dordrecht, The Netherlands.

Ishida, T. and Watanabe,Y.J. (1967) Probability computer method of determining the lattice parameters from powder diffraction data. *Journal of the Physical Society of Japan*, **23**, 556–565.

Ishida, T. and Watanabe, Y.J. (1971) Analysis of powder diffraction pattern of monoclinic and triclinic crystals. *Journal of Applied Crystallography*, **4**, 311–316.

Jenkins, R. (2000) X-ray techniques: Overview. Pp. 13269–13288 in: *Encyclopedia of Analytical Chemistry* (R.A. Meyers, editor). John Wiley & Sons Ltd, Chichester, UK.

Jenkins, R. and Snyder, R.L. (1996) *Introduction to X-Ray Powder Diffractometry*. Wiley, New York.

Jouanneaux, A., Verbaere, A., Guyomard, D., Piffard., Y, Oyetola, S and Fitch, A.N. (1991) $Sb_2(PO_4)_3$, a new mixed-valence antimony phosphate – preparation and crystal-structure. *European Journal of Solid State Inorganic Chemistry*, **28**, 755–765.

Johnson, G.G. and Vand, V., Jr. (1967) A computerized powder diffraction identification system. *Industrial & Engineering Chemistry*, **59**, 19–31.

Kabekkodu, S.N., Faber, J. and Fawcett, T. (2002) New Powder Diffraction File (PDF-4) in relational database format: advantages and data-mining capabilities. *Acta Crystallographica*, **B58**, 333–337.

Kariuki, B.M., Serrano-González, H., Johnston, R.L. and Harris, K.D.M. (1997) The applications of a genetic algorithm from solving crystal structures from powder diffraction data. *Chemical Physics Letters*, **280**, 189–195.

Keppler, H. and Frost, D.J. (2005) Introduction to minerals under extreme conditions. Pp. 1–30 in: *Mineral Behaviour at Extreme Conditions* (R. Miletich, editor): EMU Notes in Mineralogy, **7**. European Mineralogical Union, Eötvös University Press, Budapest.

Kern, A., Madsen, I.C. and Scarlett, N.V.Y. (2012) Quantifying amorphous phases. Pp. 219–231 in: *Uniting Electron Crystallography and Powder Diffraction* (K. Ute, K. Shankland, L. Meshi, A. Avilov and W. David, editors). Springer, Dordrecht, The Netherlands.

Kirkpatrick, S. (1983) Optimization by simulated annealing-quantitative studies. *Journal of Statistical Physics*, **34**, 975–986.

Klotz, S., Chervin, J.-C., Munsch P. and Le Marchand, G. (2009) Hydrostatic limits of 11 pressure transmitting media. *Journal of Physics D: Applied Physics*, **42**, 075413 (7 pp).

Langford, J.I. and Louër, D. (1996) Powder diffraction. *Reports on Progress in Physics*, **59**, 131–234.

Larson, A.C. and Von Dreele, R.B. (2004) *General Structure Analysis System (GSAS)*. Los Alamos National Laboratory, New Mexico, USA.

Le Bail A. (2008) Structure solution. Pp. 261–309 in: *Principles and Applications of Powder Diffraction* (A. Clearfield, J.H. Reibenspies and N. Bhuvanesh, editors). Wiley-Blackwell, Oxford, UK.

Le Bail, A., Duroy, H. and Fourquet, J.L. (1988) Ab-initio structure determination of LiSbWO by X-ray powder diffraction. *Materials Research Bulletin*, **23**, 447–452.

Le Meins, J.-M., Cranswick, L.M.D. and Le Bail, A. (2003) Results and conclusions of the Internet based "Search/match round robin 2002". *Powder Diffraction*, **18**, 106–113.

Loveday, J.S. (2004) Neutron diffraction studies of ices and ice mixtures. Pp. 69–80 in: *High-Pressure Crystallography* (A. Katrusiak and P. McMillan, editors). NATO Science Series II, **140**. Kluwer, Dordrecht, The Netherlands.

Madsen, I.C. and Scarlett, N.V.Y. (2008) Quantitative phase analysis. Pp. 298–331 in: *Powder Diffraction Theory and Practice* (R.E. Dinnebier and S.J.L. Billinge, editors). Royal Society of Chemistry Publishing, Cambridge, UK.

Madsen, I.C., Scarlett, N.V.Y., Cranswick, L.M.D. and Lwin, T. (2001) Outcomes of the International Union of Crystallography Commission on Powder Diffraction round robin on quantitative phase analysis: Samples 1*a* to 1*h*. *Journal of Applied Crystallography*, **34**, 409–426.

Madsen, I.C., Scarlett, N.V.Y. and Webster, N.A.S. (2012) Quantitative phase analysis. Pp. 207–218 in: *Uniting Electron Crystallography and Powder Diffraction* (K. Ute, K. Shankland, L. Meshi, A. Avilov and W. David, editors). Springer, Dordrecht, The Netherlands.

Madsen, I.C., Scarlett, N.V.Y., Riley, D.P. and Raven, M.D. (2013) Quantitative phase analysis using the Rietveld method. Pp. 285–320 in: *Modern Diffraction Methods* (E.J. Mittemejer and U. Welzel, editors). Wiley-VCH Verlag and Co., Weinheim, Germany

Makovicky, E. and Balic-Zunic, T. (1998) New measure of distortion for coordination polyhedra. *Acta Crystallographica*, B**54**, 766–773.

March, A. (1932) Mathematische Theorie der Regelun nach der Korgerstalt bei affiner Deformation. *Zeitschrift für Kristallographie*, **81**, 285–297.

Margiolaki, I., Giannopoulou, A.E., Wright, J.P., Knight, L., Norrman, M., Schluckebier, G., Fitch, A.N. and Von Dreele, R.B. (2013) High-resolution powder X-ray data reveal the T6 hexameric form of bovine insulin. *Acta Crystallographica*, **D69**, 978–990.

Markvardsen, A.J., David, W.I.F., Johnson, J.C. and Shankland, K. (2001) A probabilistic approach to space group determination from powder diffraction data. *Acta Crystallographica*, **A57**, 47–54.

McCusker, L.B., Von Dreele, R.B., Cox, D.E., Louër, D. and Scardi, P. (1999) Rietveld refinement guidelines. *Journal of Applied Crystallography*, **32**, 36–50.

McMahon, M.I. (2004) High pressure diffraction from good powders, poor powders and poor single crystals. Pp. 1–20 in: *High-pressure Crystallography* (A. Katrusiak and P. McMillan, editors). NATO Science Series II, **140**. Kluwer, Dordrecht, The Netherlands.

Miletich, R., Allan, D.R. and Kuhs, W.F. (2000) High-pressure single-crystal techniques. Pp. 445–519 in: *High-*

Temperature and High Pressure Crystal Chemistry (R.M. Hazen and R.T. Downs, editors). Reviews in Mineralogy and Geochemistry, **41**. Mineralogical Society of America and Geochemical Society, Chantilly, Virginia, USA.

Miletich, R., Hejny, C., Krauss, G. and Ullrich A. (2005) Diffraction techniques: Shedding light on structural changes.Pp. 281–338 in: *Mineral Behaviour at Extreme Conditions* (R. Miletich, editor). EMU Notes in Mineralogy, **7**. European Mineralogical Union. Eötvös University Press, Budapest.

Murnaghan, F.D. (1937) Finite deformations of an elastic solid. *American Journal of Mathematics*, **49**, 235–260.

Navias, L. (1925) Quantitative determination of the development of mullite in fired clay by an X-ray method. *Journal of the American Ceramic Society*, **8**, 296–302.

Nichols, M.C. (1966) *A Fortran II Program for the Identification of X-Ray Powder Diffraction Patterns.* UCRL-70078, Laurence Livermore Laboratory.

Nye, J.F. (1985) *Physical Properties of Crystals: Their Representation by Tensors and Matrices.* Oxford University Press, Oxford, UK, 329 pp.

Ohashi, Y. (1982) *STRAIN*: A program to calculate the strain tensor from two sets of unit-cell parameters. In: *Comparative Crystal Chemistry* (R.M. Hazen and L.W. Finger, editors). Wiley, New York, 231 pp.

Ohashi, Y. and Burnham, C.W. (1973) Clinopyroxene lattice deformations: The roles of chemical substitution and temperature. *American Mineralogist*, **58**, 843–849.

Oszlányi, G. and Sütő, A. (2004) Ab initio structure solution by charge flipping. *Acta Crystallographica*, A**60**, 134–141.

Parise, J.B. (2006) Introduction to neutron properties and applications. Pp. 1–27 in: *Neutron Studies in Earth Sciences* (H.R. Wenk, editor). Reviews in Mineralogy and Geochemistry, **63**, Mineralogical Society of America and Geochemical Society, Chantilly, Virginia, USA.

Pawley, G.S. (1981) Unit-cell refinement from powder diffraction scans. *Journal of Applied Crystallography*, **14**, 357–361.

Pawley, A.R., Redfern, S.A.T. and Holland, T.J.B. (1996) Volume behaviour of hydrous minerals at high pressure and temperature: 1. Thermal expansion of lawsonite, zoisite, clinozoisite, and diaspore. *American Mineralogist*, **81**, 335–340.

Percharsky, V.K. and Zavalij, P.Y. (2009) *Fundamentals of Powder Diffraction and Structural Characterization of Material.* Springer Science + Business Media LLC.

Peterson, R.C. and Yang, H. (2000) High-temperature devices and environmental cells designed for X-ray and neutron diffraction. Pp. 425–443 in: *High-Temperature and High-Pressure Crystal Chemistry* (R.M. Hazen and R.T. Downs, editors). Reviews in Mineralogy and Geochemistry, **41**. Mineralogical Society of America and Geochemical Society, Chantilly, Virginia, USA.

Popa, N.C. (2008) Microstructural properties: texture and microstress effects. Pp. 332–375 in: *Powder Diffraction Theory and Practice* (R.E. Dinnebier and S.J.L. Billinge, editors). Royal Society of Chemistry Publishing, Cambridge, UK.

Press, W.H., Teukolsky, S.A., Vetterling, W.T. and Flannery, B.P. (1992) *Numerical Recipes in C.* Cambridge University Press, Cambridge, UK.

Redfern, S.A.T. (2000) Order-disorder phase transitions. Pp. 105–133 in: *Transformation Processes in Minerals* (S.A.T. Redfern and M. Carpenter, editors). Reviews in Mineralogy and Geochemistry, **39**, Mineralogical Society of America and Geochemical Society, Chantilly, Virginia, USA.

Redfern, S.A.T., Henderson, C.M.B., Knight, K.S. and Wood, B.J. (1997) High temperature order-disorder in $(Fe_{0.5}Mn_{0.5})_2SiO_4$ and $(Mn_{0.5}Mg_{0.5})_2SiO_4$ olivines: an *in situ* neutron diffraction study. *European Journal of Mineralogy*, **9**, 287–300.

Redfern, S.A.T., Artioli, G., Rinaldi, R., Henderson, C.M.B., Knight, K.S. and Wood, B.J. (2000) Octahedral cation ordering in olivine at high temperature. II: An in situ neutron powder diffraction study on synthetic $MgFeSiO_4$ (Fa50). *Physics and Chemistry of Minerals*, **27**, 630–637.

Rietveld, H.M. (1969) A profile refinement method for nuclear and magnetic structures. *Journal of Applied Crystallography*, **2**, 65–71.

Rinaldi, R., Gatta, G.D., Artioli, G., Knight, K.S. and Geiger, C.A. (2005) Crystal chemistry, cation ordering and thermoelastic behaviour of $CoMgSiO_4$ olivine at high temperature as studied by in situ neutron powder diffraction. *Physics and Chemistry of Minerals*, **32**, 655–664.

Rinaldi, R., Liang, L. and Schober, H. (2009) Neutron applications in Earth, energy and environmental sciences. Pp. 1–14 in: *Neutron Applications in Earth, Energy and Environmental Sciences* (L. Liang, R. Rinaldi and H. Schober, editors). Springer Science, New York.

Rius, J. (2011) Patterson-function direct methods for structure determination of organic compounds from powder diffraction data. XV. *Acta Crystallographica*, **A67**, 63–67.

Rius, J. (2014) Application of Patterson-function direct methods to materials characterization. *IUCrJ*, **1**, 291–304.

Rius, J. and Frontera C.J. (2007) Application of the constrained S-FFT direct-phasing method to powder diffraction data. XIII. *Journal of Applied Crystallography*, **40**, 1035–1038.

Rius, J., Crespi, A. and Torrelles, X. (2007) A direct phasing method based on the origin-free modulus sum function and the FFT algorithm. XII. *Acta Crystallographica*, A**63**, 131–134.

Robinson, K., Gibbs, G.V. and Ribbe, P.H. (1971) Quadratic elongation: a quantitative measure of distortion in coordination polyhedra. *Science*, **172**, 567–570.

Sabine, T.M. (1985) Extinction in polycrystalline materials. *Australian Journal of Physics*, **38**, 507–518.

Savitzky, A. and Golay, M.J.E. (1964) Smoothing and differentiation of data by simplified least squares procedures. *Analytical Chemistry*, **36**, 1627–1639.

Scardi, P. (2008) Microstructural properties: Lattice defects and domain size effects. Pp. 376–413 in: *Powder Diffraction Theory and Practice* (R.E. Dinnebier and S.J.L. Billinge, editors). Royal Society of Chemistry Publishing, Cambridge, UK.

Scarlett, N.V.Y. and Madsen, I.C. (2006) Quantification of phases with partial or no known crystal structures. *Powder Diffraction*, **21**, 278–284.

Scarlett, N.V.Y., Madsen, I.C., Cranswick, L.M.D., Lwin, T., Groleau, E., Stephenson, G., Aylmore, M. and Agron-Oishina, N. (2002) Outcomes of the International Union of Crystallography Commission on Powder Diffraction Round Robin on quantitative phase analysis: Samples 2, 3, 4, synthetic bauxite, natural granodiorite and pharmaceuticals. *Journal of Applied Crystallography*, **35**, 383–400.

Scherrer, P. (1918) Bestimmung der Grösse und der inneren Struktur von Kolloidteilchen mittels Röntgenstrahlen. *Nachrichten von der Gesellschaft der Wissenschaften zu Göttingen*, 98–100.

Schlenker, J.L., Gibbs, G.V. and Boisen, M.B. (1975) Thermal expansion coefficients for monoclinic crystals: a phenomenological approach. *American Mineralogist*, **60**, 828–833.

Schlenker, J.L., Gibbs, G.V. and Boisen, M.B. (1978) Strain-tensor components expressed in terms of lattice-parameters. *Acta Crystallographica*, **A34**, 52–54.

Schomaker, V. and Trueblood, K.N. (1968) On the rigid-body motion of molecules in crystals. *Acta Crystallographica*, B**24**, 63–76.

Shankland, K., David, W.I.F. and Csoka, T. (1997) Crystal structure determination from powder diffraction data by the application of a genetic algorithm. *Zeitschrift für Kristallographie*, **212**, 550–552.

Sivia, D.S, and David, W.I.F. (1994) A Bayesian approach to extracting structure-factor amplitudes from powder diffraction data. *Acta Crystallographica*, **A50**, 703–714.

Smith, D.K. and Jenkins, R. (1996) The Powder Diffraction File: Past, present, and future. *Journal of research of the National Institute of Standards and Technology*, **101**, 259–271.

Smith G.S. and Snyder R.L.J. (1979) F_N, A criterion for rating powder diffraction patterns and evaluating the reliability of powder pattern index. *Journal of Applied Crystallography*, **12**, 60–65.

Snyder, R.L. (1992) The use of reference intensity ratios in X-ray quantitative analysis. *Powder Diffraction*, **7**, 186–192.

Stacey, F.D. and Davis P.M. (2004) High pressure equations of state with applications to the lower mantle and core. *Physics of Earth and Planetary Interiors*, **142**, 137–184.

Taylor, J.C. (1991) Computer programs for standardless quantitative analysis of minerals using the full powder diffraction profile. *Powder Diffraction*, **6**, 2–9.

Taylor, J.C. and Matulis, C.E. (1991) Absorption contrast effects in the quantitative XRD analysis of powders by full multiphase profile refinement. *Journal of Applied Crystallography*, **24**, 14–17.

Taylor, J.C. and Rui, Z. (1992) Simultaneous use of observed and calculated standard profiles in quantitative XRD analysis of minerals by the multiphase Rietveld method: The determination of pseudorutile in mineral sand products. *Powder Diffraction*, **7**, 152–161.

Thompson, J.B., Jr. (1969) Chemical reactions in crystals. *American Mineralogist*, **54**, 341–375.

Toby, B.H. (2001) *EXPGUI*, a graphical user interface for *GSAS*. *Journal of Applied Crystallography*, **34**, 210–213.

Toby, B.H. and Von Dreele, R.B. (2013) *GSAS-II*, the genesis of a modern open-source all purpose crystallography software package. *Journal of Applied Crystallography*, **46**, 544–549.

Toraya, H. (2016) A new method for quantitative phase analysis using X-ray powder diffraction: Direct derivation of weight fractions from observed integrated intensities and chemical compositions of individual phases. *Journal of Applied Crystallography*, **49**, 1508–1516.

Tremayne, M., Kariuki, B.M., Harris, K.D.M., Shankland, K. and Knigh, K.S. (1997) Crystal structure solution from neutron powder diffraction data by a new Monte Carlo approach incorporating restrained relaxation of the molecular geometry. *Journal of Applied Crystallography*, **30**, 968–974.

Valmas, A., Magiouf, K., Fili, S. Norrman, M., Schluckebier, G., Beckers, D., Degen, T., Wright, J., Fitch, A., Gozzo, F., Giannopoulou, A.E., Karavassili, F. and Margiolaki, I. (2015) Novel crystalline phase and first-order phase transitions of human insulin complexed with two distinct phenol derivatives. *Acta Crystallographica*, **D71**, 819–828.

Visser, J.W. (1969) A fully automatic program for finding the unit cell from powder data. *Journal of Applied Crystallography*, **2**, 89–95.

Von Dreele, R.B. (1999) Combined Rietveld and stereochemical restraint refinement of a protein crystal structure. *Journal of Applied Crystallography*, **32**, 1084–1089.

Walenta, G. and Füllmann, T. (2003) Advances in quantitative XRD analysis for clinker, cements and cementitious additions. *Advances in X-ray Analysis*, **47**, 287–296.

Werner, P.-E. (2002) Autoindexing. Pp. 118–135 in: *Structure Determination from Powder Diffraction Data* (W.I.F. David, K. Shankland, L.B. McCusker and Ch. Baerlocher, editors). Oxford University Press, Oxford, UK.

Werner, P.-E., Salomé, S. and Malmros, G. (1979) Quantitative analysis of multicomponent powders by full-profile refinement of Guinier-Hägg X-ray film data. *Journal of Applied Crystallography*, **12**, 107–109.

Werner, P.-E., Eriksson L. and Westdahl, M. (1985) *TREOR*, a semi-exhaustive trial-and-error powder indexing program for all symmetries. *Journal of Applied Crystallography*, **18**, 367–370.

Whitfield, P. and Mitchell, L. (2008). Phase identification and quantitative methods. Pp. 226–260 in: *Principles and Applications of Powder Diffraction* (A. Clearfield, J.H. Reibenspies and N. Bhuvanesh, editors). Wiley-Blackwell, Singapore.

Winchell, A.N. (1927) Notes and news – "Finger prints" of minerals. *American Mineralogist*, **12**, 261–262.

Wu, J., Leinenweber, K., Spence, J.C.H. and O'Keeffe, M. (2006) Ab initio phasing of X-ray powder diffraction patterns by charge flipping. *Nature Materials*, **5**, 647–652.

Young, R.A. (editor) (1993) *The Rietveld Method.* Oxford University Press, Oxford, UK.

Zevin, L.S. and Kimmel, G. (1995). *Quantitative X-Ray Diffractometry.* Springer-Verlag, New York.

EMU Notes in Mineralogy, Vol. 19 (2017), Chapter 3, 139–181

Electron crystallography

LUKÁŠ PALATINUS[1], MAURO GEMMI[2] and
MARIANA KLEMENTOVÁ[1,3]

[1]*Institute of Physics of the AS CR, v. v. i., Na Slovance 2, 18221 Prague 6, Czechia, e-mail: palat@fzu.cz*
[2]*Center for Nanotechnology Innovation@NEST, Istituto Italiano di Tecnologia, Piazza S. Silvestro 12, I-56127 Pisa, Italy*
[3]*Institute of Inorganic Chemistry of the AS CR, v.v.i., 250 68 Husinec-Rez 1001, Czechia*

Analysis of the structures of minerals is an important part of mineralogical investigations. Mineral assemblies are often formed by micro- or even nanocrystals, and it is most interesting to be able to shed light on the crystallography of individual grains, their structure, crystallinity and other properties. This can be done efficiently by the techniques of transmission electron microscopy. Transmission electron microscopy offers not only an ultimate spatial resolution in the imaging mode, but also the possibility of performing diffraction experiments and structure analysis of very small crystals, as well as a range of spectroscopic techniques revealing chemical composition and other properties of the crystals. This chapter reviews the basic techniques of transimssion electron microscopy and their application in mineralogical crystallography. A special emphasis is put on the methods of structure analysis of nanocrystals. This field has seen a rapid evolution in recent years, and has transformed from a being niche technique to a widely accepted and commonly used method of structure analysis. Its applications in mineralogy are especially rich and attractive.

1. Introduction

X-ray diffraction is the dominant technique for structural investigation of crystals in general and minerals in particular. Investigation of crystals by electron microscopy and diffraction – Electron crystallography – has always been a powerful alternative to X-rays or neutrons, being complementary in many aspects, and offering a series of key advantages. This chapter reviews the field of electron crystallography with special emphasis on the diffraction techniques and on the application of the described techniques to mineralogical problems.

The chapter is structured as follows: first, electron crystallography is defined, and basic characteristics of the field are discussed. In the second section the techniques used in electron crystallography are outlined, with the main focus on diffraction techniques. The final part of the chapter illustrates the potential of the techniques described for selected, mainly mineralogical applications.

DOI: 10.1180/EMU-notes.19.4

2. The scope and applications of electron crystallography

Electron crystallography can be defined broadly as a science investigating crystals with the aid of electron radiation. Although dedicated electron diffraction cameras were constructed and used in the past, today the instrument offering the most versatility and widest range of applications is the transmission electron microscope (TEM). Modern TEMs offer a wide range of imaging and diffraction techniques (described in Section 3) that enable a complex chemical and structural characterization of the material under investigation. The important feature of the TEM is the option to switch easily between imaging and diffraction modes, thus allowing easy characterization of single objects by both techniques. Modern electron microscopes are also equipped with energy-dispersive X-ray spectrometers and electron energy loss spectrometers, which provide local information about the chemical composition and other properties of the sample.

There are two main reasons why electrons are so attractive for the characterization of materials. The first reason is their wavelength, which is two orders of magnitude smaller than laboratory X-ray tubes (0.01–0.03 Å for electrons, 0.5–1.5 Å for X-rays), allowing atomic-resolution imaging. The second reason is their stronger interaction with matter that, in terms of scattering cross-section, is on average three orders of magnitude stronger than X-rays and four orders of magnitude stronger than neutrons (Vainshtein, 1964; Henderson, 1995). This stronger interaction allows measurable diffraction patterns to be obtained from very small volumes a few nanometers in diameter. In contrast, the typical size of a crystal for single-crystal X-ray diffraction experiment is 50 μm × 50 μm × 50 μm, and although much smaller crystals can be investigated under suitable circumstances, the practical limit stays well above 1 μm.

There are also two main reasons why electron crystallography is not the first method of choice for the structural studies of crystals. These are the instability of many materials, including minerals, under the intense electron beam, and the complexity of scattering effects which occur when electrons pass through the crystal, making the quantitative interpretation of the images and diffraction difficult. The problem of beam damage is a commonly occurring one, and it may thus be surprising to realize that electrons cause roughly three orders of magnitude less damage per elastic scattering event than X-rays (Henderson, 1995). The apparent paradox stems from the fact that obtaining a detectable scattering signal from a nanocrystal of a few tens of nanometers in size requires about nine orders of magnitude larger dose per unit volume than if a crystal of a few tens of micrometers in size is used.

The strong interaction of electrons with the matter means that multiple scattering occurs frequently, even for very thin samples. This is a significant hindrance in the quantitative description of the diffraction because the proper description of electron diffraction requires the dynamical theory of diffraction, while the diffraction of X-rays can be to a good approximation described by the simpler kinematical theory. A particular consequence of this problem is that the *ab initio* solution of crystal structures from electron diffraction data has traditionally been considered much more complicated than the structure solution from X-ray diffraction data. However, this problem can be largely circumvented nowadays by the use of precession electron diffraction (PED, Section 3.2.4) and tomographic methods for collecting diffraction data (Section 3.2.5). It is important to note that the dynamical nature of electron diffraction can also be useful. As an example, electron diffraction with convergent beam (CBED) can be used to determine the space-group symmetry of the crystal including the presence or absence of an inversion centre – a difficult or sometimes impossible task with X-ray or neutron diffraction (Section 4.2.3).

3. Overview of experimental techniques

The most common tool for performing experiments in electron crystallography is the transmission electron microscope (TEM). However, the TEM is not just used for obtaining electron diffraction patterns. Its power lies in the possibility of combining analyses of various signals arising from the electron beam hitting a thin specimen over an area as small as several nanometers (Fig. 1). Thus, information acquired from imaging techniques such as bright-field (BF), dark-field (DF), high-resolution (HR) TEM can be combined with structural information obtained from diffraction techniques such as selected-area electron diffraction (SAED), convergent-beam electron diffraction (CBED), precession electron diffraction (PED), as well as with chemical information obtained by spectroscopic techniques such as energy-dispersive X-ray spectroscopy (EDX) or electron energy-loss spectroscopy (EELS). All this information is retrieved from the same area of several nanometers in diameter.

A TEM is composed of several parts (Fig. 2a): (1) the source of electrons (gun), (2) electromagnetic lenses for focusing, (3) deflection coils for beam tilt/shift,

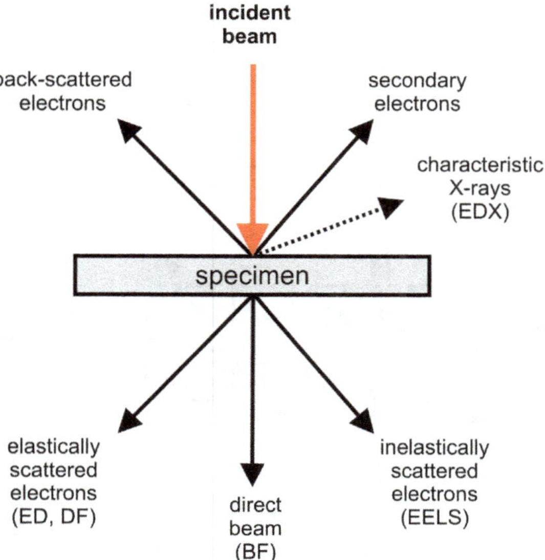

Figure 1. Various signals emerging from a thin specimen irradiated with high-energy electrons.

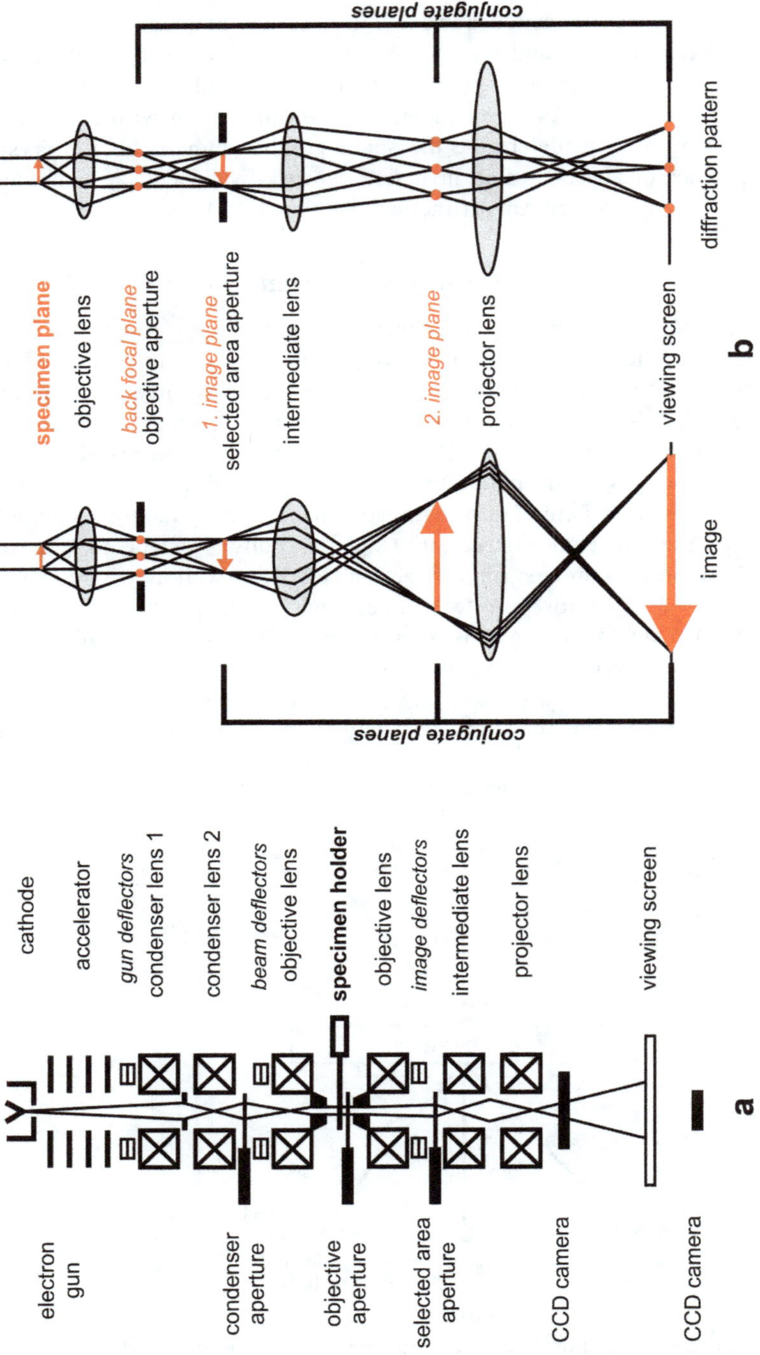

Figure 2. (a) Cross-section of a TEM, (b) ray paths for imaging and diffraction mode.

(4) selecting apertures, (5) screen or CCD for observation, and (6) specimen holder. A TEM is constructed so that it is very easy to switch between a diffraction pattern formed in the back focal plane of the objective lens and an image formed in the image plane. The switching can be performed almost instantly just by pushing a button and thus changing the focus of the intermediate lens (Fig. 2b).

Since the first TEM was built by Ernest Ruska and Max Knoll in Berlin in 1930s, there have been many textbooks covering the subject of transmission electron microscopy. Several very good texts cover the subject from the general perspective of material science (De Graef, 2003; Fultz and Howe, 2008; Williams and Carter, 2009), and a few are of special interest for mineralogy (Wenk, 1976; Buseck, 1992; McLaren, 2005).

3.1. Imaging techniques

Imaging techniques in TEM employ various types of contrast arising due to the scattering of electrons by a sample, during which the amplitude and/or the phase of the incoming electron waves are changed. Thus, the two major types of contrast are amplitude contrast and phase contrast. In most cases, both types of contrast contribute to the image; however, conditions can be selected so that only one is enhanced. The basic difference between the two types of contrast lies in the number of electron beams used for image formation. The beams are selected by the objective aperture located in the back focal plane of the objective lens (the plane of the diffraction pattern) to form the final image in the image plane (Fig. 2b). In case of amplitude contrast only one beam is used, whereas for phase-contrast imaging at least two beams are necessary to form an interference image.

3.1.1. Amplitude contrast

In crystalline samples, the most pronounced amplitude contrast is diffraction contrast which appears if only some of the elastically scattered electrons are used for the image formation. When a crystalline sample is illuminated by a parallel beam, the electrons scattered under the same angle from different parts of a single-crystal sample are focused to a single point at the back focal plane of objective lens, forming a diffracted beam. A single beam (either the direct, undiffracted beam or any diffracted beam) contains information from the whole crystal and therefore can be used to image the crystal.

To enhance the amplitude contrast only one electron beam is selected by the objective aperture located in the back focal plane of the objective lens to form the final image. When the direct beam (undiffracted beam) is selected, the resulting image is called a bright-field image (BF). If there is a diffracting crystal on a non-diffracting background, the crystal will appear dark on the bright background (field). When any of the diffracted beams is selected, the resulting image is called a dark-field image (DF). In this case, the diffracting part corresponding to the diffracted beam will appear bright on a dark background (field). In Fig. 3, BF/DF imaging of a twinned crystal is illustrated. Other types of diffraction contrast include thickness contours and bending contours.

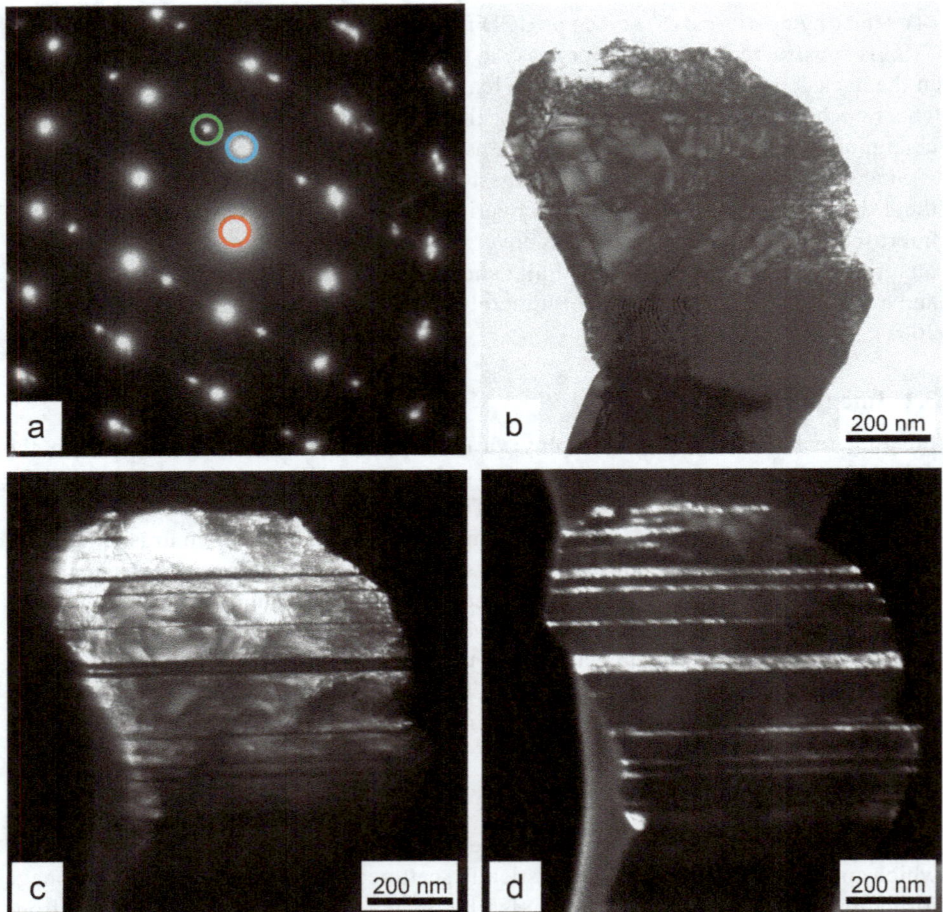

Figure 3. Diffraction contrast imaging. (a) SAED with circles representing the position of objective aperture used for imaging: red – BF image shown in (b), blue – DF image shown in (c), and green – DF image shown in (d).

Thickness contours can be used for thickness determination of a known sample. However, when thickness contours are visible, it means the sample is not thin enough for high-resolution imaging.

The contrast in BF/DF imaging is very sensitive to any particular feature that affects the local crystal structure. Therefore it is used widely to localize crystal defects such as dislocations and stacking faults and to characterize the strain fields induced by them. DF imaging becomes very effective in these applications if it is performed in weak beam mode. In this mode the diffracted beam selected by the objective aperture is far from the exact Bragg condition (large excitation error). Most of the crystal then remains almost dark, while those defects, which bend the crystal planes corresponding to the selected reflection back to Bragg condition, yield a sharp, well localized bright feature in the image. The use of a beam with a large excitation error guarantees a much sharper contrast and a better localization of the defect with respect to standard BF/DF imaging.

Mass-thickness contrast is another type of amplitude contrast which dominates in non-crystalline samples such as polymers or biological samples. It results from incoherent elastic scattering of electrons (Rutherford scattering) which is heavily dependent on the atomic number Z and the thickness. The mass-thickness contrast can be enhanced by inserting an objective aperture in the back focal plane, staining different areas of specimens with heavy metals such as Os, Pb and U, or decreasing the energy of the incident beam (lowering the accelerating voltage of the TEM).

Z-contrast is the high-resolution version of the mass-thickness contrast, which is registered by high-angle annular dark-field detectors in scanning transmission electron microscopy (STEM-HAADF). Crystalline as well as non-crystalline samples can be studied with atomic resolution, and the strong dependence of intensity on Z can be exploited. Shen *et al.* (2013) studied exsolution lamellae in Ca-rich dolomites by STEM-HAADF (Fig. 4). Intensity line profiles were taken parallel to the traces of (102) planes to estimate the variation in Ca:Mg ratio across the bright lamellae, making use of the $Z^{1.7}$ dependence of intensity in the Z-contrast image. The composition of the lamellae was estimated qualitatively to range from $Ca_{0.85}Mg_{0.15}CO_3$ to $Ca_{0.70}Mg_{0.30}CO_3$.

3.1.2. Phase contrast

The most commonly used phase contrast technique is high-resolution imaging (HRTEM), which is used for imaging details at the atomic level. The HRTEM images are in fact interference images formed by interaction of at least two beams in the

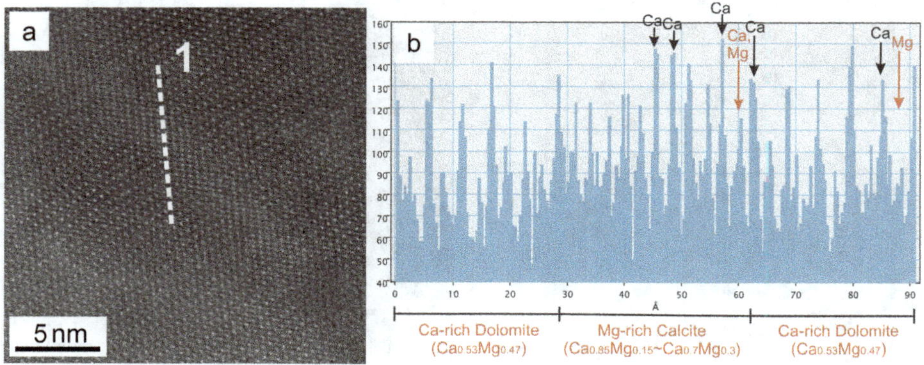

Figure 4. STEM investigation of exsolution lamellae in Ca-rich dolomite. (a) STEM-HAADF image, (b) an intensity profile of line "1" taken parallel to (102) trace to examine composition variation at atomic resolution (reproduced with the permission of the Mineralogical Society of America, from Shen *et al.*, 2013).

image plane of the objective lens. Therefore a larger objective aperture must be used to gather more than one beam (Fig. 5a). In HRTEM, we are observing a projection of the structure. We do not see individual atoms, but we observe interference fringes corresponding to the periodicity of atomic columns in a given direction. The contrast of the fringes depends on many parameters, among which the thickness and objective lens defocus are the most important (Fig. 5). In order to determine the exact positions of the atoms, simulations based on the structure model and the microscope parameters must be carried out. However, there are many applications of HRTEM where the exact positions of atoms are not necessary, such as studying dislocations or grain boundaries.

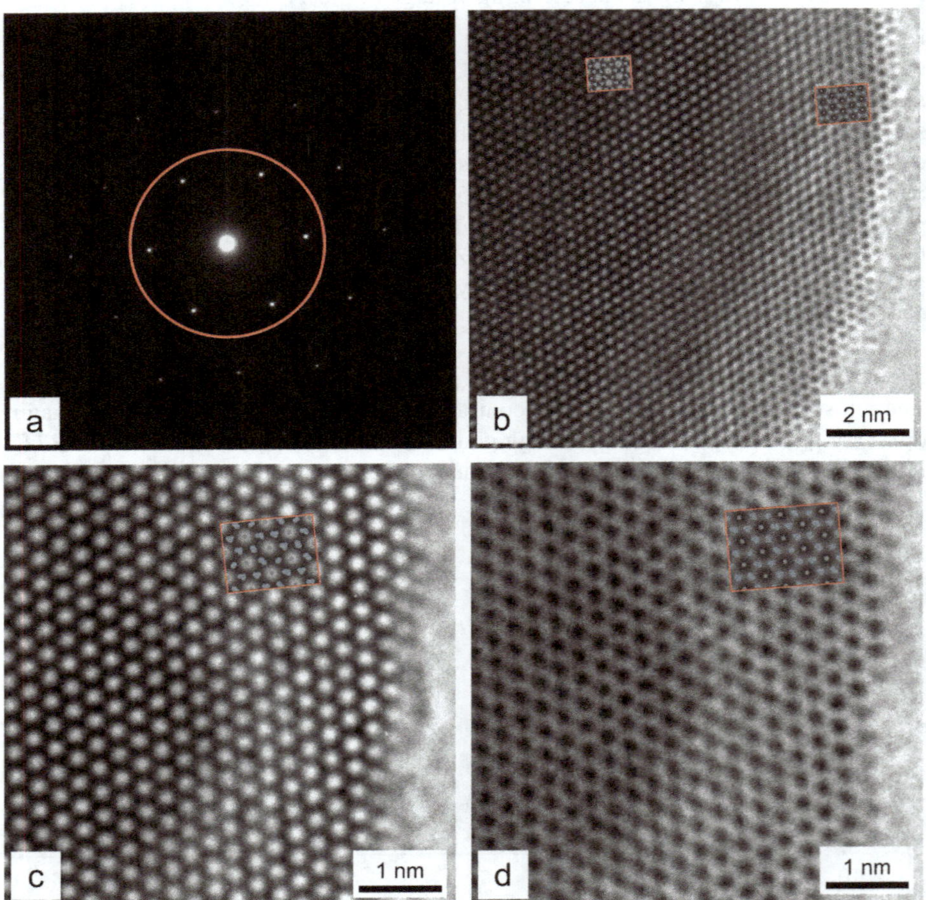

Figure 5. HRTEM images of hematite along [001] with corresponding simulations in the insets. (a) SAED with a red circle representing position of objective aperture used for HRTEM imaging; (b) variation of phase contrast with thickness, and variation of phase contrast with defocus (c) defocus 60.5 nm, thickness 5.5 nm, (d) defocus 80.5 nm, thickness 5.5 nm. Fe – grey, O – teal.

The resolution of a common HRTEM varies between 1.6 and 2.3 Å, which might seem quite poor considering the wavelength of 200 kV electrons (0.0251 Å). Such deterioration is due to the aberrations of electromagnetic lenses, mainly the spherical and chromatic aberrations. The top-end microscopes nowadays include correctors of these aberrations and achieve resolution down to 0.8 Å.

Other phase-contrast imaging techniques include moiré fringes, Fresnel fringes, electron holography, and Lorentz microscopy. Fresnel fringes are caused by an abrupt change in electrostatic potential and can be visualized by taking images with a strong defocus. They are used to study gas or liquid bubbles, voids and high-angle phase boundaries. Moire fringes result from the overlap of two or more lattice periodicities. They can be also added artificially, *e.g.* to enhance grain boundaries to measure precisely the angle between two grains (Hetherington and Dahmen, 1992). Both electron holography and Lorentz microscopy are very specialized techniques. In order to use these techniques, a TEM must be equipped with a biprism or Lorentz lens, respectively. Electron holography is used for imaging magnetic and electric fields. Lorentz microscopy is used for imaging of magnetic domains.

3.2. Electron diffraction techniques

Electron diffraction follows the same geometrical laws as X-ray or neutron diffraction. However, the availability of electromagnetic lenses for electrons makes electron diffraction a very flexible diffraction technique. In every transmission electron microscope several condenser lenses that form the illumination system allow the user to change the properties of the electron beam impinging on the sample. This flexibility gives rise to many different diffraction techniques embedded in the same instrument. The beam can be parallel or convergent, illuminating a large or small area, stationary, scanning an area or moving across different orientations, and each configuration gives rise to a specific electron diffraction technique. In the following paragraphs we will review all the main electron diffraction methods, pointing out the advantages and the peculiarities of each.

3.2.1. Selected area electron diffraction (SAED)

Selected area electron diffraction is the standard electron diffraction technique, the first electron diffraction technique we usually try on a new sample. The sample is illuminated with a parallel beam and the area from which we want to collect the diffraction pattern is selected by inserting an aperture called selected area (SA) diaphragm in the image plane of the objective lens. Depending on the dimension of the SA diaphragm the diffracting area can be varied from a few hundred nanometers to several microns in diameter. The minimum diffracting area is limited not only by the size of the SA diaphragm but also by the errors introduced by the spherical aberration of the objective lens, which brings into the SAED patterns electrons coming from regions adjacent to the area selected by the SA diaphragm (Hirsch *et al.*, 1965; Williams and Carter, 2009).

Due to the small electron wavelength the Ewald sphere has a very large radius, and it can be approximated by a plane in the vicinity of the origin of the reciprocal space.

Because the crystal is thin, the nodes of reciprocal lattice in the Ewald construction are not points, but are elongated in the direction perpendicular to the plane of the sample. As a result, if the crystal is oriented with the electron beam normal to a set of reciprocal lattice planes, all the reciprocal vectors of the plane passing through the origin are so close to the Ewald sphere that they are visible in the pattern (Fig. 6). These SAED patterns are called zone axis patterns and are used widely as they are very informative. They display undistorted representations of reciprocal lattice planes carrying information on the geometry of the crystal lattice. A set of zone axis patterns can be used for the determination of lattice parameters, point group symmetry and, to a certain extent, space-group symmetry (Section 4.2).

3.2.2. Micro- and nanodiffraction

Micro- and nanodiffraction is an alternative way of collecting an electron diffraction pattern with parallel or slightly convergent illumination. In this case the diffracting

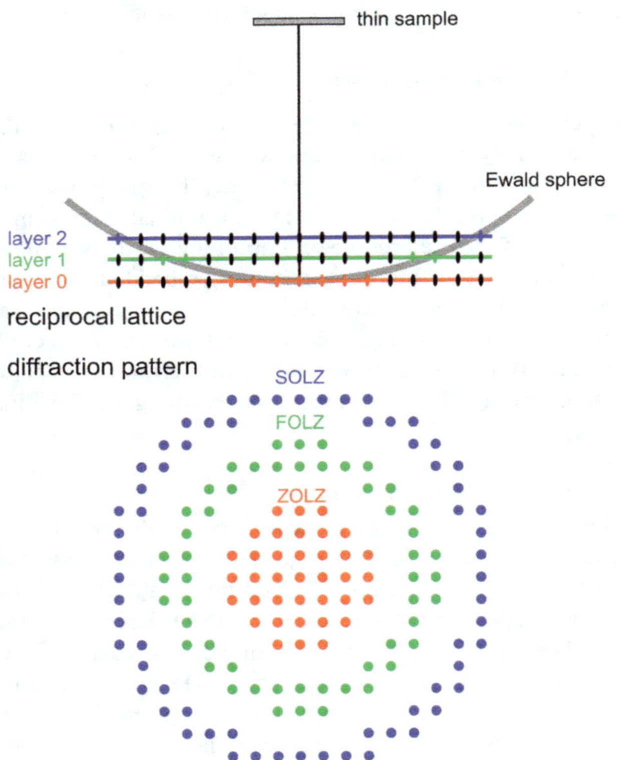

Figure 6. The geometry of electron diffraction on an oriented crystal illuminated with a parallel beam. Top: side view of the reciprocal space showing three layers of reciprocal space intersecting the Ewald sphere. Bottom: schematic representation of the resulting diffraction pattern showing the zero-order Laue zone (ZOLZ, red), first-order Laue zone (FOLZ, green) and second-order Laue zone (SOLZ, blue).

area is selected directly by forming a small parallel beam on the sample and not through an aperture. In this setting the diffraction pattern is less affected by the aberrations caused by the objective lens compared to SAED. Moreover, the beam size can be smaller than the minimum area attainable with even the smallest SA diaphragm. Depending on the electron source and on the illumination system of the microscope the minimum beam size can be as small as a few nanometers (Alloyeau *et al.*, 2008; Ganesh *et al.*, 2010). With this diffraction mode we can obtain diffraction patterns with a very high spatial resolution keeping the simple spot-like appearance of a SAED pattern. However, this technique requires a special alignment and special features of the illumination system which are not available in every microscope.

3.2.3. Convergent beam electron diffraction (CBED)

CBED is the standard technique for collecting a diffraction pattern with the maximum possible spatial resolution.

The beam used for CBED is not parallel but convergent, and as such it can be thought of as a collection of many narrow beams, each with a different propagation direction in the illumination cone. Because each direction of the incident beam generates its own set of diffracted beams, the resulting diffraction pattern does not contain spots, but discs. (Fig. 7; see Spence and Zuo, 1992 for a detailed account on the method). In a CBED pattern, we do not see diffraction effects corresponding only to reciprocal points lying on the Ewald sphere surface, but all the diffraction effects coming from the volume of reciprocal space intersected by the Ewald spheres corresponding to all the orientations of the incident beams inside the illuminating cone. A CBED pattern can carry three-dimensional information on the reciprocal space of the crystal, even if it is taken in a zone-axis orientation. The beam convergence brings the Ewald sphere to touch reciprocal lattice planes above that passing through the origin. If the field of view is sufficiently large, the diffracted discs from these lattice planes are visible in the image. The diffraction pattern can be divided in concentric zones, the innermost with discs corresponding to the plane passing through the origin called zero order Laue zones (ZOLZ) and the others, surrounding it as rings, called higher order Laue zones (HOLZ). The combination of this complicated geometry with the dynamical scattering of the electrons transforms each CBED pattern into a decorated pattern in which every disc and the space in between them show a rich contrast (Fig. 7). Because the Friedel law does not apply to the dynamical scattering (Goodmann and Lehmpfuhl, 1968), the symmetry of the pattern follows the point group symmetry of the crystal and not only the Laue class.

A CBED pattern can be recorded with an extremely high spatial resolution. Depending on the gun type of the TEM the minimum probe size for a CBED pattern can be reduced from 100 nm down to a few nm. Therefore the crystal information recorded in CBED patterns comes from an homogeneous area in terms of chemical composition, unit-cell parameters, symmetry, thickness and presence of defects that can be simulated by relatively simple models of ideal crystals because any derived quantity is a local property averaged over just a few unit cells.

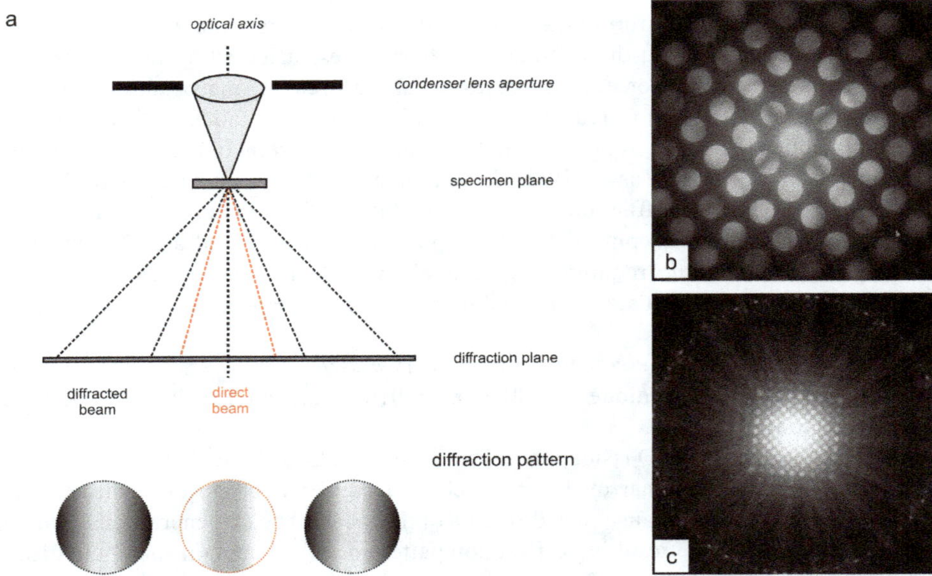

Figure. 7. (a) A sketch of the CBED geometry with a schematic depiction of the resulting CBED discs of the direct beam (lower middle) and two diffracted beams. (b) Experimental CBED pattern of the zone [001] of cassiterite showing the variation of intensity in the CBED discs and distinct Gjønnes-Moodie lines in some of the discs. (c) A CBED pattern of the same crystal as in (b), but taken with smaller magnification of the diffraction pattern (smaller camera length). A ring of FOLZ reflections surrounds the central ZOLZ pattern.

The main applications of CBED are the accurate determination of the lattice parameters, full and (almost) unambiguous characterization of the space-group symmetry, accurate refinement of structural parameters and direct determination of the amplitudes of structure factors (Sections 4.1, 4.2.3, 4.4.2).

Note that the size of the discs in the CBED patterns is restricted by the limited space between individual discs (*cf.* Figs 7 and 16), making it difficult to use the method for materials with large lattice parameters. This limitation, however, does not apply to modern ways of collecting CBED patterns by recording a consecutive set of patterns with small beam convergence, and then reconstructing the CBED discs digitally (Koch, 2011; Beanland *et al.*, 2013).

Another diffraction technique based on a convergent beam is large-angle convergent beam electron diffraction (LACBED). The electron optical configuration is the same as CBED: the beam is convergent on the object plane of the objective lens, but the convergence angle is larger than in CBED, in the range of 1 to 5°. The sample is shifted upward with respect to the object plane, therefore the convergent beam does not illuminate a small spot on the sample, as in CBED, but a wider area. As a result, electrons travelling along different incident directions impinge on different zones of the sample, and therefore there is a correlation between points in the LACBED patterns and points in the specimen. A shadow image of the sample is superimposed on the LACBED pattern and serves as a reference for this correlation. Any crystal deformation present in a specific area of the sample can be detected through a corresponding deformation of the pattern superimposed on that area. This technique has a wide application in the study of dislocations and 2D defects such as grain boundary and stacking faults, because it also has the ability to characterize the type of defect and locate its position. A thorough overview of the technique is available in Morniroli (2002).

3.2.4. Precession electron diffraction (PED)

Some electron diffraction techniques, like CBED, benefit from the dynamical diffraction effects. However, the departures from kinematical diffraction are the reason why solving crystal structures from electron diffraction data has been so difficult in the past. For many years the only possible way of avoiding strong dynamical scattering has been to reduce the crystal thickness to the minimum value permitted by thinning methods, quite a difficult task especially with powder samples. An alternative method, effective also for thicker samples, was developed when Vincent and Midgley (1994) invented the precession electron diffraction (PED) technique. This technique meant a true breakthrough for applications of electron diffraction in structure analysis.

A schematic diagram of the technique is shown in Fig. 8. Instead of keeping the incident beam static and parallel to the optical axis of the microscope, the beam is tilted away from the optical axis by a precession angle φ. Then the beam is precessed around the surface of a cone with vertex fixed on the specimen. This movement is called scan. The tilting of the beam results in a circular trace of the incident and all diffracted beams on the detector (Fig. 8a). Therefore, the movement of the beam in the diffraction plane during a precession cycle is compensated by an opposite tilt of the diffracted electrons with the post-specimen deflection coils (descan). The precession movement is very fast, usually 100 Hz, and a single exposure is an average of the diffraction signal over many precession cycles.

The resulting pattern is a spot pattern that has the same aspect as a standard SAED pattern (Fig. 9). However, a precession diffraction pattern exhibits a number of characteristics that make it different from a standard SAED or microdiffraction pattern and which make the PED technique an ideal tool for many crystallographic applications:

- The intensities are closer to kinematical reflection intensities than in a SAED pattern (Gjønnes *et al.*, 1998; Own *et al.*, 2006a; Oleynikov *et al.*, 2007). The effect is caused mainly by the integration of the intensities over many orientations of the incident beam. The dynamical effects, which cause rapid variation of the intensities with changing thickness and crystal orientation, are thus partially averaged out (Sinkler and Marks, 2010). If PED is used to record an oriented diffraction pattern, another effect contributes to the reduction of dynamical diffraction. In oriented SAED patterns, there is a large

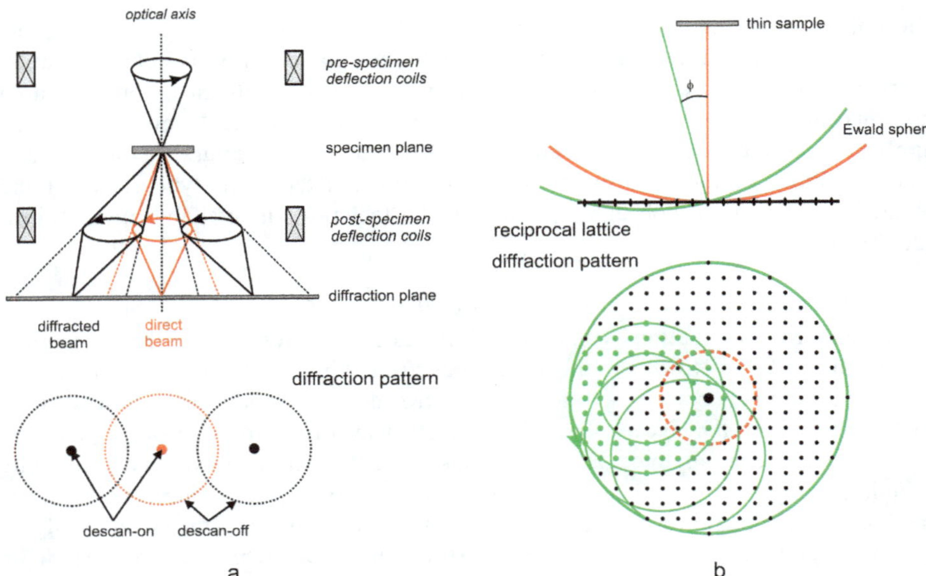

Figure 8. (a) A schematic drawing of the principle of precession electron diffraction geometry. Compare with the CBED geometry in Fig. 7a. (b) Side and top view of the diffraction during the precession movement. At each instant the reciprocal-lattice plane intersects the Ewald sphere in a circle, and reflections close to this circle appear in diffracting condition (green spots in the bottom image). During the precession, this circle of intersection is swept around the origin (two instances are shown with thin green circles), and finally all reflections within the large diameter (large green circle) appear in the diffraction pattern.

number of possible multiple diffraction paths, and this results in especially strong dynamical diffraction effects. In PED, the beam is always tilted away from the perfect orientation, and the number of possible multiple scattering paths is reduced. While the latter effect

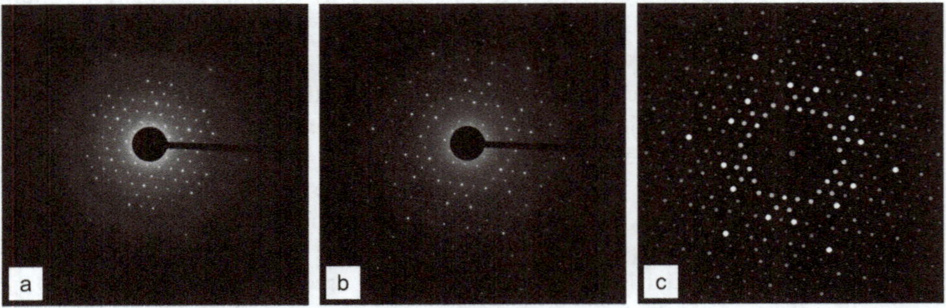

Figure 9. Diffraction pattern of almandine, zone [111] (a) standard SAED pattern. (b) PED pattern taken with precession angle $\varphi = 2.5°$. (c) Simulation of the diffraction pattern using kinematical intensities.

certainly plays an important role in oriented patterns, the former effect is the principal reason for the more kinematical character of the PED diffraction intensities. Larger precession angles result in more kinematical data (Ciston *et al.*, 2008; Eggeman *et al.*, 2010; Palatinus *et al.*, 2013), because with larger precession angles the reflection intensities are integrated over a wider range of orientations and, in the case of oriented patterns, the beam is positioned further away from the perfect on-zone orientation. Note that dynamical effects can never be avoided completely. The dynamical nature of the PED patterns is indicated, for example, by the presence of kinematically forbidden reflections. It has been demonstrated that, at least in some cases, the intensities of the forbidden reflections fall exponentially with the precession angle (Ciston *et al.*, 2008), in agreement with the general trend of more kinematical data with larger precession angle.

- Oriented zone-axis patterns recorded with PED are less sensitive to minute variations of the crystal orientation than SAED patterns. It is very difficult to obtain an oriented SAED pattern showing perfect symmetry corresponding to the point group of the crystal. On the other hand, the PED patterns average out slight deviations from the perfect orientation, and the patterns are much more symmetrical and can be better used for the determination of the Laue class (Section 4.2).
- A PED pattern contains reflections at higher scattering angles than a corresponding SAED pattern. In a SAED pattern, due to the Ewald sphere curvature, the reflections with diffraction angle higher than certain limit are too far from Bragg condition and their intensity thus vanishes. In PED, all the reflections of the ZOLZ having a Bragg angle θ smaller than φ are brought to an exact Bragg diffraction by the precessing movement.
- In a PED pattern the rocking movement of the Ewald sphere brings the reflections from the HOLZ close to the Ewald sphere similar to beam convergence in CBED. If the precession angle is large enough, the HOLZ reflections are partially superimposed on the ZOLZ reflections. Similarly to CBED, the geometry of the superposition of the HOLZ and ZOLZ two-dimensional lattices is a function of the Bravais lattice type and, together with the symmetry of the pattern, allows the determination of the crystal system, the Bravais lattice and the extinction symbol by analysing at most three different zone axis patterns (Section 4.2). HOLZ lines can be observed in standard SAED patterns too, but the PED patterns contain much larger patches of the HOLZ layers, making the interpretation easier.

While the superposition of ZOLZ and HOLZ can be useful for symmetry determination, it can be dangerous for an accurate determination of the reflection intensities. Depending on the crystal geometry and on the zone axis this superposition can cause a complete overlapping between reflections of different Laue zones. Therefore, if we are interested in collecting reliable diffraction intensities, the best precession angle should be as large as possible, but small enough to avoid overlapping of different Laue zones.

- PED patterns are useful also if dynamical diffraction theory is used for calculating diffraction intensities. PED patterns are less sensitive than SAED patterns to the variation of crystal thickness and orientation, and more sensitive to the parameters of the crystal structure, such as atomic positions and occupancies. For this reason, it is easier to obtain good fits to PED patterns than to SAED patterns, and PED patterns can also be used more easily for structure refinement using dynamical diffraction theory (Section 4.4; Palatinus *et al.*, 2013).

The PED technique is not a standard diffraction method embedded in every TEM. The first PED patterns were obtained by sending external signals to the magnetic lenses using homemade devices (Vincent and Midgley, 1994; Gemmi *et al.*, 2003; Own *et al.*, 2004) but nowadays a commercial device is obtainable for performing the technique and is adaptable to most commercial TEMs (Gemmi *et al.*, 2013).

3.2.5. Electron diffraction tomography (EDT)[1]

All the diffraction techniques reviewed so far are based on data collections on a definite crystal orientation, usually with an important zone axis oriented parallel to the incident beam. Even if, because of the convergence or the precession of the beam, a portion of the reciprocal lattice larger than a single plane is explored, the coverage of the reciprocal space through a collection of zone axes is far from complete. A set of thin planar slices will always miss some reflections in between the recorded planes. If the crystal system has a high symmetry this problem can be overcome partly by collecting symmetry-equivalent reflections in different zone-axis patterns, but if the symmetry is low, as in the triclinic and monoclinic systems, the diffraction data will always be more or less incomplete. This problem is not critical if the diffraction data are just used for the determination of the crystal symmetry or for measuring the unit-cell parameters. However, if we want to investigate the crystal structure of the material, the completeness of the data is of utmost importance for our ability to solve or refine the crystal structure. In 2007 Ute Kolb and coworkers (Kolb *et al.*, 2007) of Mainz University proposed the collection of a sequence of non-oriented diffraction patterns while the sample is rotated around the axis of the goniometer. The physical rotation of the sample is usually performed in angular steps of 1° within the maximum tilt range

[1] In some literature this method is referred to as ADT for automated diffraction tomography. The word automated does not refer to the data collection, but to the data processing, which has to be performed automatically by a computer. We prefer EDT for electron diffraction tomography as a more general and less confusing term.

allowed by the sample holder. In the case of special tomographic holders a tilt range of 150° can be achieved, while standard sample holders allow for tilt ranges typically between 90° and 120°. Because such a fine mapping of the diffraction space resembles a tomography of the reciprocal space, the technique is called electron diffraction tomography (EDT). An X-ray crystallographer will notice immediately, however, that it is equivalent to a standard single-crystal diffraction experiment performed on a modern single-crystal X-ray diffractometer equipped with an area detector. During each tilt, unavoidable mechanical instabilities of the TEM goniometer displace the crystal slightly and it must be recentred every time. In the original EDT procedure developed in Mainz this is done automatically with a cross correlation routine on STEM images that moves the beam back into its original position (Kolb *et al.*, 2007). The technique can also be used without the automated procedure. The patterns can be collected in SAED mode and the crystal can be tracked manually in image mode (Palatinus *et al.*, 2011; Gemmi *et al.*, 2012). However, the crystal tracking using the STEM mode rather than the standard TEM imaging reduces significantly the dose on the sample, avoiding sample contamination and radiation damage. With the STEM tracking routine, samples that are beam sensitive have a better chance of remaining crystalline for the entire EDT experiment.

The sampling of the reciprocal space obtained by EDT is much finer than collecting patterns in a zone axis only, but a one degree step is still quite coarse. If the Bragg condition for a reflection falls between the steps, its intensity may be severely underestimated. This undersampling problem can be solved in two different ways: by precession EDT (PEDT; Mugnaioli *et al.*, 2009a) and rotation EDT (REDT; Zhang *et al.*, 2010).

In PEDT the patterns are collected in PED mode with φ comparable with the angles between the patterns. PED plays two roles here. First, the precessing movement integrates the intensities in the range $\pm\varphi$ and thus fills the gaps between the recorded sections. Second, integrating the diffracted intensities makes them closer to the kinematical limit and thus makes the structure solution process easier and more straightforward. In REDT the missing wedges are avoided by a finer sampling of the reciprocal space. This is achieved by a combination of relatively large mechanical tilts (>1°) and very small electronic beam tilts (<0.1°). The final data collection results in a number of patterns about ten times larger than in PEDT. Both techniques give as output a set of patterns that must be processed further with dedicated software. The individual patterns can be combined to give a three-dimensional reconstruction of the reciprocal space (Fig. 10). Just to illustrate the current rapid development of the fields, we note that only 5 years from the first EDT data collection, four different programs have been developed to perform a complete data reduction of an EDT experiment: *ADT3D* (Kolb *et al.*, 2011), *PETS* (Palatinus, 2011), *EDT-PROCESS* (Oleynikov, 2011a) and *RED* (Wan *et al.*, 2013). The principal output from all these programs is the lattice parameters of the crystal, and a list of indexed reflections with their intensities. EDT data sets have reciprocal space coverage usually >70%. The reflection intensities have a reduced dynamical character, as they are collected in a random orientation usually far

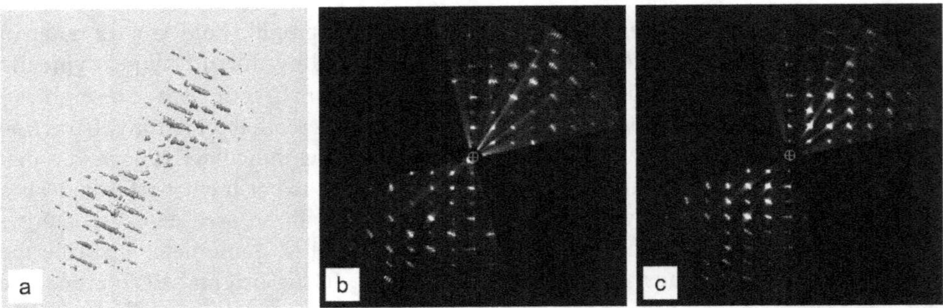

Figure 10. Reconstruction of the reciprocal space of widenmannite from PEDT data (93 patterns, tilt step 1°, φ = 1.2°). (a) 3D reconstruction of the reciprocal space viewed along the tilt axis, which is approximately parallel to the c-axis of the crystal. (b) Reconstruction of the reciprocal layer *hk*0. (c) Reconstruction of the reciprocal layer *hk*1. Comparison of the two sections shows that in the *hk*0 the reflections of the type $h+k=2n+1$ are systematically weaker than the rest. No such pattern is visible in the *hk*1 section. This is an indication of the presence of an *n*-glide perpendicular to **c**. For more information on widenmannite see Plášil *et al.* (2013).

away from a perfect zone axis condition. The dynamical character is further suppressed by using PED in the PEDT technique. These data can be used for *ab initio* structure solution of an unknown crystal structure with a high degree of success (Section 4.3), and also for structure refinement (Section 4.4).

3.2.6. Phase and orientation mapping

We conclude the review of the available diffraction techniques with a recently developed method in which nanodiffraction is exploited to obtain orientation and phase maps with nanometric resolution. It is an equivalent of the EBSD technique (electron back-scatter diffraction), but performed in a TEM at higher resolution. Although, to the authors' knowledge, it has not been applied to mineralogical problems so far, we think that, like EBSD, its application in mineralogy and geology is just a matter of time.

In this method a small (few nm), parallel beam is scanned over an area of a few square microns. The TEM is kept in nanodiffraction mode and while the beam is scanning, the patterns that appear on the fluorescent screen are recorded by an external CCD at the speed of up to 100 frames per second. The collected patterns are then compared with a database of precomputed diffraction patterns (called templates) corresponding to all the possible crystal orientations for each crystal phase that might be present in the sample. By matching the experimental and computed patterns the crystal phase and the crystal orientation is determined for each pattern (Rauch and Dupuy, 2005). The entire procedure is more effective if the patterns are collected in PED with a precession angle between 0.5° and 1°. PED increases the number of reflections in the patterns and makes the patterns more kinematical, thus improving the matching procedure (Portillo *et al.*, 2010). The output of the procedure comprises several maps: an index map where for every point the grey level is proportional to the quality of the match (Fig. 11a); a phase map where different colours are assigned to different phases; and an orientation map

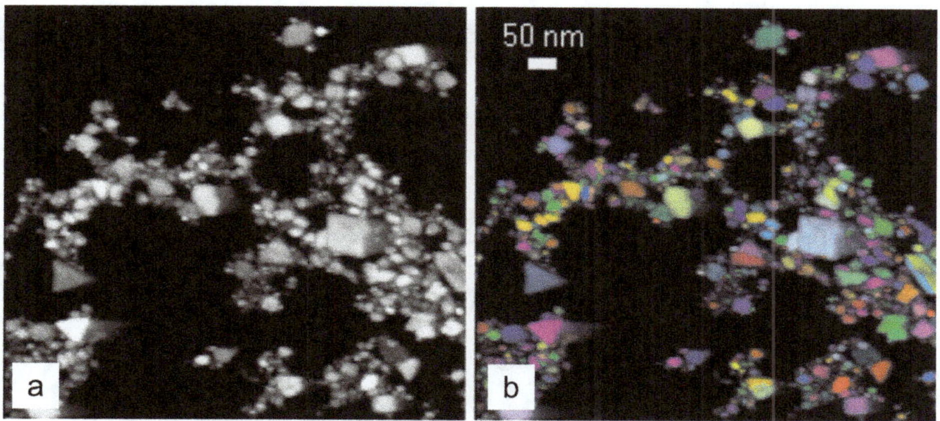

Figure 11. Index (a) and orientation (b) maps of CeO_2 nanoparticles collected on a Zeiss Libra 120 TEM working at 120 kV.

where different colours represent different crystallographic orientations (Fig. 11b). As already mentioned, the results are similar to EBSD but on a much smaller scale. Due to the nanodiffraction mode, the resolution can go beyond 5 nm in the most favourable cases (Ganesh *et al.*, 2012), whereas the analysed area is much smaller than in EBSD, only a few square microns. No special sample preparation is necessary. The only requirement is that the area of interest is electron transparent.

3.3. Spectroscopic techniques

Spectroscopic techniques do not belong, strictly speaking, to experimental techniques of electron crystallography. However, they are so important for crystallographic analysis that a brief description is appropriate here.

3.3.1. Energy dispersive X-ray spectroscopy (EDX)

When an electron beam hits a specimen, some of the incoming electrons can knock out electrons from inner atomic energy levels creating a hole. This hole can be filled with an electron from a higher level. As this electron fills the hole, it loses energy, which is emitted in the form of characteristic X-rays. The energy of the X-ray generated thus corresponds to the energy difference between the two atomic levels and therefore is characteristic for a given element. In EDX spectrometry, the energy of the characteristic X-rays emitted is detected and thus the elemental composition of the sample can be determined. The basic principle and instrumentation of EDX are common to both TEM and scanning electron microscopy (SEM). However, there is one significant advantage of EDX in a TEM – the spatial resolution of the analysis. Due to the thin samples which are used in TEM the interaction volume of the electrons basically equals a cylinder defined by the diameter of the beam and the sample thickness (Fig. 12). In a thin specimen, X-ray absorption and fluorescence can, to a first

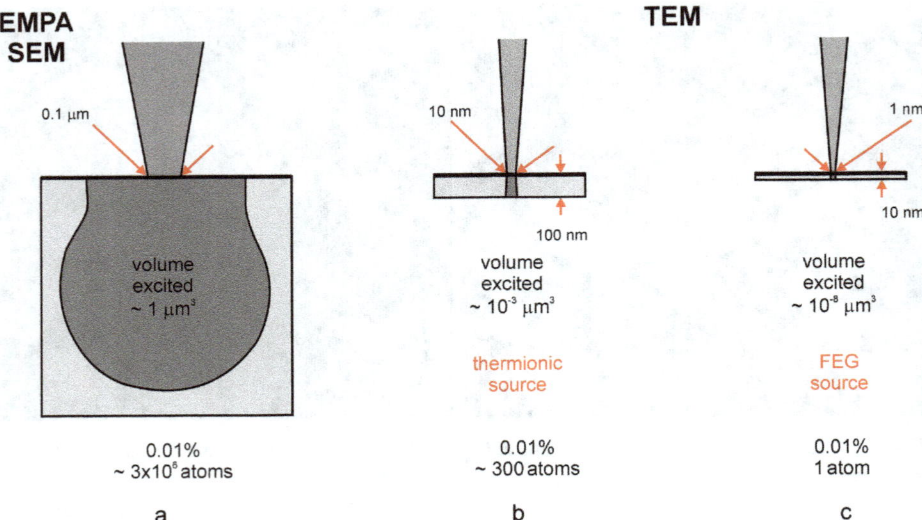

Figure 12. Comparison of interaction volume in a bulk sample used in SEM or EMPA (a); with a thin sample used in TEM with a thermionic source (b); and a FEG source (c) (after Williams and Carter, 2009).

approximation, be ignored and the X-ray intensity ratios, I_A/I_B, observed can be converted into weight fraction ratios, C_A/C_B, by multiplying by a constant k_{AB}, which can be calculated or determined experimentally (Lorimer, 1987).

3.3.2. Electron energy-loss spectroscopy (EELS) and energy-filtered TEM (EFTEM)
During the process of X-ray generation described in Section 3.3.1, the incident electron loses some energy depending on the type of atom and interaction. Such electrons can be detected by a spectrometer and the technique is called electron energy-loss spectroscopy (EELS). The most interesting parts of the EELS spectrum are the core-loss edges. Energy of an edge is characteristic for a given element. EELS is preferable especially for the analysis of light elements. In addition to the determination of elemental composition, the shape of an edge can provide more detailed information about valence or type of bonding. In energy-filtered TEM (EFTEM), a certain energy window is selected to image elemental composition as well as valence state. As an example, valence state mapping of Fe^{2+} and Fe^{3+} in exsolution intergrowth of hematite-ilmenite in a natural crystal of titanohematite from a granulite is shown in Fig. 13 (Golla and Putnis, 2001).

4. Applications

4.1. Phase identification

As mentioned previously, the strength of TEM lies in the possibility of combining information from the different types of signal generated by the interaction of electron

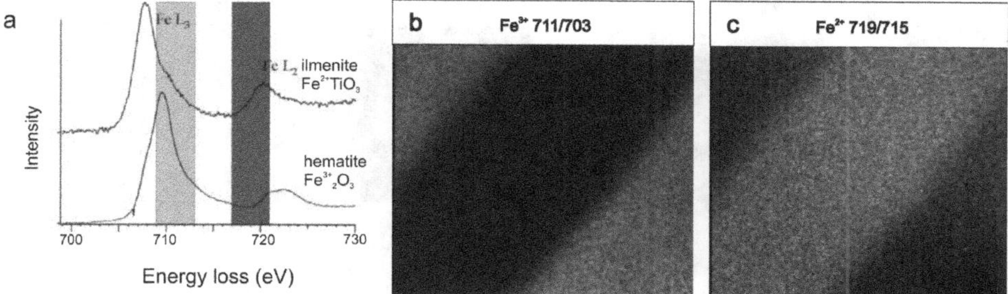

Figure 13. EELS and EFTEM observations of hematite-ilmenite exsolution intergrowth in natural titanohematite crystal. (a) Fe $L_{2,3}$-edge EELS near-edge structure and energy shift of the ilmenite and hematite exsolution lamellae, and post-edge energy window positions used for Fe^{3+}, Fe^{2+} imaging; (b) Fe^{3+} valence imaging; and (c) Fe^{2+} valence imaging with the indication of post/pre-edge energy window positions (reproduced with the permission of Springer from Golla and Putnis, 2001).

beam with sample. Combination of structural information derived from electron diffraction together with chemical information obtained by EDX can unequivocally identify phases at the nanometer scale. Phase identification together with observation of morphology of individual phases is the most common task in TEM, especially in the initial stages of sample investigation. It is especially important in mineralogy where samples commonly contain many different phases which are difficult to separate during sample preparation.

Two types of SAED patterns can be used to identify the mineral structure in combination with the composition determined by EDX spectroscopy. In the case of the ring pattern from diffraction on a powder sample, the pattern can be transformed into a typical powder diffractogram (intensity *vs.* diffraction angle) and the structure can be search-matched against a suitable database, *e.g.* the PDF-database (ICDD, 2003) using dedicated software (*e.g. ProcessDiffraction* – Lábár, 2005), similarly to powder XRD analysis. In the case of oriented single-crystal patterns, the patterns must be indexed satisfactorily by matching the observed reflection positions to the lattice parameters of a known phase (*e.g.* Process Diffraction). Oriented patterns of known structures can be simulated in various programs to compare the intensities as well as positions of reflections (*e.g. JEMS* – Stadelmann, 2004). In addition, in *JEMS* it is also possible to search-match against several pre-selected structures to find the matching zone axis.

A complex case illustrating all of the above methods is the phase identification in Pb smelter fly ash (Fig. 14). The TEM-EDX-SAED observations indicated that formation of secondary mimetite $Pb_5(AsO_4)_3Cl$ (identified from ring diffraction pattern) and Sn-bearing bindheimite $Pb_2(Sb,Sn)_2O_6(O,OH)$ (determined from single-crystal diffraction) was the most important mineralogical control of Sb and As release from the Pb smelter fly ash (Ettler *et al.* 2012).

In the case of an unknown crystal structure, the unit cell can be determined manually by collecting several SAED patterns from the same crystal while tilting the crystal around a specific reciprocal-space direction and measuring the angles between the patterns. In this way all the patterns will have one row of reflections in common, and, by

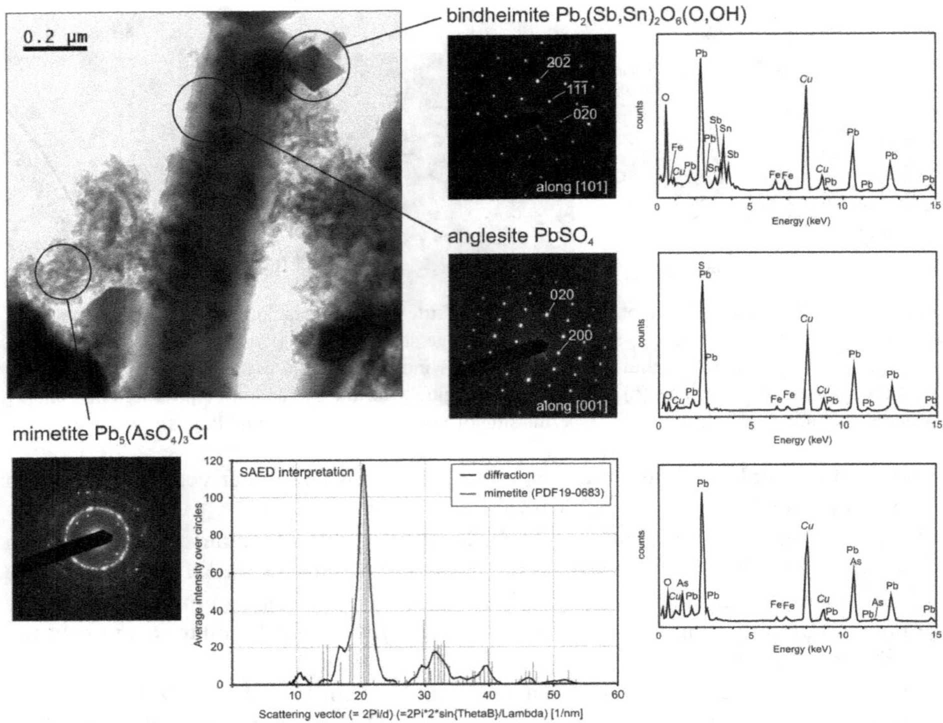

Figure 14. Phase identification of the main solubility-controlling phases for Pb, As and Sb in a weathered Pb smelter fly ash based on SAED patterns and EDX spectra (Cu originated from the supporting grid).

knowing the angles between the patterns, a three-dimensional reciprocal lattice can be reconstructed and indexed. The same procedure can be carried out *via* software by the *TRICE* module of the *ELD* program (Zou *et al.*, 1993). Indexation is also possible if the patterns of the unknown phase are collected from different crystal grains and the relative orientation between them is unknown (*QED* – Belletti *et al.*, 2000; *EDIFF* – Jiang *et al.*, 2011a). The unit-cell parameters can be determined also with EDT (Kolb *et al.*, 2008; Schlitt *et al.*, 2012), through the 3D reconstruction of the reciprocal space. This method is based on the exploration of a fraction of the reciprocal space in a semi-automatic way and does not require orientation of the crystal along a particular direction (Section 3.2.5). For unit-cell determination the EDT data collection can be short – 60° is more than enough, and often 20° or 30° is sufficient for at least an approximate estimation. Note however, that the unit-cell parameters determined from spot electron diffraction patterns, are, in general, much less accurate than parameters based on X-ray diffraction data, with errors of up to 5%. The main reason for the inaccuracy is the geometrical distortions introduced by the microscope's optics. Accuracy can be improved by a careful calibration of the patterns (Mugnaioli *et al.*, 2009b). Unit-cell parameters can also be determined to a high precision with CBED.

The fine details inside the central CBED disc are related directly to the unit-cell geometry and the simulation of their relative positions allows the determination of the unit-cell parameters with a high accuracy (0.01% in the best case) (Brunetti *et al.*, 2010) and also the local strain in the crystal (Armigliato *et al.*, 2006; Béché *et al.*, 2013).

4.2. Determination of the crystal symmetry

Determination of the symmetry of crystal structures is one of the first steps in crystal structure determination, but is also an important part of phase identification. In mineralogy, symmetry determination is an especially important subject as, on many occasions, subtle differences in the symmetry and hence in the structure of the mineral grain may reflect important details about the growth conditions and the history of the embedding rock.

A TEM offers three different approaches for the determination of the symmetry that will be described here ordered by the increasing difficulty of the experiment, but also by the increasing accuracy and amount of extractable information.

4.2.1. Determination of the Laue class and extinction symbol by diffraction tomography

3D electron diffraction tomography is the simplest and fastest way of obtaining lattice parameters and an estimate of the Laue class of the crystal. The data collection described in Section 3.2.5 is fast and does not require the tedious and time-consuming search for oriented zone-axis patterns. This is especially important if the material studied is not stable in the beam. The raw diffraction data can be processed with one of several available software packages (Section 3.2.5) to obtain a three-dimensional representation of the reciprocal space (Fig. 10a), or an arbitrary 2D-section of the diffraction pattern (Fig. 10b,c). Appropriate processing of the recorded reflection positions allows for the determination of the lattice parameters (Section 4.1). Once the lattice parameters are determined, the positions of the reflections on the diffraction patterns can be predicted, and intensities of individual reflections can be integrated. The match between symmetry-related intensities can be used to estimate the Laue class in a way similar to the procedure used in X-ray crystallography. The match is commonly expressed in terms of the so-called internal R value, R_{int}, defined as:

$$R_{\text{int}} = \frac{\sum |I_i - \langle I \rangle|}{\sum I_i}$$

where I_i is the experimental intensity of a reflection i, $\langle I \rangle$ is the mean intensity of all reflections equivalent by symmetry for a given Laue class and the summation runs over all experimentally determined reflection intensities. R_{int} equals zero for a perfect, noise-free dataset and for the correct choice of the Laue class or one of its subgroups. In practice, positive R_{int} values are obtained as a consequence of experimental noise and systematic errors. In EDT data, the discrepancies between symmetry-related reflections are also caused by deviations from the kinematical diffraction. This deviation is so strong that standard EDT data cannot be used efficiently for Laue class estimation, with the possible exception of very thin crystals or crystals with only light elements and large mosaicity (typically organic crystals). The situation can be improved by using the precession electron diffraction. As mentioned in Section 3.2.4, PED partly averages out the effects of multiple scattering, hence making the intensities of the symmetry-related reflections closer to each other. Despite this improvement, the R_{int} obtained typically from EDT data is much larger than values obtained from X-ray or neutron diffraction data. Moreover, the differences between the correct and incorrect Laue classes are less clear-cut. In cases of doubt, reconstructed sections through reciprocal space can be of great help.

Often the indication obtained by R_{int} in combination with the lattice metrics and reciprocal-space reconstructions is sufficient to estimate reliably the Laue class. In many cases, however, the estimation has to be confirmed by other means – either by a successful structure solution, or by using an alternative symmetry-determination method.

Similar problems can be encountered in the determination of the extinction symbol. The determination of lattice centering is straightforward because the corresponding reflection absences cannot be violated by the dynamical scattering effects. However, zonal (in one plane) and axial (along one axis) systematic absences can be violated and non-zero intensities observed at kinematically forbidden positions. Here too, PED can help a lot in suppressing the dynamical diffraction effects, but cannot eliminate them altogether (Fig. 10b). Hence, the systematic absence conditions can be estimated only by the relatively low intensities of the forbidden reflections compared to other reflections in the same plane/axis. The systematic absences are generally most prominent at high-angle regions of the diffraction patterns.

4.2.2. Determination of the Laue class and extinction symbol by the collection of oriented spot diffraction patterns

The flat Ewald sphere in electron diffraction allows direct experimental recording of almost undistorted sections of the reciprocal space. To record an oriented diffraction pattern, the crystal must be oriented with a principal zone axis parallel to the incident electron beam. For a long time measurement and analysis of oriented diffraction patterns has been the principal method of symmetry analysis of crystals in TEM. Oriented zone-axis patterns have one principal advantage over three-dimensional EDT data. Although the dynamical effects are still present in such patterns, the symmetry of the diffraction is preserved and therefore the point-group symmetry of such a diffraction pattern corresponds to the point-group symmetry of the crystal projected on the recorded plane. Oriented diffraction patterns can thus be used more reliably for the determination of the Laue class than computer-reconstructed reciprocal-space sections obtained from EDT data.

The disadvantage of oriented patterns is that only sections passing through the origin of reciprocal space are accessible, while the 3D data sets allow for the reconstruction of arbitrary sections. This drawback can be compensated partially by recording higher-order Laue zones in the diffraction pattern (Fig. 15). These circular intersections of the Ewald sphere with reciprocal lattice planes contain information about systematic absences out of the zero plane, and allow also, albeit approximately, the determination of all six lattice parameters from only one diffraction pattern. However, measurement of several zone-axis patterns provides more accurate and more reliable results.

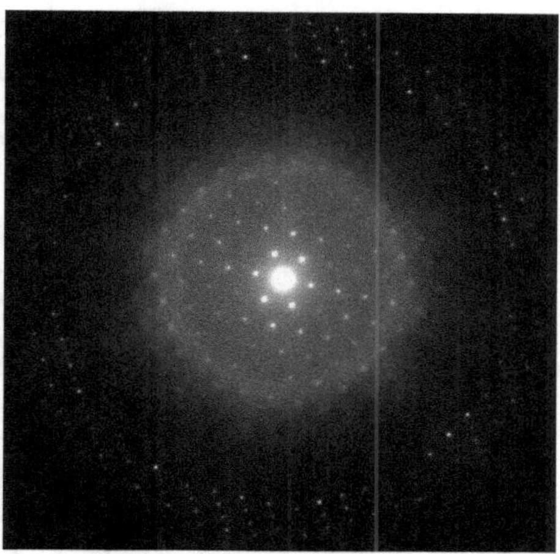

Figure 15. [001] zone axis pattern of coesite, showing the first-order Laue zone (FOLZ) as a ring of spots around the zero-order Laue zone (ZOLZ) reflections. Data courtesy of Damien Jacob, University of Lille.

To simplify the practical application of the method, the properties of patterns belonging to different symmetries have been tabulated. A classic tabulation is that by Buxton *et al.* (1976). The recently released *Atlas of Electron Zone-axis Patterns* produced by Jean-Paul Morniroli (Jacob *et al.*, 2012; Morniroli and Jacob, 2012; Morniroli, 2013) builds on these tables, augments them with additional information, and as a result contains all the information necessary for the extraction of the extinction symbol from oriented spot diffraction patterns (and also space group from CBED patterns, see Section 4.2.3). This work represents an important contribution to the practical applicability of the method. The atlas can be downloaded from http://www.electron-diffraction.fr). The web site also contains several worked examples.

Here again, the power of the method can be greatly enhanced by using precession electron diffraction. As mentioned in Section 3.2.4, oriented PED patterns usually contain many more reflections, especially in HOLZ, are less sensitive to the exact orientation of the crystal and make the kinematically forbidden (*i.e.* systematically absent) reflections generally weaker than is observed in non-precessed patterns.

4.2.3. Space-group determination by CBED

The diffraction symmetry of the spot diffraction patterns is, in principle, given by the point group of the crystal, and not the Laue class. In practice, however, the deviations are either too small or too sensitive to crystal orientation, so that it is impossible to use them to determine the true point-group symmetry, and only the Laue class can be determined reliably. The situation is different in convergent beam electron diffraction patterns. In these patterns, each reflection forms a disc, and each point in the disc corresponds to one ray within the cone of convergent incident beam (Section 3.2.3). In such discs, small variations in intensity can be observed reliably and reproducibly. As a result, the correct point-group symmetry can be determined from such patterns relatively easily (Fig. 16). In particular, the presence or absence of the inversion centre

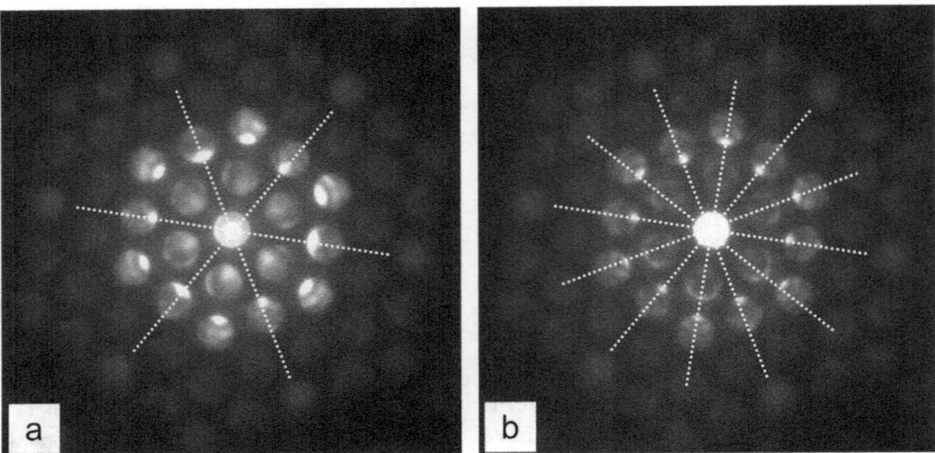

Figure 16. CBED pattern of the [001] zone axis of α-quartz (a) and β-quartz (b). The absence of the inversion centre is clearly visible in the pattern of α-quartz (space group *P*312). Data courtesy of Damien Jacob, University of Lille, France.

in the structure can be detected from the CBED patterns (Fig. 16a). The accurate determination of the point group combined with the extinction symbol allows for an unambiguous determination of the space-group symbol in almost all cases[2]. The *Atlas of Electron Zone-axis Patterns* contains information about the CBED symmetries of special zone axis patterns for each point group.

CBED patterns can also be used to confirm if certain reflections are systematically absent. Gjønnes and Moodie (1965) have shown, using beautifully simple geometrical arguments, that the multiple scattering paths in dynamical diffraction cancel so that the CBED discs of kinematically forbidden reflections contain typical dark regions in the form of a line or a cross (Gjønnes-Moodie lines, Fig. 7b). These lines can be used as reliable indicators of forbidden reflections. However, not all forbidden reflections show Gjønnes-Moodie lines. The *Atlas of Electron Zone-axis Patterns* contains for each pattern an indication of which CBED discs should exhibit Gjønnes-Moodie lines.

Unfortunately, not all materials can be analysed with this technique. Focusing the beam on a small area exposes the sample to a large electron dose that can damage the sample, masking the fine details of the CBED pattern or, in the most beam-sensitive samples, destroying the crystallinity of the sample. Moreover, as mentioned in Section 3.2.3, useful CBED patterns can be collected only on structures with small to medium unit-cell dimensions, unless special sequential recording techniques are used to circumvent the problem of disc overlap.

[2] An exception to this rule are the 11 enantiomorphic space-group pairs, and two special pairs of space groups, $I222/I2_12_12_1$ and $I23/I2_13$, which have the same set of symmetry operators, only arranged differently in the unit cell.

4.3. *Ab initio* structure solution

Dynamical scattering has always been an important obstacle to any *ab initio* structure solution based on electron diffraction data only. Although several successes were obtained using conventional electron diffraction data (see for example the books by Pinsker (1953), Vainshtein (1964), Zvyagin (1967), Dorset (1995) and also Gemmi *et al.* (2000), Zhukhlistov *et al.* (2001), Zhukhlistov (2001), Zvyagin (2001), Weirich *et al.* (1998, 2000, 2002), Albe and Weirich (2003)), and HREM imaging (Hovmöller *et al.* (1984), Weirich *et al.* (1996), Zou and Hovmöller (2008)), the use of electron diffraction for solving crystal structures has traditionally been the method of last resort. The invention of PED with the possibility of obtaining more kinematical data gave a boost to the method. In PED it is possible to record several zone axes with a reduced contribution of the dynamical scattering. The intensities extracted from the different patterns can be merged in a three dimensional set using the common reflections. Although the residual dynamical scattering is responsible for errors in calculating the scaling factors between intensities of different patterns and the final data sets show high $R_{\rm int}$ values, at least the hierarchy of the reflections is retained. The strong reflections remain strong and those that are weak remain weak even if the corresponding ratio is probably wrong by >30%. This is enough for a structure solution done with direct methods to succeed (Marks and Sinkler, 2003; Klein and David, 2011). Several unknown structures were solved using collections of PED zone-axis patterns (*e.g.* Gjønnes *et al.*, 1998; Gemmi *et al.*, 2003, 2010; Dorset, 2006; Own *et al.*, 2006b; Weirich *et al.*, 2006; Boullay *et al.*, 2009; Hadermann *et al.*, 2010; Meshi *et al.*, 2010; White *et al.*, 2010, Klein, 2011). However, the probability of failure remains quite high, as the method provides an incomplete coverage of the reciprocal space (see Section 3.2.5) and relies on a merging procedure that is not always reliable. The appearance of the EDT method was a veritable breakthrough in the field. EDT guarantees almost full reciprocal-space coverage, and does not suffer from merging problems, as the data are collected on one crystal under constant illumination conditions. Recording the patterns in random orientations and not in zone axis condition means further reduction of dynamical scattering effects. In both its PEDT or REDT version (see Section 3.2.5) an integration of each reflection intensity is assured by the scanning of the reciprocal space done by precession in PEDT or by the fine sampling in REDT. The resulting three-dimensional sets of intensities usually have a smaller internal *R* value, $R_{\rm int}$, than zone-axis PED data (still between 20% and 30% higher than X-ray), are usually quite complete (70–100% coverage), and provide structure solutions with high success rates. Structure solution with EDT data is usually carried out with direct methods or with charge flipping and also with a new procedure called δ recycling (Rius *et al.*, 2013). All these methods, built originally for X-ray diffraction, are supposed to deal with completely kinematical data, a situation which is not true for EDT. In spite of this, most of the time the structure is recovered. Because of the neglected dynamical effects, any structure refinement based on the kinematical approximation may improve the model, but the accuracy of the refinement is limited (Section 4.4.1). For more accurate refinement, dynamical diffraction has to be taken into account (see Section 4.4.2).

The PEDT technique, in particular, is quite productive in solving new structures, and several unknown structures are solved every year (*e.g.* Birkel *et al.*, 2010; Rozhdestvenskaya *et al.*, 2010; Andrusenko *et al.*, 2011; Gemmi *et al.*, 2011, 2012; Jiang *et al.*, 2011b; Palatinus *et al.*, 2011; Bellussi *et al.*, 2012; Gorelik *et al.*, 2012; Mugnaioli *et al.*, 2012; Boullay *et al.*, 2013). Several very interesting results have also been obtained by the REDT technique (Willhammar *et al.*, 2012; Martínez-Franco *et al.*, 2013; Wan *et al.*, 2013; Zhang *et al.*, 2013).

But when is EDT the technique of choice for *ab initio* structure solution? Every time a single-crystal X-ray diffraction is not available and powder X-ray diffraction is limited by peak overlapping or poor crystallinity. Analysis of unknown phases in multiphase samples is the most common case. In mineralogy and geology there are plenty of situations that fall into this category and the application of EDT in this field has already given several impressive results, some of which we review in the following paragraphs.

4.3.1. EDT on nanocrystalline minerals: the vaterite crystal structure

Vaterite, together with calcite and aragonite, is a polymorph of $CaCO_3$. The determination of its crystal structure is controversial as, unlike calcite and aragonite, which form large euhedral single crystals easily, vaterite crystallizes in polycrystalline spherulites composed of thousands of nanometric (10–50 nm) crystallites. Before the advent of EDT, all the proposed structural models were derived from X-ray diffraction data which, in the case of nanocrystalline materials, can be quite poor. In PEDT it is possible to collect the patterns in nanodiffraction mode with a beam-size comparable with the size of the vaterite crystals. Mugnaioli *et al.* (2012) performed this experiment with a 50 nm beam on ten different crystals of synthetic vaterite. In all these data sets the three dimensional reciprocal space of the crystals could be reconstructed and the reflections could be indexed with the same monoclinic unit cell $a = 12.17$ Å, $b = 7.12$ Å $c = 9.47$ Å, $\beta = 118.9°$. The single-crystal data allowed a reliable intensity extraction. A reproducible model could be derived from these data using direct methods. The correctness of the structure was confirmed by the stability of the kinematical structure refinement, although restraints had to be imposed on the C–O and O–O distances. The vaterite structure comprises alternating layers of Ca atoms and CO_3^{2-}groups as for the other polymorphs, but the anion groups are aligned orthogonally to the Ca planes. Each CO_3^{2-} anion layer is stacked on top of a Ca layer with a shift of $1/3a$ that destroys the hexagonal symmetry along the stacking direction and gives to the vaterite structure a 2-layer sequence unit cell. The way in which the anion and cation layers are superimposed leaves open the possibility of stacking polymorphism. Together with the 2-layer structure already described, PEDT has identified small ordered domains of a triclinic 6-layer polymorph (Fig. 17), while recent investigations have detected in HREM images an intergrowth of two vaterite structures, one corresponding to the hexagonal model of Kamhi (1963) and another still unknown (Kabalah-Amitai *et al.*, 2013). The investigations of vaterite are far from being concluded, and theoretical computations propose several other possible stable structures for vaterite that have not

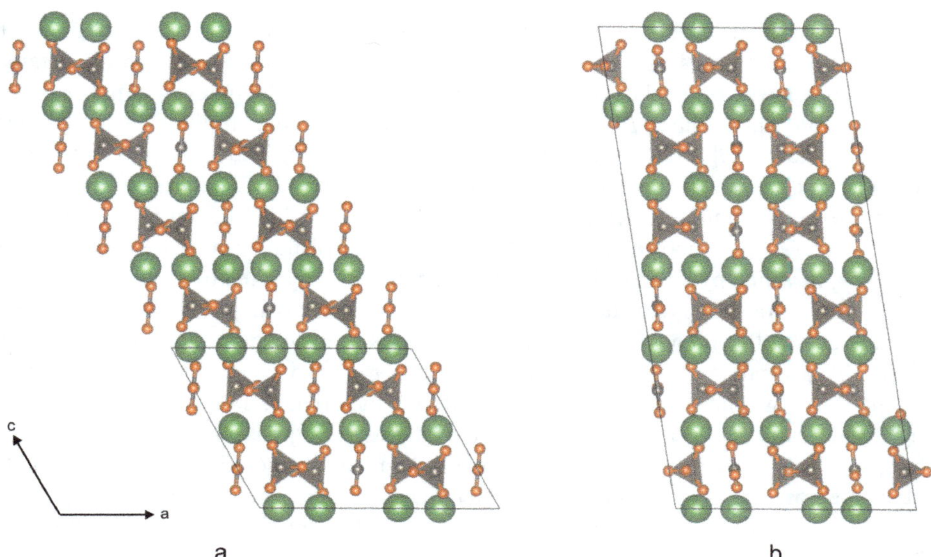

Figure 17. Crystal structures of the 2-layer (left) and 6-layer vaterite (right) structures viewed along the *b* axis. Ca is green, C brown and O red.

yet been observed (Demichelis *et al.*, 2012, 2013). In the future we may expect several surprising results on the subject and EDT will definitely play an important role in obtaining these results.

4.3.2. EDT in experimental petrology: the HAPY phase

Experimental petrology is a very promising field for EDT. A typical experimental petrology sample is a result of a high-pressure, high-temperature synthesis, in which the amount of sample is a few mm^3 and the number of phases is greater than one, as in a rock sample. The phases in the sample are usually identified by their chemical composition using electron microprobe analysis (EMPA), or by their crystal structure from an X-ray powder pattern. However, if the grain size is smaller than the spatial resolution of an EMPA analysis (~1 μm), X-ray powder diffraction remains the only available technique. If all the phases in the sample are known, X-ray powder diffraction can identify them with a phase-match procedure, and Rietveld refinement can furnish a quantitative analysis. But what happens if one or more phases are unknown? Some of the peaks in the pattern will remain unindexed and the peaks belonging to other phases will partially mask the peaks belonging to the unknown phase(s). Therefore X-ray powder diffraction gives the telltale but not the final answer. TEM is perfect for these situations. In a TEM we can literally sail through the powder, checking the chemical composition of every grain with EDX and the crystal structure with electron diffraction and once we find something new, we can use EDT for the structure solution. The discovery of the HAPY phase follows exactly this path. During a study of the stability field of chlorite in the MgO–Al_2O_3–SiO_2–H_2O (MASH) system in an experiment at 5.4 GPa and 720°C, researchers at the Earth Science Department of Milan University found that some peaks of the X-ray powder pattern could not be indexed with the usual phases stable under these *PT* conditions. TEM revealed that in addition to pyrope, forsterite and spinel a new phase with the Mg:Al:Si cation ratio close to 2:2:1 was present in the sample. Its crystal structure was solved using PEDT data as monoclinic in space group *C*2/*c* and unit-cell parameters a = 9.88 Å, b = 11.7 Å, c = 5.11 Å and β = 109°. The crystal structure, displayed in Fig. 18, corresponds to a new inosilicate topology, in which pyroxene-like tetrahedral chains are separated by octahedral layers which, in contrast to pyroxenes, host three rather than two independent cation sites (Gemmi *et al.*, 2011). The larger octahedral layer has some oxygen sites that are not part of the SiO_4 tetrahedra. These oxygen sites have a charge compatible with an (OH) group; the new phase is, therefore, a hydrated inosilicate. Following its peculiar structural features the authors decided to call this new phase Hydrous Al-bearing Pyroxene (HAPY). Once the new phase and its chemical composition were determined, it became possible to program an experiment having bulk composition exactly that of HAPY. The result of the experiment was a sample having HAPY as the main phase (>85 wt.%) and therefore a Rietveld refinement on synchrotron radiation powder diffraction data could refine the structure confirming all the structural features from the direct methods solution with PEDT data.

4.3.3. EDT on new minerals: the sarrabusite case

New minerals are always a hot topic in mineralogy and every year ~100 new species are discovered. One of the most critical aspects in describing a new species is finding a large crystal or enough microcrystalline powder to characterize the mineral with the standard X-ray diffraction techniques. Sarrabusite is an example in which a new phase

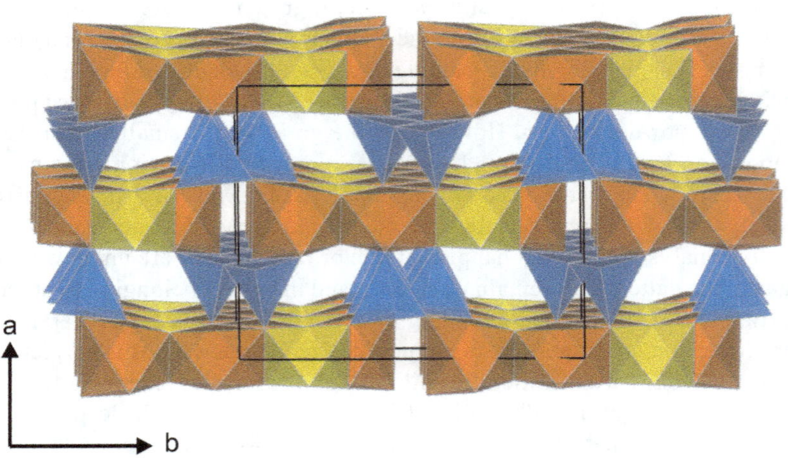

Figure 18. The structure of the HAPY phase viewed along *c*. Si tetrahedra are blue, Mg octahedra are orange and Al octahedra are yellow.

was identified through its chemical analysis performed in a SEM, but due to its particular crystallographic aspect it was impossible to obtain reliable crystallographic information by X-ray techniques. Sarrabusite crystalizes in lemon yellow spherical aggregates of <100 μm, formed by crystals <10 μm. It was originally discovered in 1999 (Campostrini *et al.*, 1999) but, because it is extremely difficult to separate from other minerals in quantities large enough for a powder X-ray diffraction experiment, its crystallographic data remained unknown and it was not recognized as a new mineral species. In a TEM experiment, on the contrary, it is relatively easy to isolate a single crystal and to perform a PEDT experiment on it. Because the crystals have a micrometric size, PEDT could be done in SAED mode rather than in nanodiffraction. A PEDT data collection of 93° was sufficient for the determination of the unit cell and the crystal structure (Gemmi *et al.*, 2012). Sarrabusite is monoclinic with a C-centred unit cell a = 24.7 Å, b = 5.50 Å, c = 14.2 Å and β = 102°. Its structure was solved in space group *C2/c* by direct methods on a data set of 915 independent reflections (corresponding to 67% coverage of reciprocal space at 1 Å resolution) collected in PEDT mode. Sarrabusite is a selenite of copper and lead that also contains Cl anions. Despite the large difference in atomic number (Z_{Pb} = 82, Z_O = 8), all the atomic positions of the different chemical species could be located correctly, and the sarrabusite chemical formula was determined to be $Pb_5CuCl_4(SeO_3)_4$. The sarrabusite crystal structure is constructed from distorted 8-fold coordination polyhedra centred by Pb, which coordinates both O and Cl atoms, $(SeO_3)^{2-}$ pyramidal groups and CuO_4Cl_2 square bipyramids (Fig. 19a). After the crystal structure determination by PEDT, sarrabusite was recognized by the IMA Commission on New Minerals, Nomenclature and Classification as a new mineral species.

4.3.4. EDT on finely intergrown polytypes: the charoite case

Charoite has been known as a new mineral since 1978 (Rogova *et al.*, 1978) but its crystal structure remained enigmatic for years. The reason the structure solution is so difficult is the dimension of well ordered domains which are smaller than few hundred nm. A study with standard SAED revealed that at least three different polytypes coexist. Among the polytypes identified are two ordered polytypes, one with a monoclinic cell (β = 96°, Charoite96) one with a metrically orthorhombic unit cell (Charoite90), a rare ordered polytype with doubled a axis with respect to the other two, and a completely disordered polytype. All these polytypes are finely intergrown. Working in nanodiffraction it has been possible to isolate ordered domains of Charoite90 and Charoite96 and to perform a complete PEDT experiment on them. From the three-dimensional reconstruction of the reciprocal space, EDT allows an unambiguous identification of the unit cell of Charoite 90 (a = 31.96 Å, b = 19.64 Å, c = 7.09 Å and β = 90.0°) and Charoite96 (a = 32.11 Å, b = 19.77 Å, c = 7.23 Å and β = 95.9°) and the inspection of the intensities revealed that both polytypes had a monoclinic symmetry compatible with a 2_1 screw axis parallel to b. In both cases the intensity data set collected by PEDT had a completeness of 97% up to a resolution of 1.1 Å for a total number of ~3000 independent reflections. The completeness and the kinematical

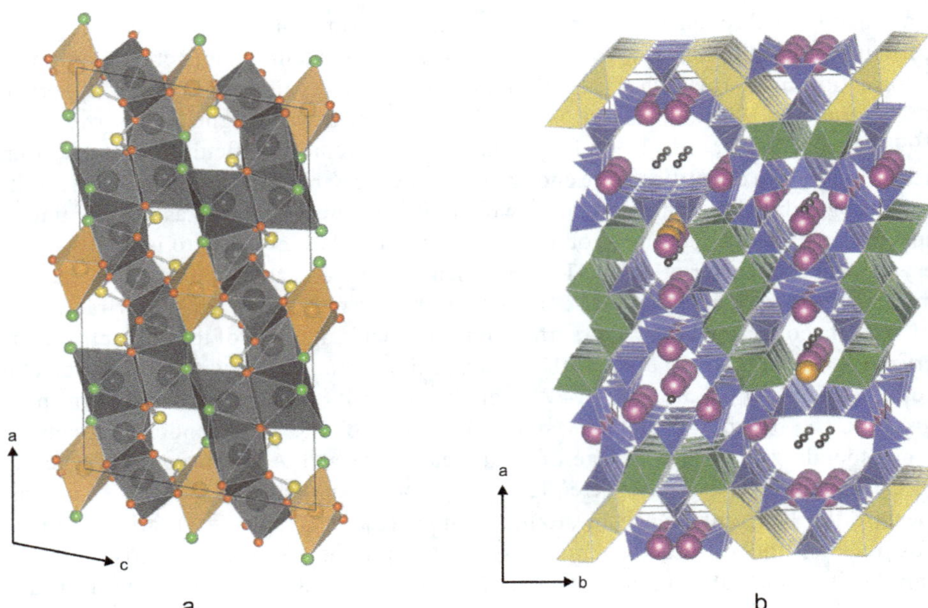

Figure 19. (a) The structure of sarrabusite viewed along *b*. The Pb polyhedra are grey, the Cu polyhedra are orange, Se is yellow, O red and Cl green. (b) The structure of charoite viewed along *c*. Ca octahedra are green, Na octahedra are yellow, Si tetrahedra are blue, K is violet, Sr is orange and water molecules are grey.

character of the PEDT data allowed the structure solution in the space group $P2_1/m$ in both cases (Rozhdestvenskaya *et al.*, 2010, 2011). Despite the complexity of these structures, direct methods allowed the retrieval of the complete framework structure which is formed by 2 Na, 6 Ca, 20 Si, 50 O and one OH in the asymmetric unit. The extraframework cation (7 K, 1 Sr and 3 H_2O sites) can be located through difference-Fourier maps. The crystal-structure determination also yielded a reliable chemical formula for the mineral: $(K,Sr,Ba,Mn)_{15-16}(Ca,Na)_{32}[(Si_{70}(O,OH)_{180}](OH,F)_4\ .nH_2O$. The charoite structure is a demonstration that PEDT can be used successfully even in the case of a very complex mineral structure (90 atoms in the asymmetric unit). The structure is shown in Fig. 19b.

4.4. Structure refinement

Structure refinement is the final step of structure determination. Unlike structure solution, which results in an approximate structure model, structure refinement aims to find the best set of structural parameters given the available diffraction (and other) data. A refined structure is the best available model, which we should use for any subsequent interpretation. The most frequently used refinement method for relatively small crystal structures is the full matrix least-squares refinement. The method is based on the linearization of the expression for the diffracted intensities as a function of structural parameters, and solving the set of linear least-squares equations. Two principal assumptions underlie the method. First, it is assumed that the starting structure model is close enough to the correct model so that the linearized least-squares procedure converges to the global and not local minimum. Second, it is assumed that the errors in the data are independent and random. The second condition can also be stated slightly differently: the model should describe all features in the data except for the independent, random experimental errors.

4.4.1. Kinematical structure refinement

It is common practice to refine the structures against electron diffraction data using the kinematical approximation, *i.e.* assuming that the intensities can be calculated as the squares of the structure-factor amplitudes. This assumption is, however, only a very rough approximation. As a result, these kinematical refinements must be understood only as a tool for completion and validation of the structure solution and for obtaining approximate structural information, but the accuracy of the refined parameters is much lower, sometimes by orders of magnitude, than the precision indicated by the estimated standard deviations following from the least-squares procedure. It is especially problematic to draw any conclusions about the partially occupied atomic sites or relative occupancies of mixed atomic positions because these quantities are more sensitive to accurate intensities than the atomic positions. Nevertheless, the kinematical refinement is still a very useful tool – in many cases there is no other approach available for at least an approximate optimization of the structural model. The limitation of this method must, however, be borne in mind. Kolb *et al.* (2011) gave an interesting overview of several structures refined against electron diffraction data with kinematical approximation, often with comparison to the structural models refined against X-ray diffraction data. This publication confirms the general experience that a typical average error on atomic positions tends to be ~0.1 Å (less for heavy elements, more for lighter), while the largest positional error can easily exceed 0.3 Å.

Many mineral structures, especially clay and related minerals, were studied by this method by the group of Vainshtein, Pinsker, Zvyagin and coworkers mainly in the 1950s and 1960s. This 'Soviet school' worked out a complete instrumentation and methodology for collecting electron diffraction patterns from single crystals, powder samples and also patterns collected on partially oriented (textured) samples. Among minerals studied by this approach, in particular, were layered silicates (kaolinite, palygorskite, nacrite, celadonite, muscovite, *etc.*; Zvyagin, 1967), but also other minerals and materials (Vainshtein, 1964; Vainshtein *et al.*, 1992; Zvyagin, 2001).

In the first decade of the current century interest in this method was awakened with the advent of precession electron diffraction and electron diffraction tomography. Several groups solved and refined mineral structures by this technique. Most, if not all structures mentioned in Section 4.3 were in the final stages verified and optimized with kinematical structure refinement.

4.4.2. Dynamical structure refinement

In spite of the successes achieved by the combination of direct methods for structure solution and kinematical refinement for structure completion and validation, the approach cannot be considered the final answer to the problem of structure analysis by electron diffraction. The limitations of structure refinement using kinematical theory cannot be overcome by any experimental setup or optimization of data acquisition because it is intrinsic to the interaction of electrons with the matter. Extracting accurate structural information with reliable statistical estimation of accuracy almost always requires the use of dynamical diffraction theory. The theory of dynamical diffraction is well known and two principal methods have been developed for the calculation of the diffracted intensities from the structure model: Bloch-wave method (Bethe, 1928; Humphreys, 1979) and the multislice method (Cowley and Moodie, 1957). Both methods have been used successfully to refine crystal structures from electron diffraction data.

The refinements can be divided into two broad classes: refinements against spot diffraction patterns, and refinements against CBED patterns. The CBED patterns contain a wealth of information and individual structure-factor amplitudes can be obtained by fitting the fine structure of the CBED discs, allowing, for example, a detailed study of the valence-electron distribution in the material (Zuo *et al.*, 1999). The method is, however, both experimentally and computationally demanding. A review of the early works with this method can be found *e.g.* in Cowley, 1992). Recently, very accurate structure analyses by CBED refinement were obtained by K. Tsuda and coworkers on silicon (Ogata *et al.*, 2008), $LaCrO_3$ (Tsuda *et al.*, 2002), hematite (Theissmann *et al.*, 2012), perovskite-like $SmBaMn_2O_6$ (Morikawa *et al.*, 2012) and other materials.

An alternative approach is the refinement against spot diffraction patterns. The amount of information in these patterns is less than in the CBED patterns, but in general, is sufficient for extraction of the structural parameters, and they are experimentally easier to obtain. Dynamical refinement against one or several oriented diffraction patterns is available in several software packages, using either the multislice method (Marks *et al.*, 1993; Jansen *et al.*, 1998) or the Bloch-wave method (Dudka, 2007; Oleynikov, 2011b; Palatinus *et al.*, 2015a,b). The calculations using dynamical diffraction theory are much more time consuming than the kinematical calculations, but with modern desktop computers one cycle of refinement takes only seconds for moderately complicated structures.

Successful dynamical refinement requires that the intensities in the diffraction pattern are sufficiently sensitive to the structural parameters. This is true if the sample is thin. As the sample gets thicker, the redistribution of intensities due to the multiple scattering is such that essentially all structure-sensitive information is lost (*cf.* Fig. 9). Moreover, with increasing thickness the calculated intensities are very sensitive to the orientation of the crystal – a misfit of a small fraction of a degree can significantly compromise the quality of the fit (Fig. 20). For all these reasons the samples investigated by the dynamical refinement against standard spot diffraction patterns

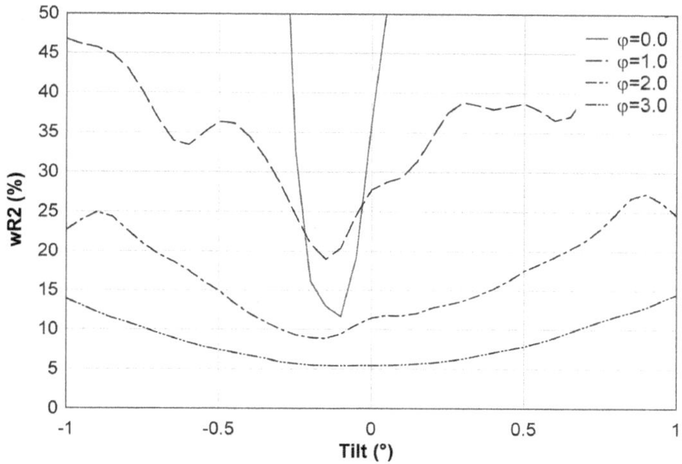

Figure 20. Plot of *R* value *vs*. crystal orientation for silicon [110] zone axis with various precession angles using dynamical diffraction calculations (from Palatinus *et al*., 2013).

should be very thin. As an example, the structure of the intermetallic Mg-Si alloy could be refined to an *R* factor of <5% from samples with the thickness between 2.7 and 17 nm (Andersen *et al*., 1998).

The problems just described can be circumvented to a large extent by using precession electron diffraction. As described above, this technique has several advantages if used for the analysis of symmetry and for structure solution. It is also very beneficial for accurate structure refinement, but for different reasons. As illustrated in Fig. 20, the PED intensities are much less sensitive to the crystal orientation than diffraction patterns obtained without precession. At the same time their sensitivity to thickness variation is also decreased and the sensitivity to structural parameters is increased. As a result, the refinements are more stable and more accurate, and it is easier to find a good starting point for the refinement. The samples can also be much thicker. The accessible thickness depends on the particular crystal and its composition, but in general samples with thickness up to 100 nm (and sometimes even more) can be treated. The price to pay is increased computing time, because the precession diffraction patterns must be modelled as a sum of many (between 100 and 500 depending on the crystal thickness) individual diffraction patterns with different orientations of the incident beam along the precession circuit.

The refinement against PED data is a recent development (Dudka *et al*., 2008; Palatinus *et al*., 2013, 2015a,b), and practical, applicable software has also been developed only recently (Petříček *et al*., 2014). Nevertheless, it has already found interesting mineralogical applications.

Jacob, Palatinus and coworkers (Jacob *et al*., 2013; Palatinus *et al*., 2013) studied the cation distribution in the crystal structure of orthopyroxenes with the general composition $(Fe,Mg)_2Si_2O_6$. This mineral has a structure with two octahedrally coordinated cationic sites (M1 and M2, Fig. 21). Both sites can host variable amounts

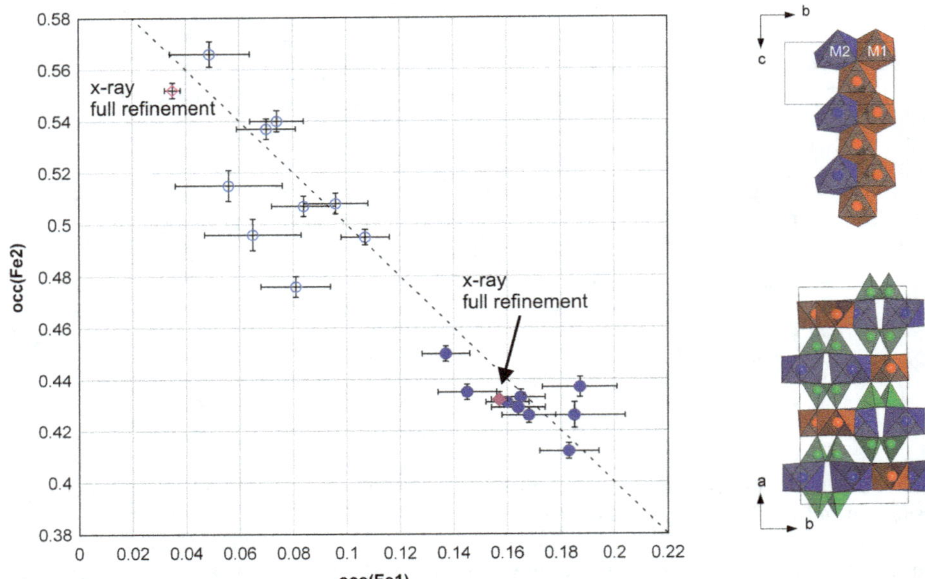

Figure 21. The site occupancies of Fe in positions M1 and M2 as refined against X-ray diffraction data (purple diamonds) and several PED data sets (blue circles). Empty symbols refer to the natural, ordered sample, full symbols to the heat-treated disordered sample (from Jacob *et al.*, 2013).

of Fe and Mg, and the distribution of these two chemical elements can be related to the closing temperature and hence the rate of cooling of the specific mineral grain. The cationic distribution is commonly determined by single-crystal X-ray diffraction. However, orthopyroxenes frequently occur as microcrystalline aggregates and intergrowths with other minerals. In such cases X-ray diffraction cannot be used and electron diffraction is the only available technique.

To test the applicability of the method, a single crystal of orthopyroxene was first analysed by X-ray diffraction, and then a lamella ~30–40 nm thick was cut from the very same grain by focused ion beam, and this lamella was used for electron diffraction. Two grains with different cationic distribution were analysed in this way. A number of diffraction patterns along zone axis [001] were recorded from several spots on the sample and with several precession angles ranging from 1.6° to 2.8°. The atomic positions and the atomic displacement parameters from the X-ray refinement were used, and only the occupancies of the mixed atomic sites were refined. The results are shown graphically in Fig. 21. For one of the samples, the refined occupancies match very well the reference value obtained from X-ray diffraction analysis. The match for the second sample is poorer, although still not too far from the correct value. A definitive explanation for the worse results from the second sample is not available yet, but one possible reason could be the disordering due to the interaction with the intense electron beam.

In another application Gemmi *et al.* (2016) studied phases in the MASH (MgO–Al_2O_3–SiO_2–H_2O) system synthesized at high pressure and temperature, matching deep subduction conditions in the upper mantle. Among the different phases a new hydrous magnesium aluminium silicate $Mg_3Al\,(OH)_3(Si_2O_7)$ was identified, called the Hyso phase. It could be solved and refined kinematically from PEDT data. The structure could be identified as a layered sorosilicate. The structure could be refined using the dynamical structure refinement. The refinement converged to very acceptable figures of merit and revealed an intricate disorder. One Mg position was occupied only partially, and the difference Fourier map revealed two additional partially occupied tetrahedral groups (Fig. 22). Detailed analysis of the structure model and the refinement of occupancies at several cationic positons revealed that the true chemical formula is $Mg_{3.30}Al((OH)_{2.40}O_{0.60})(Si_2O_7)$, *i.e.* there is a freedom in exchange between $Mg^{2+}\ 2O^{2-} \leftrightarrow 2OH^-$. The mineral can thus adopt variable amounts of water.

5. Conclusion

Electron crystallography is a rich subject offering a wide range of techniques that are able to cover essentially all aspects of crystallographic work, from phase identification or determination of lattice parameters to symmetry determination, structure solution and refinement up to the extraction of very accurate information on valence-electron distribution. Its greatest advantage is the ability to analyse routinely samples too small for any other diffraction technique. This advantage is, however, also its main drawback.

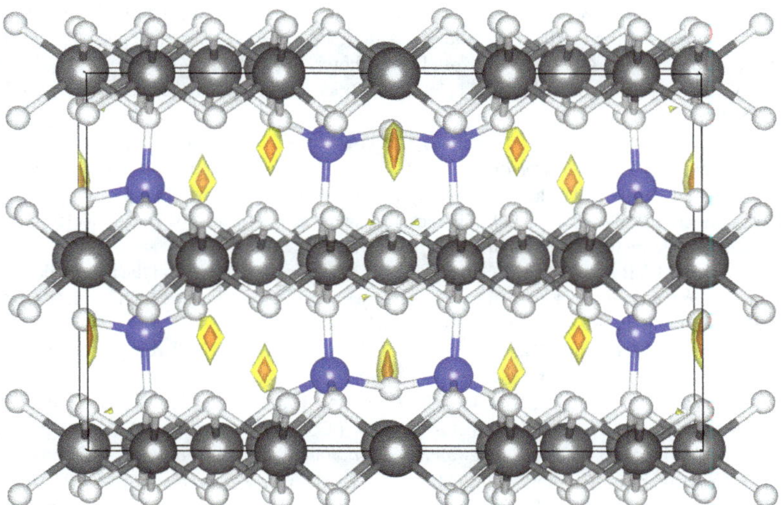

Figure 22. Difference Fourier map calculated from the dynamical refinement of the Hyso phase superimposed on the structural model viewed along **c**. Oxygen atoms are white, octahedrally coordinated Mg and Al are light grey, tetrahedral Si is blue. Yellow and orange colours represent two different isosurface levels in the difference map. Two maxima clearly indicate two new tetrahedral sites (from Gemmi *et al.*, 2016).

Electron diffraction is a very local technique, and it is very difficult to prove that the small part of the sample being analysed is indeed representative of the whole sample. It is therefore always desirable to combine electron crystallography with other diffraction techniques. Indeed, electron and powder X-ray diffraction are wonderfully complementary techniques. While the former provides information on the finest scale, but extracting accurate structural parameters remains challenging, the latter provides information on the bulk, possible mixture of several phases, but once approximate lattice parameters and structural models are available, it provides easier access to accurate structural parameters.

We believe that electron crystallography is a very interesting field with potential to grow. The developments of the past few years, be they the advent of corrected microscopes providing true atomic-resolution images, the advent of new diffraction techniques, or the amazing increase in computing power, boost the power of established techniques and open up entirely new areas of application. Mineralogy and petrology already benefit from these developments and will certainly make use of electron crystallography in the future.

References

Albe, K. and Weirich, T.E. (2003) Structure and stability of α- and β-Ti_2Se. Electron diffraction versus density-functional theory calculations. *Acta Crystallographica A*, **59**, 18–21.

Alloyeau, D., Ricolleau, C., Oikawa, T., Lanlois, C., Le Bouar, Y. and Loiseau, A. (2008) STEM nanodiffraction technique for structural analysis of CoPt nanoparticles. *Ultramicroscopy* **108**, 656–662.

Andersen, S.J., Zandbergen, H.W., Jansen, J., Traeholt, C., Tundal, U. and Reiso, O. (1998) The crystal structure of the beta'' phase in Al-Mg-Si alloys. *Acta Materialia*, **46**, 3283–3298.

Andrusenko, I., Mugnaioli, E., Gorelik, T.E., Koll, D., Panthöfer, M., Tremel, W. and Kolb, U. (2011) Structure analysis of titanate nanorods by automated electron diffraction tomography. *Acta Crystallographica B*, **67**, 218–225.

Armigliato, A., Spessot, A., Balboni, R., Benedetti, A., Carnevale, G., Frabboni, S., Mastracchio, G. and Pavia, G. (2006) Convergent beam electron diffraction investigation of strain induced by Ti self-aligned silicides in shallow trench Si isolation structures. *Journal of Applied Physics*, **99**, 064504.

Beanland, R., Thomas, P.J., Woodward, D.I., Thomas, P.A. and Roemer, R.A. (2013) Digital electron diffraction – seeing the whole picture. *Acta Crystallographica A*, **69**, 427–434.

Béché, A., Rouvière, J.L., Barnes, J.P. and Cooper, D. (2013) Strain measurement at the nanoscale: Comparison between convergent beam electron diffraction, nano-beam electron diffraction, high resolution imaging and dark field electron holography. *Ultramicroscopy*, **131**, 10–23.

Belletti, D., Calestani, G., Gemmi, M. and Migliori, A. (2000) QED V 1.0: a software package for quantitative electron diffraction data treatment. *Ultramicroscopy*, **81**, 57–65.

Bellussi, G., Montanari, E., Di Paola, E., Millini, R., Carati, A., Rizzo, C., O'Neil Parker Jr., W., Gemmi, M., Mugnaioli, E., Kolb, U. and Zanardi, S. (2012) ECS-3: a crystalline hybrid organic-inorganic aluminosilicate with open porosity. *Angewandte Chemie International Edition*, **51**, 666–669.

Bethe, H. (1928) Theorie der Beugung von Elektronen an Kristallen. *Annalen der Physik*, **392**, 55–129.

Birkel, C.S., Mugnaioli, E., Gorelik, T., Kolb, U., Panthöfer, M. and Tremel, W. (2010) Solution synthesis of a new thermoelectric $Zn_{1+x}Sb$ nanophase and its structure determination using automated electron diffraction tomography. *Journal of the American Chemical Society*, **132**, 9881–9889.

Boullay, P., Dorcet, V., Pérez, O., Grygiel, C., Pellier, W., Mercey, B. and Hervieu, M. (2009) Structure determination of a brownmillerite $Ca_2Co_2O_5$ thin film by precession electron diffraction. *Physics Review B*, **79**, 184108.

Boullay, P., Palatinus, L. and Barrier, N. (2013) Precession electron diffraction tomography for solving complex modulated structures: the case of $Bi_5Nb_3O_{15}$. *Inorganic Chemistry*, **52**, 6127–6135.

Brunetti, G., Bouzy, E., Fundenberger, J.J., Morawiec, A. and Tidu, A. (2010) Determination of lattice parameters from multiple CBED patterns: a statistical approach. *Ultramicroscopy*, **110**, 267–277.

Buseck, P.R. (editor) (1992) *Minerals and Reactions at the Atomic Scale: Transmission Electron Microscopy*. Reviews in Mineralogy, **27**. Mineralogical Society of America, Chantilly, Virginia, USA.

Buxton, B.F., Eades, J.A., Steeds, J.W. and Rackham, G.M. (1976) The symmetry of electron diffraction zone axis patterns. *Philosophical Transactions*, **281**, 171–194.

Campostrini, I., Gramaccioli, C.M. and Demartin, F. (1999) Orlandiite, $Pb_3Cl_4(SeO_3).H_2O$, a new mineral species, and an associated lead-copper selenite chloride from the Baccu Locci mine, Sardinia, Italy. *The Canadian Mineralogist*, **37**, 1493–1498.

Ciston, J., Deng, B., Marks, L.D., Own, C.S. and Sinkler, W. (2008) A quantitative analysis of the cone-angle dependence in precession electron diffraction. *Ultramicroscopy*, **108**, 514–522.

Cowley, J.M. (editor) (1992) *Electron Diffraction Techniques, Vol. 1*. Oxford University Press, New York.

Cowley, J.M. and Moodie, A.F. (1957) The scattering of electrons by atoms and crystals. I. A new theoretical approach. *Acta Crystallographica*, **10**, 609–619.

De Graef, M. (2003) *Introduction to Conventional Transmission Electron Microscopy*. Cambridge University Press. Cambridge, UK.

Demichelis, R., Raiteri, P., Gale, J.D. and Dovesi, R. (2012) A new structural model for disorder in vaterite from first-principles calculations. *CrystEngComm*, **14**, 44–47.

Demichelis, R., Raiteri, P., Gale, J.D. and Dovesi, R. (2013) The multiple structures of vaterite. *Crystal Growth and Design*, **13**, 2247–2251.

Dorset, D.L. (1995) *Structural Electron Crystallography*. Plenum Press, New York.

Dorset, D.L. (2006) The crystal structure of ZSM-10, a powder X-ray and electron diffraction study. *Zeitschrift für Kristallographie* **221**, 260–265.

Dudka, A. (2007) ASTRA – a program package for accurate structure analysis by the intermeasurement minimization method. *Journal of Applied Crystallography*, **40**, 602–608.

Dudka, A.P., Avilov, A.S. and Lepeshov, G.G. (2008) Crystal structure refinement from electron diffraction data. *Crystallography Reports* **53**, 530–536.

Eggeman, A.S., White, T.A. and Midgley, P.A. (2010) Is precession electron diffraction kinematical? Part II: A practical method to determine the optimum precession angle. *Ultramicroscopy*, **110**, 771–777.

Ettler, V., Mihaljevic, M., Sebek, O., Valigurova, R. and Klementova, M. (2012) Differences in antimony and arsenic releases from lead smelter fly ash in soils. *Chemie der Erde*, **72**, 15–22.

Fultz, B. and Howe, J.M. (2008) *Transmission Electron Microscopy and Diffractometry of Materials* (First edition 2001, second edition 2002, third edition 2008). Springer, Berlin.

Ganesh, K.J., Kawasaki, M., Zhou, J.P. and Ferreira, P.J. (2010) D-STEM: A parallel electron diffraction technique applied to nanomaterials. *Microscopy and Analysis* **16**, 614–621.

Ganesh, K.J., Darbal, A.D., Rajasekhara, S., Rohrer, G.S., Barmak, K. and Ferreira, P.J. (2012) Effect of downscaling nano-copper interconnects on the microstructure revealed by high resolution TEM-orientation-mapping. *Nanotechnology*, **23**, 135702.

Gemmi, M., Righi, L., Calestani, G., Migliori, A., Speghini, A., Santarosa, M. and Bettinelli, M. (2000) Structure determination of phi-$Bi_8Pb_5O_{17}$ by electron and powder X-ray diffraction. *Ultramicroscopy*, **84**, 133–142.

Gemmi, M., Zou, X.D., Hovmöller, S., Migliori, A., Vennström, M. and Andersson, Y. (2003) Structure of Ti_2P solved by three-dimensional electron diffraction data collected with the precession technique and high-resolution electron microscopy. *Acta Crystallographica A*, **59**, 117–126.

Gemmi, M., Klein, H., Rageau, A., Strobel, P. and Le Cras, F. (2010) Structure solution of the new titanate $Li_4Ti_8Ni_3O_{21}$ using precession electron diffraction. *Acta Crystallographica B*, **66**, 60–68.

Gemmi, M., Fischer, J., Merlini, M., Poli, S., Fumagalli, P., Mugnaioli, E. and Kolb, U. (2011) A new hydrous Al-bearing pyroxene as a water carrier in subduction zones. *Earth and Planetary Science Letters*, **310**, 422–428.

Gemmi, M., Campostrini, I., Demartin, F., Gorelik, T.E. and Gramaccioli, C.M. (2012) Structure of the new mineral sarrabusite, $Pb_5CuCl_4(SeO_3)_4$, solved by manual electron-diffraction tomography. *Acta*

Crystallographica B, **68**, 15–23.

Gemmi, M., Galanis, A., Karavassili, F., Das, P.P., Calamiotou, M., Gantis, A., Kollia, M., Margiolaki, I. and Nicolopoulos, S. (2013) Structure determination of nano-crystals with precession 3D electron diffraction tomography in the transmission electron microscope. *Microscopy and Analysis*, **27**, 24–29.

Gemmi, M., Merlini, M., Palatinus, L., Fumagalli, P. and Hanfland, M. (2016) Electron diffraction determination of 11.5 Å and HySo structures: Candidate water carriers to the Upper Mantle. *American Mineralogist*, **101**, 2645–2654.

Gjønnes, J. and Moodie, A.F. (1965) Extinction conditions in the dynamic theory of electron diffraction. *Acta Crystallographica*, **19**, 65–67.

Gjønnes, J., Hansen, W., Berg, B.F., Runde, P., Cheng, Y.F., Gjønnes, K., Dorset, D.L. and Gilmore, C.J. (1998) Structure model for the phase AlmFe derived from three-dimensional electron diffraction intensity data collected by a precession technique. Comparison with convergent-beam diffraction. *Acta Crystallographica A*, **54**, 306–319.

Golla, U. and Putnis, A. (2001) Valence state mapping and quantitative electron spectroscopic imaging of exsolution in titanohematite by energy filtered TEM. *Physics and Chemistry of Minerals* **28**, 119–129.

Goodman, P. and Lehmpfuhl, G. (1968) Observation of the breakdown of Friedel's law in electron diffraction and symmetry determination from zero-layer interactions. *Acta Crystallographica A*, **24**, 339–347.

Gorelik, T.E., Van De Streek, J., Kilbinger, A.F.M., Brunklaus, G. and Kolb, U. (2012) Ab-initio crystal structure analysis and refinement approaches of oligo p-benzamides based on electron diffraction data. *Acta Crystallographica B*, **68**, 171–181.

Hadermann, J., Abakumov, A.M., Tsirlin, A.A., Filonenko, V.P., Gonnissen, J., Tan, H., Verbeeck, J., Gemmi, M., Antipov, E.V. and Rosner, H. (2010) Direct space structure solution from precession electron diffraction data: resolving heavy and light scatterers in $Pb_{13}Mn_9O_{25}$. *Ultramicroscopy*, **110**, 881–890.

Henderson, R. (1995) The potential and limitations of neutrons, electrons and X-rays for atomic resolution microscopy of unstained biological molecules. *Quarterly Reviews in Biophysics*, **28**, 171–193.

Hetherington C.J.D. and Dahmen U. (1992) An optical moire technique for the analysis of displacements in lattice images. *Scanning Microscopy Supplement*, **6**, 405–414.

Hirsch, P.B., Howie, A., Nicholson, R.B., Pashley, D.W. and Whelan, M.J. (1965) *Electron microscopy of thin crystals*. Butterworths, London.

Hovmöller, S., Sjögren, A., Farrants, G., Sundberg, M. and Marinder, B.-O. (1984) Accurate atomic positions from electron microscopy. *Nature*, **311**, 238–241.

Humphreys, C.J. (1979) The scattering of fast electrons by crystals. *Reports on Progress in Physics*, **42**, 1825–1887.

ICDD (2003) *PDF-2 Database, Release 2003*. International Centre for Diffraction Data, Newton Square, PA.

Jacob, D., Ji, G. and Morniroli, J.P. (2012) A systematic method to identify the space group from PED and CBED patterns part II – practical examples. *Ultramicroscopy*, **121**, 61–71.

Jacob, D., Palatinus, L., Cuvillier, P., Leroux, H., Domeneghetti, C. and Cámara, F. (2013) Ordering state in orthopyroxene as determined by precession electron diffraction. *American Mineralogist*, **98**, 1526–1534.

Jansen, J., Tang, D., Zandbergen, H.W. and Schenk, H. (1998) MSLS, a least-squares procedure for accurate crystal structure refinement from dynamical electron diffraction patterns. *Acta Crystallographica A*, **54**, 91–101.

Jiang, L., Georgieva, D. and Abrahams, J.P. (2011a) EDIFF: a program for automated unit-cell determination and indexing of electron diffraction data. *Journal of Applied Crystallography*, **44**, 1132–1136.

Jiang, J., Jorda, J.L., Yu, J., Baumes, L.A., Mugnaioli, E., Diaz-Cabanas, M.J., Kolb, U. and Corma, A. (2011b) Synthesis and structure determination of the hierarchial meso-microporous zeolite ITQ-43. *Science*, **333**, 1131–1134.

Kabalah-Amitai, L., Mayzel, B., Kauffmann, Y., Fitch, A.N., Bloch, L., Gilbert, P.U.P.A. and Pokroy, B. (2013) Vaterite crystals contain two interspersed crystal structures. *Science*, **340**, 454–457.

Kamhi, S. (1963) On the structure of vaterite $CaCO_3$. *Acta Crystallographica*, **16**, 770–772.

Klein, H. (2011) Precession electron diffraction of Mn_2O_3 and $PbMnO_{2.75}$: solving structures where X-rays fail. *Acta Crystallographica A*, **67**, 303–309.

Klein, H. and David, J. (2011) The quality of precession electron diffraction data is higher than necessary for structure solution of unknown crystalline phases. *Acta Crystallographica A*, **67**, 297–302.

Koch, C.T. (2011) Aberration-compensated large-angle rocking-beam electron diffraction. *Ultramicroscopy*, **111**, 828–840.

Kolb, U., Gorelik, T., Kübel, C., Otten, M.T. and Hubert, D. (2007) Towards automated diffraction tomography: Part I – Data acquisition. *Ultramicroscopy*, **107**, 507–513.

Kolb, U., Gorelik, T. and Otten, M.T. (2008) Towards automated diffraction tomography. Part II – Cell parameter determination. *Ultramicroscopy*, **108**, 763–772.

Kolb, U., Mugnaioli, E. and Gorelik, T.E. (2011) Automated electron diffraction tomography – a new tool for nano crystal structure analysis. *Crystal Research and Technology*, **46**, 542–554.

Lábár, J.L. (2005) Consistent indexing of a (set of) SAED single crystal pattern(s) with the process diffraction program. *Ultramicroscopy*, **103**, 237–249.

Lorimer, G.W. (1987) Quantitative X-ray microanalysis of thin specimens in the transmission electron microscope; a review. *Mineralogical Magazine*, **51**, 49–60.

Marks, L.D., Xu, P. and Dunn, D.N. (1993) UHV transmission electron microscopy of Ir(001?) II. Atomic positions of the (5 × 1) reconstructed surface from HREM and R-factor refinements. *Surface Science*, **294**, 322–332.

Marks, L.D. and Sinkler, W. (2003) Sufficient conditions for direct methods with swift electrons. *Microscopy and Microanalysis*, **9**, 399–410.

Martínez-Franco, R., Moliner, M., Yun, Y., Sun, J., Wan, W., Zou, X. and Corma, A. (2013) Synthesis of an extra-large molecular sieve using proton sponges as organic structure-directing agents. *Proceedings of the National Academy of Sciences of the U.S.A.*, **110**, 3749–3754.

McLaren, A.C. (2005) *Transmission Electron Microscopy of Minerals and Rocks* (First edition 1991, second edition 2005). Cambridge University Press, Cambridge, UK.

Meshi, L., Ezersky, V., Kapush, D. and Grushko, B. (2010) Identification of a new hexagonal phase in the Al-Cu-Re system. *Journal of Alloys and Compounds*, **496**, 208–211.

Morikawa, D., Tsuda, K., Maeda, Y., Yamada, S. and Arima, T.H. (2012) Charge and orbital order patterns in an A-site ordered perovskite-type manganite $SmBaMn_2O_6$ determined by convergent-beam electron diffraction. *Journal of the Physical Society of Japan*, **81**, 093602.

Morniroli, J.P. (2002) *Large Angle Convergent-beam electron diffraction (LACBED)*. Société Française des Microscopie, Paris.

Morniroli, J.P. (2013) *Atlas of Electron Zone-Axis Paterns*. http://www.electron-diffraction.fr.

Morniroli, J.P. and Jacob, D. (2012) A systematic method to identify the space group from PED and CBED patterns part I – theory. *Ultramicroscopy*, **121**, 42–60.

Mugnaioli, E., Gorelik, T. and Kolb, U. (2009a) "Ab initio" structure solution from electron diffraction data obtained by a combination of automated diffraction tomography and precession technique. *Ultramicroscopy*, **109**, 758–765.

Mugnaioli, E., Capitani, G., Nieto, F. and Mellini, M. (2009b) Accurate and precise lattice parameters by selected-area electron diffraction in the transmission electron microscope. *American Mineralogist*, **94**, 96–104.

Mugnaioli, E., Andrusenko, I., Schuler, T., Loges, N., Dinnebier, R.N., Panthofer, M., Tremel, W. and Kolb, U. (2012) Ab initio structure determination of vaterite by automated electron diffraction. *Angewandte Chemie International Edition*, **51**, 7041–7045.

Ogata, Y., Tsuda, K. and Tanaka, M. (2008) Determination of the electrostatic potential and electron density of silicon using convergent-beam electron diffraction. *Acta Crystallographica A*, **64**, 587–597.

Oleynikov, P. (2011a) *EDT-PROCESS software package*. Analitex, Sweden, http://www.edt3d.com.

Oleynikov, P. (2011b) eMap and eSlice: a software package for crystallographic computing. *Crystal Research & Technology*, **46**, 569–579.

Oleynikov, P., Hovmöller, S. and Zou, X. (2007) Precession electron diffraction: Observed and calculated intensities. *Ultramicroscopy*, **107**, 523–533.

Own, C.S., Subramanian, A.K. and Marks, L.D. (2004) Quantitative analyses of precession diffraction data for a large cell oxide. *Microscopy and Microanalysis*, **10**, 96–104.

Own, C.S., Marks, L.D. and Sinkler, W. (2006a) Precession electron diffraction 1: multislice simulation. *Acta*

Crystallographica A, **62**, 434–443.
Own, C.S., Sinkler, W. and Marks, L.D. (2006b) Rapid structure determination of a metal oxide from pseudo-kinematical electron diffraction data. *Ultramicroscopy* **106**, 114–122.
Palatinus, L. (2011) *PETS – program for analysis of electron diffraction data*. Institute of Physics of the CAS, Prague, Czechia.
Palatinus, L., Klementová, M., Dřínek, V., Jarošová, M. and Petříček, V. (2011) An incommensurately modulated structure of η′-phase of $Cu_{3+x}Si$ determined by quantitative electron diffraction tomography. *Inorganic Chemistry*, **50**, 3743–3751.
Palatinus, L., Jacob, D., Cuvillier, P., Klementová, M., Sinkler, W. and Marks, L.D. (2013) Structure refinement from precession electron diffraction data. *Acta Crystallographica A*, **69**, 171–188.
Palatinus, L., Petříček, V. and Corrêa, C.A. (2015a) Structure refinement using precession electron diffraction tomography and dynamical diffraction: theory and implementation. *Acta Crystallographica A*, **71**, 235-244.
Palatinus, L., Corrêa, C.A., Steciuk, G., Jacob, D., Roussel, P., Boullay, P., Klementová, M., Gemmi, M., Kopeček, J., Domeneghetti, M.C., Cámara, F. and Petříček, V. (2015b) Structure refinement using precession electron diffraction tomography and dynamical diffraction: tests on experimental data. *Acta Crystallographica B*, **71**, 740–751.
Petříček, V., Dusek, M. and Palatinus, L. (2014) Crystallographic computing system JANA2006: general features. *Zeitschrift für Kristallographie*, **229**, 345–352.
Pinsker, Z.G. (1953) *Electron Diffraction*. Butterworths Scientific Publications, London.
Plášil, J., Palatinus, L., Rohlíček, J., Houdková, L., Goliáš, V. and Škácha, P. (2013) Crystal structure of lead uranyl carbonate mineral widenmannite: Precession electron diffraction and synchrotron powder diffraction study. *American Mineralogist*, **79**, 276–282.
Portillo J., Rauch, E.F., Nicolopoulos, S., Gemmi, M. and Bultreys, D. (2010) Precession electron diffraction assisted orientation mapping in the transmission electron microscope. *Materials Science Forum*, **644**, 1–7.
Rauch, E.F. and Dupuy, L. (2005) Rapid spot diffraction pattern identification through template matching. *Archives of Metallurgy and Materials*, **50**, 87–99.
Rius, J., Mugnaioli, E., Vallcorba, O. and Kolb, U. (2013) Application of delta recycling to electron automated diffraction tomography data from inorganic crystalline nanovolumes. *Acta Crystallographica A*, **69**, 536–546.
Rogova, V.P., Rogov, Yu.P., Drits, V.A. and Kuznetsova, N.N. (1978) Charoite, a new mineral, and a new jewelry stone. *Zapiski Vsesoyuznogo Mineralogicheskogo Obshchestva*, **107**, 94–100.
Rozhdestvenskaya, I., Mugnaioli, E., Czank, M., Depmeier, W., Kolb, U., Reinholdt, A. and Weirich, T. (2010) The structure of charoite, $(K,Sr,Ba,Mn)_{15-16}(Ca,Na)_{32}[(Si_{70}(O,OH)_{180})](OH,F)_{4.0}.nH_2O$, solved by conventional and automated electron diffraction. *Mineralogical Magazine*, **74**, 159–177.
Rozhdestvenskaya, I., Mugnaioli, E., Czank, M., Depmeier, W., Kolb, U. and Merlino, S. (2011) Essential features of the polytypic charoite-96 structure compared to charoite-90. *Mineralogical Magazine*, **75**, 2833–2846.
Schlitt, S., Gorelik, T.E., Stewart, A.A., Schomer, E., Raasch, T. and Kolb, U. (2012) Application of clustering techniques to electron-diffraction data: determination of unit-cell parameters. *Acta Crystallographica* A, **68**, 536–546.
Shen, Z., Konishi, H., Brown, P.E. and Xu, H. (2013) STEM investigation of exsolution lamellae and "c" reflections in Ca-rich dolomite from the Platteville Formation, western Wisconsin. *American Mineralogist*, **98**, 760–766.
Sinkler, W. and Marks, L.D. (2010) Characteristics of precession electron diffraction intensities from dynamical simulations. *Zeitschrift für Kristallographie*, **225**, 47–55.
Spence, J. and Zuo, J.M. (1992) *Electron Microdiffraction*. Plenum Press, New York.
Stadelmann, P. (2004) *JEMS, Electron Microscopy Software*. CIME-EPFL, Lausanne, Switzerland.
Theissmann, R., Fuess, H. and Tsuda, K. (2012) Experimental charge density of hematite in its magnetic low temperature and high temperature phases. *Ultramicroscopy*, **120**, 1–9.
Tsuda K., Ogata Y., Takagi K., Hashimoto T. and Tanaka, M. (2002) Refinement of crystal structural parameters and charge density using convergent-beam electron diffraction – the rhombohedral phase of $LaCrO_3$. *Acta Crystallographica A*, **58**, 514–525.

Vainshtein, B.K. (1964) *Structure Analysis by Electron Diffraction.* Pergamon Press, Oxford, UK.

Vainshtein, B.K., Zvyagin, B.B. and Avilov, A.S. (1992) *Electron diffraction structure analysis.* Pp. 216–312 in: *Electron Diffraction Techniques, Vol.* **1** (J.M. Cowley, editor). IUCr Monographs on Crystallography, Oxford Science Publications, New York.

Vincent, R. and Midgley, P.A. (1994) Double conical beam-rocking system for measurement of integrated electron diffraction intensities. *Ultramicroscopy*, **53**, 271–282.

Wan, W., Sun, J., Su, J., Hovmöller, S. and Zou, X. (2013) Three-dimensional rotation electron diffraction: software RED for automated data collection and data processing. *Journal of Applied Crystallography*, **46**, 1863–1873.

Weirich, T.E., Ramlau, R., Simon, A., Hovmöller, S. and Zou, X. (1996) A crystal structure determined with 0.02 Å accuracy by electron microscopy. *Nature*, **382**, 144–146.

Weirich, T.E., Hovmöller, S., Kalpen, H., Ramlau, R. and Simon, A. (1998) Electron diffraction versus X-ray diffraction – a comparative study on the structure of the Ta_2P structure. *Crystallography Reports*, **43**, 956–967.

Weirich, T.E., Zou, X., Ramlau, R., Simon, A., Cascarano, G.L., Giacovazzo, C. and Hovmöller, S. (2000) Structures of nanometer-size crystals determined from selected-area electron diffraction data. *Acta Crystallographica A*, **56**, 29–35.

Weirich, T.E., Winterer, M., Seifried, S. and Mayer, J. (2002) Structure of nanocrystalline anatase solved and refined from electron powder data. *Acta Crystallographica A*, **58**, 308–315.

Weirich, T., Portillo, J., Cox, G., Hibst, H. and Nicolopoulos, S. (2006) Ab initio determination of the framework structure of the heavy-metal oxide $Cs_xNb_{2.54}W_{2.46}O_{14}$ from 100 kV precession electron diffraction data. *Ultramicroscopy*, **106**, 164–175.

Wenk, H.R. (editor) (1976) *Electron Microscopy in Mineralogy* (First edition 1976, reprint 2011). Springer, Berlin.

White, T.A., Moreno, M.S. and Midgley, P.A. (2010) Structure determination of the intermediate tin oxide Sn_3O_4 by precession electron diffraction. *Zeitschrift für Kristallographie*, **225**, 55–66.

Willhammar, T., Sun, J., Wan, W., Oleynikov, P., Zhang, D., Zou, X., Moliner, M., Gonzalez, J., Martínez, C., Rey, F. and Corma, A. (2012) Structure and catalytic properties of the most complex intergrown zeolite ITQ-39 determined by electron crystallography. *Nature Chemistry*, **4**, 188–194.

Williams, D.B. and Carter, C.B. (2009) *Transmission Electron Microscopy* (First edition 2004, second edition 2009). Springer, New York.

Zhang, D., Oleynikov, P., Hovmöller, S. and Zou, X. (2010) Collecting 3D electron diffraction data by the rotation method. *Zeitschrift für Kristallographie*, **225**, 94–102.

Zhang, Y., Su, J., Furukawa, H., Yun, Y., Gándara, F., Duong, A., Zou, X. and Yaghi, O.M. (2013) Single-crystal structure of a covalent organic framework. *Journal of the American Chemical Society*, **135**, 16336–16339.

Zhukhlistov, A.P. (2001) Crystal structure of lepidocrocite FeO(OH) from the electron-diffractometry data. *Crystallography Reports*, **46**, 805–808.

Zhukhlistov, A.P., Zvyagin, B.B. and Getmanskaya, T.I. (2001) Three-layer monoclinic $3M_2$ polytype of muscovite revealed from oblique-texture electron diffraction patterns. *Crystallography Reports*, **46**, 932–936.

Zou, X. and Hovmöller, S. (2008) Electron crystallography: imaging and single-crystal diffraction from powders. *Acta Crystallographica A*, **64**, 149–160.

Zou, X., Sukharev, Y. and Hovmöller, S. (1993) ELD – a computer program system for extracting intensities from electron diffraction patterns. *Ultramicroscopy*, **49**, 147–158.

Zuo, J.M., Kim, M., O'Keeffe, M. and Spence, J.C.H. (1999) Direct observation of d-orbital holes and Cu–Cu bonding in Cu_2O. *Nature*, **401**, 49–52.

Zvyagin, B.B. (1967) *Electron Diffraction Analysis of Clay Mineral Structures*. Plenum Press, New York.

Zvyagin, B.B. (2001) From electron diffraction by clays to modular crystallography. *Crystallography Reports*, **46**, 550–555.

EMU Notes in Mineralogy, Vol. 19 (2017), Chapter 4, 183–211

Environmental mineralogical applications of total scattering and pair distribution function analysis

F. MARC MICHEL

Department of Geosciences, Virginia Tech, Blacksburg, Virginia 24061 USA, e-mail: mfrede2@vt.edu

Total scattering experiments using high-energy synchrotron X-rays and spallation neutrons are providing new insights into the structures of nanoscale and poorly crystalline materials of environmental and mineralogical relevance. The pair distribution function (PDF) derived from these total scattering data is a real-space depiction of the atomic arrangements over short (<3–5 Å), intermediate (up to ~20 Å), and even longer length scales. Structural information can be extracted both directly from the PDF and through modelling. PDF analysis approaches are described using selected examples of natural and synthetic nanoparticles as well as a sample that is a mixture of amorphous and crystalline structural phases. Several applications include combined analysis of the real- and reciprocal-space forms of the scattering data. Greater application of the total scattering and PDF methods to environmental minerals that are nanoscale and poorly crystallized will provide new insight to structure, including structural disorder at different length scales, and help to develop further structure-property relationships.

1. Introduction

Determining the atomic structure of natural crystalline solids is of critical importance to modern mineralogy. The arrangement and type of atoms determine the properties of minerals and often provide important insights into the physical and chemical conditions of the geological process by which they formed. Indeed, mineralogical studies of Earth materials carried out using modern crystallographic methods have played a central role in progress made by the geosciences community in unravelling the complexity of geological phenomena involved in Earth's history. For long-range ordered (crystalline) solids, the determination of atomic structure has relied mainly on X-ray diffraction (XRD) using single-crystal and powder specimens.

Work by geoscientists over the past two decades or so has also shown that a large number of inorganic Earth materials exist between crystals and glassy or amorphous solids. In particular, mineral weathering and a variety of other (bio)geochemical processes result in secondary mineralization products that include a vast array of inorganic solids with less than perfect crystallinity (poorly crystalline). Among these, clay minerals and poorly crystalline silicates are common and influential in geochemical processes such as the cycling of carbon in terrestrial soils (*e.g.* Mikutta *et al.*, 2005). Other important and abundant secondary mineralization products include

DOI: 10.1180/EMU-notes.19.5

an assortment of nanoscale transition metal oxide, (oxy-)hydroxide, phosphate, sulfide, sulfate, *etc.*, compounds. These materials fall generally into one of two categories: mineral nanoparticles and nanominerals, with the distinction being that nanominerals exist only with nanometric particle sizes, whereas mineral nanoparticles also crystallize in larger sizes (Hochella *et al.*, 2008). In both cases, extreme small particle size and poor crystallinity correlate commonly with high reactivity in the environment, making these phases important in many (bio)geochemical processes (see Banfield and Zhang, 2001 for review). This is in part because particles with nanosized dimensions typically exhibit large surface area relative to microcrystalline solids. In addition, the chemical reactivity of the surfaces, which can be significantly enhanced in nanoparticles, can affect the biogeochemical cycling of nutrients and contaminants in the environment through various surface chemical processes (*e.g.* surface complexation, precipitation, dissolution, redox transformations, adsorption and desorption). Importantly, and as in the case of long-range ordered minerals, the arrangement and type of atoms in nanoscale and poorly crystalline minerals largely determine their physicochemical properties.

Despite their importance, understanding atomic structure in nanoscale and poorly crystalline materials has proved to be more challenging than for most crystalline solids. This is due mainly to size-induced broadening and suppression of the Bragg diffraction spots or lines (peaks), and is a well known effect of extreme small particle size on diffraction data. Variations from periodicity (*i.e.* structural disorder) further complicate the development of a complete representation of nanoparticle structure. As discussed in the following sections, various types of structural disorder spanning different length scales are often present at the nanoparticle surfaces and interiors. Short-range structural variations are difficult to detect in standard laboratory diffraction data, as they do not result in sharp peaks, but rather manifest as an underlying diffuse scattering. This diffuse component is often difficult to detect and separate from other parasitic forms of diffuse scattering (background). As such, sample diffuse scattering is usually ignored or discarded as background during analysis of powder or single-crystal diffraction data by conventional crystallographic methods (*e.g.* Rietveld). Yet, the diffuse scattering in cases of nanoscale and poorly crystalline minerals contains important structural information over interatomic distances of short (<3–5 Å) and intermediate (up to ~20 Å) length scales. Unlike a conventional diffraction experiment, the primary goal of a 'total scattering' experiment is to collect both the diffuse and Bragg scattering and treat these on an equal basis. Fourier transformation of the total scattering after a number of corrections including subtraction of the measured parasitic intensity results in the pair distribution function, or PDF. The PDF represents a 'roadmap' of the real-space distribution of interatomic distances in the solid. The length scales over which interatomic distances are observed in the experimental (real) PDF vary significantly for amorphous, poorly crystalline and disordered, and crystalline materials. Interatomic distances in the real PDF for crystalline materials can be visible in the experimental PDF over length scales extending up to tens of nm, and are limited only by the resolution of the instrument. In cases of nanoscale, poorly crystalline, and amorphous solids, the ranges of interatomic distances are typically much shorter (*e.g.* see Egami and Billinge, 2003 and references therein) due to the effects of structural disorder and finite particle size effects on the PDF.

Structural information for nanoparticles can be extracted from both the real and reciprocal-space forms of the scattering data (X-ray or neutron). Interatomic distances, including bond lengths, coordination numbers, and indications of static and dynamic disorder of atoms are obtainable in a model-independent way directly from the PDF. In terms of structure modelling, the method of fitting a single unit-cell to the PDF, a practice sometimes referred to as 'real-space Rietveld', is a common and straightforward first approach. Although useful and appropriate for nanocrystalline materials, the periodic model approach alone is often inadequate, however, in cases of poorly crystalline nanoscale solids due to limitations in fully describing structural disorder. Fortunately, recent advances in PDF analysis and simulation methods are leading to improvements in our understanding of nanoparticle structure, in particular the nature of disorder. Some recent studies of nanoparticle structure, several of which are reviewed below, achieve this by combining the periodic model approach in real-space with analysis of reciprocal-space form of the total scattering (diffraction) data. The reverse Monte Carlo (RMC) method is one alternative to the periodic model approach that is attracting attention in the analysis of nanoparticle structure, and we review here a recent study using a whole-nanoparticle RMC simulation. Other analysis methods are available (Egami and Billinge, 2003), and their usefulness will depend ultimately on what particular structural information is required.

In this chapter, we review mainly studies that have considered (or reconsidered) the structures of nanoscale and poorly crystalline solids with environmental and mineralogical relevance (environmental minerals). All of the studies included use total scattering and PDF methods as part of the structure analysis approach. Although application of these methods in structural studies of nanoparticles continues to grow, it is important to bear in mind that these methods are not new and have been used for many decades in studies of short-range order in atomically disordered systems such as liquids and amorphous solids (*cf.* Wright, 1998). The goal of this contribution is to provide an introduction to the total scattering and PDF methods, as well as several currently available approaches to structure analysis. We illustrate some of the advantages of these different – but often complementary – approaches to analysing total scattering and PDF data using selected studies of nanoscale materials (nanominerals, mineral nanoparticles) and a study of a mixture of crystalline and amorphous structural phases. Some of the studies included evaluate materials formed in nature, while others evaluate only laboratory-prepared samples that are often considered to be synthetic analogues of natural nanomaterials.

2. Effects of size and structural disorder on diffraction data

Crystalline materials exhibit sharp well defined Bragg diffraction spots or lines (peaks) and background intensity that is low and relatively featureless. The diffraction profiles obtained from nanoscale (and amorphous) materials, compared with those obtainable from crystalline phases, are much degraded. Crystallite size broadening of Bragg diffraction maxima and the build up of diffuse scattering becomes increasingly significant as the particle size and/or coherent scattering domain size approaches the coherence length of the radiation used to study these materials. These changes impede efforts to use the reciprocal-space form of the diffraction data to discriminate between competing structural models and interpret structural changes under both ambient and non-ambient environmental conditions, including pressure, temperature, humidity, *etc*.

To illustrate further the effects of size-induced broadening on powder XRD data, we will consider three zinc sulfide (ZnS) samples with different average particle size and degree of crystallinity. As shown in Fig. 1, the XRD profile (CuKα, λ = 1.5405 Å) observed for a natural crystalline sphalerite (ZnS) with micron-sized particles is well defined and indexed readily using the known sphalerite structure. The full-width-half-maximum (FWHM) of the indexed (111) reflection for the sample is ~0.18°2θ. For comparison, the FWHM of the (111) reflection for a crystalline quartz standard measured on the same instrument under identical conditions is ~0.14°2θ (data not shown). Due to high purity and crystallinity, the quartz peak profile in this example indicates essentially the effects of instrumental broadening on the sample (Klug and Alexander, 1954). Note that the slightly larger FWHM value in the case of natural sphalerite is probably not size-related, but may be due to strain (*e.g.* from impurities) or some other type of structural disorder. XRD profiles for two synthetic ZnS samples of different sized nanoparticles are also shown in Fig. 1. The smallest nanoparticles are for the 'fresh' ZnS sample that was formed at ambient temperature by rapid mixing of Na_2S and $ZnSO_4$ aqueous solutions. Transmission electron microscopy (TEM) imaging

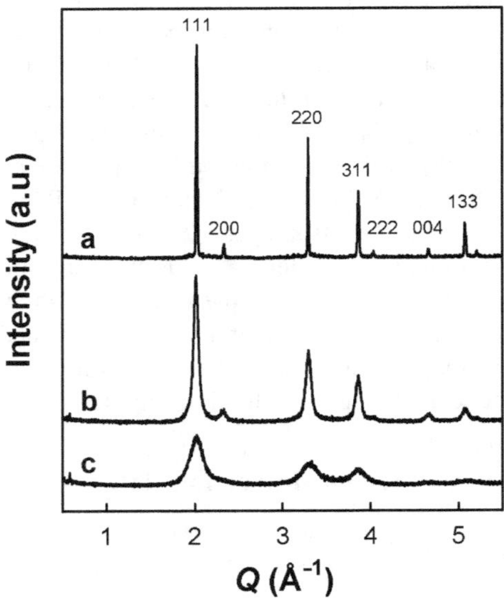

Figure 1. Comparison of XRD intensities for (a) natural crystalline sphalerite, (b) synthetic 8-nm-diameter ZnS, and (c) synthetic 2.6-nm-diameter ZnS collected using a CuKα (8.05 keV; λ = 1.5405 Å) laboratory X-ray source.

indicates that the size of particles is of the order of 3–5 nm. XRD data for an 'aged' ZnS nanoparticle sample that was precipitated using the same method, but was subsequently aged at 70°C for ~142 h (see Zhang *et al.*, 2006 for additional details regarding sample preparations) is also included. TEM suggests greater polydispersity in the aged ZnS, with particle sizes of ~10 nm and larger. After correction for instrumental broadening using an average FWHM obtained from the crystalline quartz data, the (111) reflections for the fresh and aged ZnS nanoparticles have FWHM of ~2.74° and 1.11°2θ, respectively. These values are considerably larger than those observed for natural crystalline sphalerite and are due to the effects of size-induced broadening.

The average particle sizes of the ZnS nanoparticle samples can be estimated from the XRD peak broadening using the Scherrer equation (Zsigmondy, 1920)

$$L = \frac{\lambda K}{\beta \cos \theta} \tag{1}$$

where L is the mean crystallite dimension in Å normal to the reflecting plane, λ is the wavelength, K is a constant of 0.9, and β is the FWHM expressed in radians of 2θ. Using the average FWHM based on several reflections for each sample, equation 1 yields particle sizes of 8.0 ± 0.4 nm and 2.6 ± 0.4 nm for the aged and fresh synthetic ZnS, respectively. Although both the 8-nm-diameter (aged) and 2.6-nm-diameter (fresh) ZnS exhibit their main diffraction features in positions similar to crystalline sphalerite, the size-induced line broadening of the 2.6-nm-diameter ZnS sample is so severe (Fig. 1) that several of the weaker reflections (*e.g.* (200), (222), (004)) appear absent or buried in the background intensity.

Although particle size clearly has a significant effect on diffraction data, it is important to bear in mind that structural variations such as increased strain and the presence of defects often accompany decreasing size, particularly in cases of nanoparticles with dimensions <~10 nm. Strain, defects and other types of structural disorder also contribute to the suppression of Bragg intensity, as well as an increase in diffuse scattering. As mentioned earlier, because the diffuse scattering intensity is not normally considered in conventional crystallographic methods, the degraded diffraction data in nanoscale materials often present a challenge to structure determination, and alternative approaches are required.

3. Overview of total scattering methodology

A powder diffraction experiment (X-rays or neutrons) consists of measuring the scattered intensity from a sample as a function of scattering angle, usually expressed as 2θ. Normally applied to crystalline materials, the primary goal of the experiment is to measure accurately the position, shape and intensity of the Bragg diffraction spots or lines (peaks). These sharp and often intense features contain important information regarding the long-range order of the structure. Other forms of scattered intensity such as inelastic (Compton) also occur and are usually corrected for empirically. As mentioned earlier, diffuse scattering intensity, which lies between and beneath the

Bragg features, and which contains information regarding short-range order in a structure, is usually discarded as background during analysis (*e.g.* Rietveld). This is not an issue for determining long-range order, but sensitivity to short-range order is essential for understanding structures of materials that are nanoscale or poorly crystalline.

The primary goal of a 'total' scattering experiment is to measure accurately the elastic Bragg and diffuse scattering intensities, thereby providing information on short-range order and long-range order (if present) in a sample. One issue is that the measured total scattering intensity will contain a diffuse scattering component that is parasitic and not sample-related (*i.e.* background). Fortunately, this issue can usually be resolved easily. In practice, this normally involves also measuring an empty sample container under identical conditions in the same instrument and subtracting this dataset directly from the sample. A primary consideration in the experiment is also the range of scattering angles over which data will be collected, more commonly expressed as momentum transfer, $Q = (4\pi\sin\theta)\lambda^{-1}$, where θ is the scattering angle and λ is radiation wavelength. A wide Q range is important because the real-space resolution of the PDF can be approximated as $\delta r = \pi/Q_{max}$, where Q_{max} corresponds to the maximum Q-value. The range of scattering data used in high-resolution PDF studies involving nanoparticles typically exceeds ~25 Å^{-1}, significantly higher than what is normally achievable using a laboratory diffractometer using a conventional source of X-rays (*e.g.* Cu). As such, X-rays from high-energy synchrotron sources with high photon flux such as the Advanced Photon Source (APS), Argonne National Laboratory, are often preferred. An example of the difference between the laboratory and synchrotron scattering data is illustrated in Fig. 2. Note that wide Q range is important using either X-rays or neutrons.

The Debye-Scherrer-type scattering geometry with a 2D area detector to measure the scattered intensity (Fig. 3) is often preferred for a total scattering experiment using high-energy synchrotron X-rays. This setup is amenable to rapid acquisition times

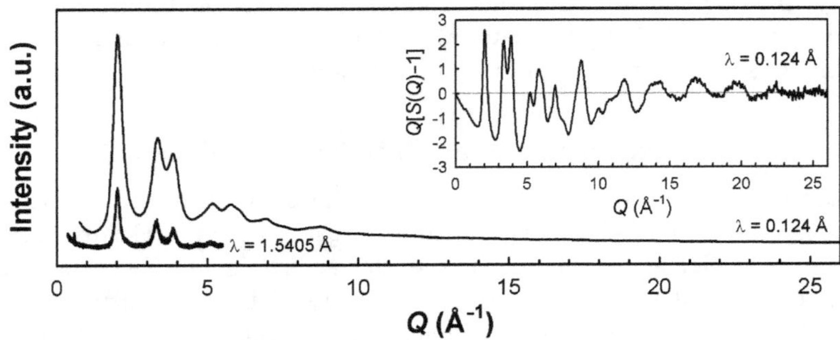

Figure 2. Comparison of raw scattering intensities from synthetic 2.6-nm-diameter ZnS collected using a high-energy synchrotron X-ray source (100 keV; $\lambda = 0.124$ Å) and a CuKα laboratory source (8.05 keV; $\lambda = 1.5405$ Å). Inset shows the reduced structure function ($f(Q)$ or $Q[S(Q)-1]$. Fourier transform of the reduced structure function results in the PDF.

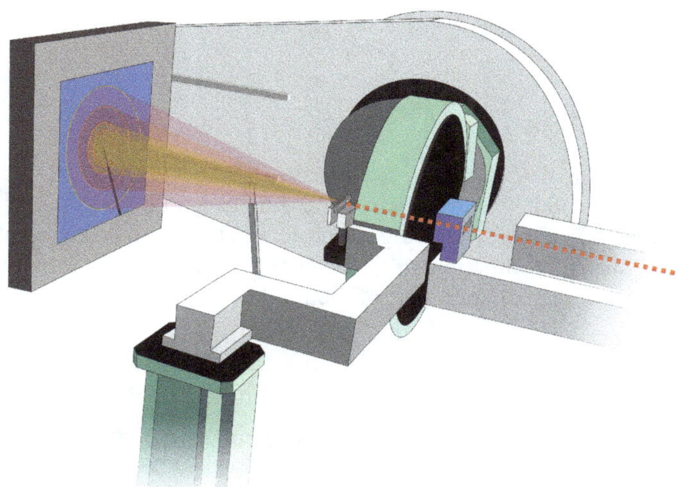

Figure 3. Experimental setup with Debeye-Scherrer-type geometry and a 40 cm × 40 cm area detector on dedicated PDF beamline 11-ID-B at the APS. The incident high-energy monochromatic X-ray beam (dashed red line) scatters from the sample and the 2-D area detector measures the scattered intensity. Within certain limits, decreasing the sample-to-detector distance increases the available Q_{max} of a measurement. Reprinted from Reeder and Michel (2013), with permission from Elsevier.

(Chupas *et al.*, 2003) and rapid sample throughput. The Bragg-Brentano geometry typical of powder diffractometers equipped with laboratory sources of X-rays is also a suitable option, although drawbacks typically include substantially longer exposure times, smaller signal/noise, and less accessible Q range. The reader should refer to Egami and Billinge (2003) for a thorough review of total scattering data collection and analysis. Most examples discussed in this review are of PDF studies carried out at high-energy synchrotron X-ray experimental beamlines that are dedicated or adaptable for PDF studies, such as 11-ID-B, 11-ID-C and 1-ID at the APS. Also considered is a study based on neutron total scattering data collected at the Neutron Pair Distribution Function (NPDF) beamline at the Lujan Center, Los Alamos National Laboratory.

4. The pair distribution function

The PDF gives the probability of finding two atoms separated by a distance, r. The atomic PDF, $G(r)$, is defined as

$$G(r) = 4\pi r[\rho(r) - \rho_0]$$

where r is radial distance (*e.g.* from a central atom to a neighbouring atom), ρ_0 is the average atomic number density, $\rho(r)$ is the atomic pair-density. The PDF contains the real-space distribution of interatomic distances in a material and is obtained from the Fourier transform of the measurable Bragg and/or diffuse scattering intensities. The relationship of $G(r)$ to the measured scattering pattern (X-rays or neutrons) is through a Fourier transform

$$G(r) = \frac{2}{\pi} \sum_{Q=0}^{Q_{max}} Q[S(Q) - 1] \sin(Qr) dQ$$

where $S(Q)$ is the total scattering structure function and contains the measured scattering intensity. The total scattering structure function $S(Q)$ is commonly displayed as the reduced structure function $f(Q)$, or $Q[S(Q)-1]$. The experimental structure function is related to the coherent part of the total diffracted intensity,

$$S(Q) = \frac{I^{coh}(Q) - \sum c_i |f_i(Q)|^2}{|\sum c_i f_i(Q)|^2}$$

where $I^{coh}(Q)$ is the corrected measured scattering intensity from a powder sample. In this equation, c_i and f_i are the atomic concentration and X-ray atomic form factor, respectively, for the atomic species of type i (Egami and Billinge, 2003).

As mentioned earlier, the PDF is like a map of interatomic distances in a solid. To illustrate this further, we will now consider a total X-ray scattering experiment of the nanosized ZnS discussed above in Section 2. Total scattering for both samples was collected at APS 1-ID at an X-ray energy of ~100 keV (λ = 0.124 Å) using a standard experimental setup (Fig. 3). Total scattering intensity data were collected from ~$0.5 < Q < 26$ Å^{-1}, and reduced structure functions ($Q[S(Q)-1]$, Fig. 2, inset) were obtained following subtraction of the measured background intensity and a number of other standard corrections (*e.g.* Compton). Figure 4 presents the PDFs calculated from the Fourier transform of the reduced structure function truncated at $Q_{max} = 25$ Å^{-1} for each sample. The intense, sharp peaks that are clearly observed from $2 < r < 30$ Å in the PDFs for the samples are related to the different atom-atom pairs (Zn–S, Zn–Zn, S–S) in the ZnS nanoparticle structures.

The position, intensity, and extent of peaks in the PDFs for the 2.6-nm and 8-nm-diameter ZnS samples contain information that can be used in qualitative analysis and for comparison of their structures. In sphalerite, the Zn and S atoms are tetrahedrally coordinated and the average Zn–S bond is ~2.342 Å (Skinner, 1961). The ZnS_4 tetrahedra are connected *via* corner-sharing linkages that result in Zn–Zn and S–S distances of ~3.83, and longer Zn–Zn distances of 4.49 Å. The positions of the first few correlations in the PDFs for both samples agree well with this structural topology (Fig. 4 inset). The most striking difference between the two samples is the overall reduction of amplitudes in the PDF with increasing r (Fig. 4) which is related to differences in the size of the coherent scattering domain (CSD). As shown, the signal in the PDF for the 2.6-nm-diameter ZnS sample is attenuated rapidly and is truncated (signal reduces to approximately zero) at r ~20 Å, whereas the signal for the 8-nm-diameter ZnS extends to r >30 Å (full extent of data not shown). This indicates a difference in crystallinity, due to hydrothermal aging in the case of the 8-nm-diameter ZnS, and was expected for these samples based on differences in XRD peak widths

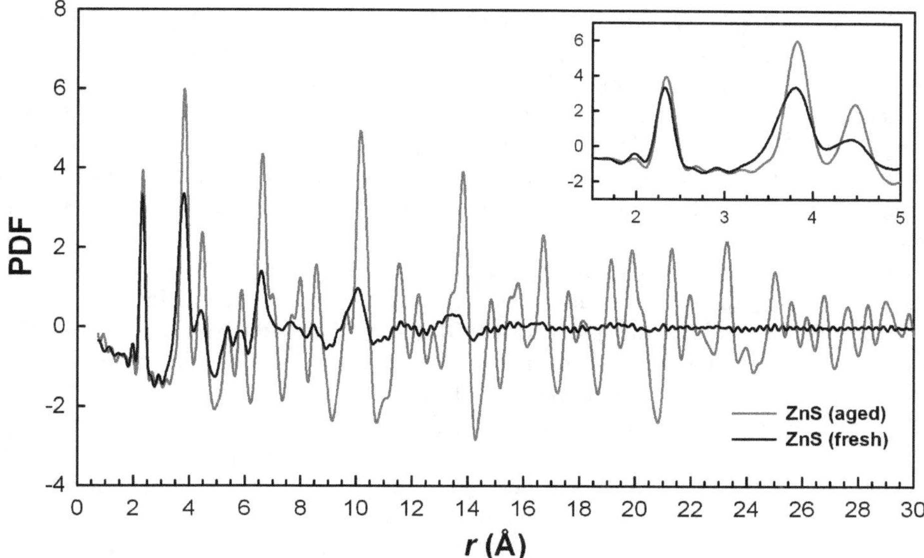

Figure 4. Comparison of X-ray PDFs for the synthetic 8-nm-diameter ("aged") ZnS and 2.6-nm-diameter ("fresh") ZnS nanoparticles. Expanded inset shows the short-range order (<5 Å) of the same ZnS samples. Note that these samples are similar but not identical to the samples measured by Gilbert *et al.* (2004).

(Fig. 1) and the TEM characterization discussed earlier. Along these lines, another interesting difference is noted for the 'fresh' ZnS sample between the CSD size estimates from PDF attenuation (~2.2 nm) and peak broadening in powder XRD (~2.6 nm, calculated using equation 1), and particle size estimated from direct microscopic imaging (3–5 nm, Zhang *et al.*, 2006). Other differences in the positions, intensities and widths and shapes of peaks are also apparent in the PDFs for the two samples, and suggest size-dependent differences in the ZnS structure. These aspects are discussed further in the next section.

5. Extracting information from the PDF: Examples from selected studies of environmental minerals

5.1. Direct structural information from the PDF

5.1.1. Peak position, intensity, width and shape

The position, integrated intensity and width and shape of peaks in the PDF reveal certain structural information directly. The peak positions in the 1–3 Å region of the PDF yield bond lengths directly for the nearest neighbours to each origin atom. Peaks at >~3 Å correspond typically to second, third or higher coordination shells. It is common practice to obtain bond lengths and also longer atom–pair distances by fitting one or more Gaussian lineshapes to peaks in the PDF. However, the interpretation of peaks at larger r is usually more difficult due to peak overlap, and model fitting is often required, as discussed in the next section. Nevertheless, the integrated area of a resolvable peak can be used to calculate the average coordination number for a particular pair of atoms and the width and shape of the peak contain information regarding static and dynamic structural disorder (*cf.* Egami and Billinge, 2003). Numerous PDF studies have used structural information obtained directly from the PDF of atomically disordered systems such as glasses, liquids and gasses, disordered crystalline materials, and nanoscale and poorly crystalline solids. Only examples of the latter are included in the remaining sections.

One clear example where direct structural information from the PDF proved useful is in understanding the structural differences between 3.4-nm-diameter zinc monosulfide nanoparticles and crystalline sphalerite (ZnS). As noted previously, in sphalerite the Zn and S atoms are tetrahedrally coordinated and the ZnS_4 tetrahedra are connected *via* corner-sharing linkages. Crystalline sphalerite forms a long-range-ordered crystal structure, and the peaks in the PDF are sharp in crystalline sphalerite because the long-range periodicity of the structure means that all neighbours at all lengths are well defined. As also noted previously, reducing particles to the nanometric size regime can result in increased structural modification. In addition to size, the surroundings of nanoparticles can also drive substantial structural changes and increased disorder (Zhang *et al.*, 2003).

Gilbert *et al.* (2004) showed that differences in crystallinity and disorder between nanoparticles and bulk ZnS can be resolved in a high-resolution PDF measurement. Analysis of the real X-ray PDF for 3.4-nm-diameter ZnS nanoparticles showed that it is distinct from that of real crystalline sphalerite, as well as from that of a simulated PDF for nanocrystalline 3.4-nm-diameter ZnS nanoparticles. The latter was calculated by truncating the simulated PDF of bulk ZnS with a real-space curve associated with a 3.4-nm-diameter sphere. Note that the structure of the simulated 3.4-nm-diameter is fully ordered and the PDF differs from crystalline sphalerite only in that the peak amplitudes diminish more rapidly with increasing r and that no peaks are observed beyond 3.4 nm in the PDF due to the imposed finite particle size. More interestingly, a number of differences in the PDFs are evident when comparing the simulated and real 3.4-nm-diameter nanoparticles (see fig. 2A from Gilbert *et al.* 2004). For example, lower intensity and increased broadening for the real *vs.* the ideal nanoparticles is seen in the first peak in the PDF, representing the Zn–S bonds throughout the nanoparticle. The amplitudes of peaks at higher r also become attenuated more rapidly and shift in position more for the real *vs.* the ideal nanoparticles. These changes were evaluated at 50 K and at 298 K and showed only weak temperature dependence.

Analysis of peak positions and widths obtained directly from the PDF through fitting a Gaussian lineshape to the major resolvable peaks suggested the presence of internal static structural disorder within the real nanoparticles. Further, the reduction in peak amplitudes with increasing r for the real nanoparticles, which truncates the PDF at ~2.0 nm rather than the expected 3.4 nm, and the correlated shifts in the positions of atoms relative to the perfect sphalerite lattice (Fig. 5), suggested strain within the nanoparticle. These conclusions were based mainly on information extracted directly

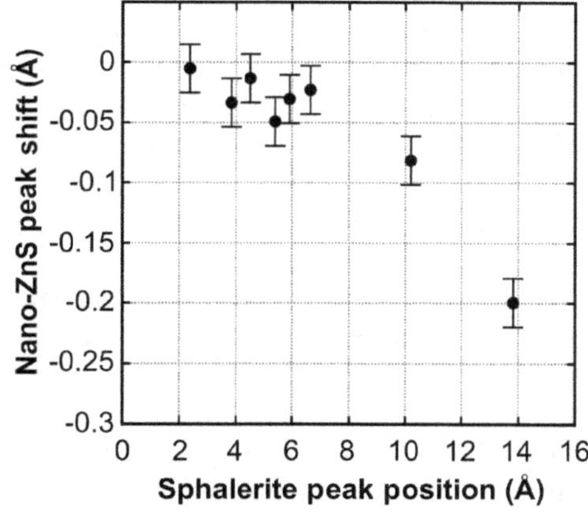

Figure 5. The shift in PDF peak positions observed in ZnS nanoparticles plotted against the positions of the equivalent peaks in the bulk sphalerite PDF. Reprinted from Gilbert *et al.* (2004), with permission from *Science*.

from the PDF and were supported by a variety of other measurements including high-resolution TEM, small-angle X-ray scattering and X-ray absorption fine structure spectroscopy.

For other examples of structural data extracted directly from the PDF, the reader should refer to studies of natural and synthetic (analogue) nanoparticles by Waychunas *et al.* (1996), Cismasu *et al.* (2011, 2012), Harrington *et al.* (2011), Xu *et al.* (2011), Aimoz *et al.* (2012), Dyer *et al.* (2012) and Toner *et al.* (2012).

5.1.2. Peak attenuation and PDF truncation

The amplitudes of correlations in all PDFs for real samples demonstrate a gradual falloff with increasing r due to resolution effects from the experimental setup. This is a limitation of the spatial coherence of the measurement and is different for each beamline or endstation (see Billinge and Egami (2003) for review). Yet there are certain instances where the attenuation in peak amplitudes with increasing r and truncation of the PDF is related directly to finite particle size effects from the sample of nanoparticles. In such cases, the PDF attenuation and truncation can be used to estimate scattering body size (*i.e.* CSD) and shape characteristics. Two methods are commonly used for estimating CSD size from PDF data for nanoparticles. In one approach the CSD size is assigned based on the radial distance of the final (largest r) feature observed in the experimental PDF. In practice, it is sometimes advantageous to scale the PDF, *e.g.* as r^2, in order to amplify artificially the PDF signal and discern the final real, albeit weak, features in the PDF from other oscillations due to termination or normalization errors. This method is clearly qualitative and subjective, and an uncertainty of ±3 Å based on Hall *et al.* (2000) is sometimes given with the estimated CSD size. A second and more quantitative approach is to incorporate a form factor in PDF simulations that relates the amplitude decay to nanoparticle size and shape. This method usually requires assumption of a particular domain shape and, while this assumption is unlikely to be rigorously correct in all cases, it can be constrained further by additional information, *e.g.* from small-angle scattering measurements or TEM imaging. It is important to bear in mind that PDF-derived size (and shape) estimates can deviate significantly from the actual physical dimensions of the particles, and that the effects of structural disorder and particle size on PDF data may be difficult to separate because they are sometimes highly correlated (Gilbert, 2008). The reader should refer to Hall *et al.* (2000) and Gilbert (2008) for additional information.

Using the amplitude attenuation and truncation in the PDF to estimate nanoparticle CSD size is possible only if that size is within the instrumental envelope of the PDF, as determined by instrument resolution. This is a limitation of the spatial coherence of the measurement and is different for each beamline or endstation (see Billinge and Egami (2003) for review). In practice, to evaluate whether the decay of the amplitudes with increasing *r* reflects the structural coherence of the sample or the limitation in the resolution of the instrument, one must measure the instrumental envelope as seen in the PDF for a crystalline reference material. Figure 6 presents an instrument envelope (shaded region) for APS beamline 11-ID-B that was measured using the experimental PDF for crystalline CeO_2. Also included in the figure are PDFs for two hypothetical

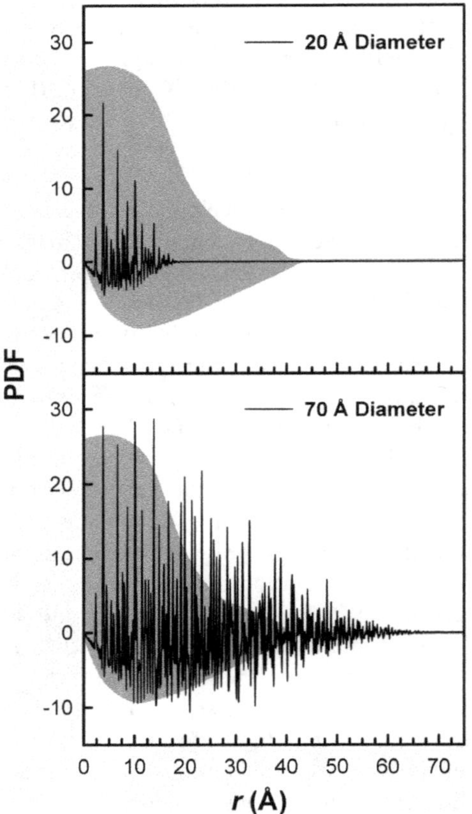

Figure 6. Calculated PDFs for spherical CeO_2 nanoparticles with diameters of 20 Å (black, top pane) and 70 Å (black, bottom pane). The CSD size of the 20-Å nanoparticles can be estimated at APS beamline 11-ID-B because the attenuation of correlations in the PDF is within the instrument envelope (shaded region). Correlations for the larger 70-Å nanoparticles extend beyond the instrument envelope and therefore would not be apparent beyond ~45 Å for the data from this same beamline. Reprinted from Reeder and Michel (2013), with permission from Elsevier.

spherical CeO_2 nanoparticles with CSD diameters of 20 Å and 70 Å. The PDFs were calculated assuming infinite instrument resolution (*i.e.* PDF attenuation and truncation is due only to imposed finite particle size effects). It is clear in this example that the peaks in the PDF for the 20-Å particle fall entirely within the instrument envelope. As such, the reduction in amplitudes and truncation of the PDF would not be affected by the instrument resolution of this particular beamline and experimental setup. The larger *r* correlations in the calculated PDF for the 70-Å particle, however, extend outside and beyond the instrument envelope. As such, the full extent of correlations in the PDF for a real 70-Å sample would not be observed due to instrument resolution and an accurate estimate of CSD size would not be possible.

Studies of various natural and synthetic (analogue) nanoparticles by the present author and others have reported PDF-derived CSD sizes (*e.g.* Michel *et al.*, 2005, 2007a, 2010; Scheinost *et al.*, 2008; Cismasu *et al.*, 2011, 2012; Dyer *et al.*, 2012; Toner *et al.*, 2012). In many cases, CSD size was of interest as it gave some indication of the effects of different synthesis procedures, compositional changes across a series or treatment (*e.g.* hydrothermal) on sample crystallinity. As mentioned previously, the reader should be cautious in using these and other estimated CSD sizes as actual estimates of particle size, because these are not always interchangeable (Gilbert, 2008).

5.2. Real-space fitting with periodic models

As described above, useful structural information can be obtained in a model-independent way directly from the PDF. However, the unit cell approach traditionally used in crystallography is proving to be useful for interpreting PDFs from materials, including nanoparticles, which exhibit some degree of periodicity. Sometimes called real-space Rietveld, this approach involves calculating a PDF based on a structure model composed of a small unit cell with atom types delineated and positions specified in terms of fractional coordinates. The various structure parameters for the model are then refined against the experimental (real) PDF using least-squares refinement procedures to minimize the residuals function. PDFFIT (Proffen and Billinge, 1999) is one of the most commonly used programs to perform such refinements, although home written refinement programs are also sometimes used (*e.g.* Zhu *et al.*, 2012).

It is important to bear in mind that the approach to fitting the PDF in real space is sample-dependent and no single strategy works in all cases. The range of PDF data used in a refinement and the number and type of structural parameters and PDF-dependent parameters can vary significantly. In general, however, it is important to start a refinement with reasonable values for the different parameters including unit-cell dimensions, atomic positions and atomic (thermal) displacement. In cases of mineral nanoparticles, these values can often be taken from the known structure of a crystalline counterpart. In other instances it may be appropriate to build a model from scratch or test a number of competing models, perhaps taken from different chemical systems. The order in which parameters are allowed to move from their starting values (*i.e.* refine) is not the same in every case, but often mimics the standard Rietveld method to some degree.

The following should be considered as an example strategy for fitting a model to the PDF in real space. In the first iteration, only a scale factor and a PDF-dependent parameter (σ_Q) (for modelling the decay in the amplitudes of the PDF with increasing r) are allowed to vary for the model. Upon convergence, visual comparison is used to check that the peak positions and relative intensities of the model PDF are reasonably similar to the experiment. Unit-cell parameters are then added to the refinement and allowed to vary in accordance with the restrictions of the space-group symmetry of the model. Atom coordinates are added next and refined, again in accordance with space-group symmetry, followed by atomic displacement (thermal) parameters usually starting with the highest-z atoms and finishing with the lowest-z atoms. Site occupancies are also refined for certain structures. It is also standard practice to refine the PDF-dependent parameter δ to model sharpening of PDF near-neighbour peaks due to correlated motion between atom pairs. In general, care must be taken when initially refining terms as correlation between parameters is high in certain cases. In the end, the quality of the fit and suitability of the model is judged usually by the general aesthetics of the fit, a 'goodness-of-fit' parameter, and by assessing if the refined values are chemically plausible. The reader should refer to Billinge and Egami (2003) for a more thorough review and additional references regarding modelling the PDF. Three examples of studies using the unit-cell approach to fitting the PDF in real space are reviewed below. In addition to these examples, there are several other studies that take a similar approach and focus on nanoparticles of environmental and mineralogical significance, both under ambient (*e.g.* Bell *et al.*, 2008; Gualtieri *et al.*, 2008) and non-ambient (*e.g.* Ehm *et al.*, 2008) conditions. Only their references are included here. The reader should refer also to Dove (2002) for additional information and references regarding applications of neutron total scattering methods in mineral sciences.

5.2.1. Nanosized mackinawite

One straightforward example where real-space fitting of a periodic model to the PDF proved useful is in understanding the structural differences between iron monosulfide nanoparticles and crystalline mackinawite (FeS). In mackinawite the Fe and S are tetrahedrally coordinated and the average Fe–S bond is ~2.256 Å (Lennie *et al.*, 1995). Mackinawite possesses a layer structure with individual layers formed from edge-sharing FeS_4 tetrahedra. The sheets are stacked normal to the c axis and relatively weak van der Waals forces are thought to hold the layers together. The tetragonal mackinawite unit cell with space group *P4/nmm* contains two S and two Fe atoms.

Michel *et al.* (2005) used synchrotron PDF measurements to evaluate and compare the structures of iron monosulfide nanoparticles with different particle sizes in the range 2–5 nm and crystalline mackinawite (FeS). Each PDF was calculated from the Fourier transform of the corresponding experimental reduced structure function truncated at Q_{max} = ~20 $Å^{-1}$. The model for crystalline mackinawite was fitted in real space to the experimental PDFs ($1.5 \leqslant r \leqslant 20$ Å) and structural parameters including unit-cell dimensions, atomic coordinates and isotropic displacement (thermal) parameters were refined from initial values based on Lennie *et al.* (1995). Two additional PDF-dependent parameters δ and σ_Q were incorporated in each refinement to model sharpening of PDF near-neighbour peaks due to correlated motion between atom pairs and the exponential decay of the PDF, respectively. These parameters were added sequentially during the refinement process in a manner analogous with the Rietveld methodology. An example fit of the refined mackinawite model to the real (experimental) PDF is included as Fig. 7a. Comparison of the partial PDFs calculated from the refined mackinawite model demonstrates the sensitivity of the X-ray PDF technique with respect to different atoms within the unit cell. Figure 7b shows the contributions of the different atoms in the mackinawite unit cell to the calculated (total) PDF.

The unit-cell parameters, atomic coordinates (S *z*-position) and calculated bond lengths and angles derived from real-space refinement of the PDFs for FeS nanoparticles and bulk mackinawite samples were in good agreement with those reported by Lennie *et al.* (1995) based on Rietveld refinement of crystalline mackinawite. Only slight structure relaxation in the FeS nanoparticles was suggested by differences in FeS bond lengths (Δ_{max} = 0.5%), S–Fe–S–Fe bond angles (Δ_{max} = −0.7%), unit-cell dimensions (Δ_{max} = −1.1%), and S *z*-position (Δ_{max} = 3%). Michel *et al.* (2005) established from real-space fitting of the PDF that the FeS nanoparticles, which were often described as 'amorphous' in past literature, were in fact highly ordered over length scales comparable to the particle size. Hence, the authors concluded that the FeS nanoparticles were described better as a nanocrystalline material with mackinawite structure.

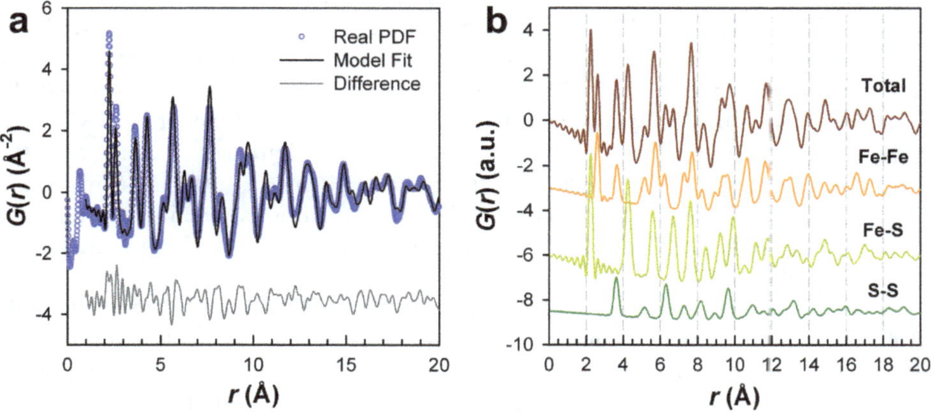

Figure. 7. (a) Experimental PDF data (blue circles) fitted by the refined model PDF (black line). The difference plot (grey line) is included just below. Reprinted from Michel *et al.* (2005), with permission from *Chemistry of Materials*. (b) Simulated refined model PDF of mackinawite with its partial PDFs for the Fe–Fe, Fe–S and S–S atom–atom pairs. Reprinted from Michel *et al.* (2005), with permission from *Chemistry of Materials*.

5.2.2. Ferrihydrite

Another case in which real-space fitting of a periodic model to the PDF proved useful is in understanding the structure of ferrihydrite, a poorly crystalline ferric oxy-hydroxide phase that is common in many natural environments due to the variety of (bio)geochemical processes. Both natural and laboratory ferrihydrite occur usually as nanoparticles with CSD and particles sizes that are in the range 1–7 nm. However, unlike the cases of FeS or ZnS discussed above, ferrihydrite has no microcrystalline counterpart that can be synthesized in the laboratory or found in nature. As such, a long-standing controversy has existed within the mineralogical community concerning the correct initial structure model for ferrihydrite. Due largely to varying interpretations or misinterpretations of powder XRD and an assortment of spectroscopic data, there has been debate over the most basic structural characteristics of ferrihydrite, including the coordination environment of Fe^{3+}, FeO_x polyhedral connectivity and the nature and abundance of structural disorder. These topics are reviewed elsewhere in the literature (*cf.* Jambor and Dutrizac, 1998; Xu *et al.*, 2011) and will not be restated here. Instead, this section will focus on use of the unit-cell approach to test structural models, evaluate size-dependent changes and detect structural order in ferrihydrite.

Michel *et al.* (2007a,b) used synchrotron PDF measurements to evaluate and compare the structures of three laboratory-prepared ferrihydrite samples with particle and CSD sizes in the range 2–7 nm. A new single-phase structure model was proposed for ferrihydrite based on real-space fitting of a periodic model (space group $P6_3mc$) to the PDFs for these three samples. The structure was based on a lesser-known aluminium hydroxide mineral named akdalaite ($Al_{10}O_{14}(OH)_2$) and its synthetic counterpart, tohdite (Yamaguchi *et al.*, 1969). Unlike the prior ferrihydrite structure model proposed by Drits *et al.* (1993), the Michel *et al.* (2007b) model was single-phase and required Fe^{3+} in both octahedral (80%) and tetrahedral (20%) coordination with oxygen. Subsequent work showed that the model for ferrihydrite proposed by Michel *et al.* could also be applied successfully to PDFs calculated from neutron scattering from ferrihydrite (Harrington *et al.*, 2011). However, the validity of the model and the structural aspects of ferrihydrite continue to be debated (*e.g.* Rancourt and Meunier, 2008; Manceau, 2009, 2011, 2012; Michel *et al.*, 2010; Maillot *et al.* 2011; Mikutta, 2011; Guyodo *et al.*, 2012; Harrington *et al.*, 2011; Hiemstra, 2012; Hocking *et al.*, 2012; Peak and Regier, 2012; Gilbert *et al.*, 2013). Additional details regarding the application of total scattering in studies of natural and synthetic ferrihydrite can be found in Michel (2014).

Michel *et al.* (2010) showed that synchrotron total scattering and PDF methods could also be used to evaluate changes in ferrihydrite structure during its hydrothermal transformation to hematite. Previous work indicated that the transformation of ferrihydrite under certain conditions involved the formation of an intermediate phase that was strongly ferrimagnetic (Barrón *et al.*, 2003). This intermediate, initially designated "hydromaghemite" but later renamed "ferrimagnetic ferrihydrite" (Barrón *et al.*, 2012), was shown to occur in abundance if the precursor ferrihydrite was co-precipitated with a strongly adsorbing surface ligand *e.g.* citrate and aged at elevated temperature. The presence of citrate during aging retarded the fast transformation, *via* aggregation, of ferrihydrite to hematite.

In Michel *et al.* (2010) a ferrihydrite/citrate suspension was aged hydrothermally at <175°C and samples collected hourly from 0–14 h were evaluated *ex situ* by total scattering/PDF, XRD and TEM, as well as by a number of complementary methods not discussed in detail here. Changes in the XRD profiles over time suggested progressive transformation of the precursor ferrihydrite (*Fh*) to the ferrimagnetic ferrihydrite intermediate (*Ferrifh*), and finally to hematite (*Ht*). As in the previous cases, reciprocal-space structural analysis was problematic due to the extreme small sizes of the particles. Additional complications arose for the samples near the middle of the aging series, as the solids were mixtures of up to three structural phases (*Fh*, *Ferrifh*, *Ht*).

Michel *et al.* (2010) used chemometric analysis methods combined with real-space model-fitting of the PDFs to evaluate the structure in a series of transformation products and address several fundamental questions regarding the PDF data: How many distinct solids phases are present in the aging series, how do they vary in abundance with aging time, what are the structures of the different phases and how are they related? In short, chemometric analysis results showed that only three distinct solid phases were required to satisfy 99.8% of the variance of the experimental PDFs. Multivariate curve resolution (MCR), which is a well established technique for resolving mixtures in an evolving system by determining the number of constituents and their response profiles, revealed the individual PDFs for the three structural phases. One of the structural phases corresponded with the precursor *Fh* (with citrate) and a second corresponded with the final *Ht* transformation product. The third was the individual PDF assigned to the ferrimagnetic intermediate. Chemometrics also provided an indication of the percent phase abundances for the series, and showed an ~85% decrease in the concentration of the precursor *Fh* during the first 8 h of aging time. This decrease was concomitant with an increase in the *Ferrifh* concentration, followed by the complete conversion to *Ht* after 14 h.

Real-space analyses of the raw (total) PDFs and the MCR-derived PDFs for the individual phases were performed using the Michel *et al.* (2007b) ferrihydrite structure model and the known structure model for hematite. A single unit-cell approach was used for the MCR-derived PDFs for the individual phases. Interestingly, analysis of the individual phases showed that only one structure model, the single-phase structure proposed by Michel *et al.* (2007b), was needed to fit the precursor *Fh* and intermediate *Ferrifh*. In addition to describing the real-space data, the PDF-refined model also predicted the positions and intensities of maxima corresponding to the *Ferrifh* phase in the corresponding XRD data. Additional real-space analysis of the raw PDFs, which included dual unit-cell fits for the samples containing both ferrihydrite and hematite, revealed additional changes in ferrihydrite. Among other findings, fitting results indicated systematic changes in the ferrihydrite structure were occurring during aging and particle growth. For example, size-dependent structural relaxation was indicated by systematic and anisotropic variations in the refined a and c unit-cell parameters (see fig. 5 in Michel *et al.*, 2010). Increases in the refined occupancies of two of the three Fe sites in the ferrihydrite structure also indicated filling of vacancies during aging and particle growth. Michel *et al.* attributed these and other structural, chemical and magnetic changes to an increase in ferrihydrite crystallinity that resulted from hydrothermal treatment in the presence of citrate.

5.2.3. Manganese oxide

Yet another example where the unit-cell approach to fitting the PDF in real space proved advantageous is in understanding the crystal structures of poorly crystalline manganese (Mn) oxides. Naturally occurring Mn oxides such as birnessite and birnessite-like minerals are common in oceanic and terrestrial environments and can play important roles in elemental cycling. Both layer and tunnel structures of natural biotic and abiotic Mn oxides phases have been identified, although the former type consisting of birnessite and birnessite-like minerals is considered more abundant. Like a number of other important environmental nanominerals, Mn oxides present a substantial challenge to accurate crystal-structure determination using conventional analysis by XRD because of poor crystallinity and extreme small particle size. In layer Mn oxide structures, Mn and O are only octahedrally coordinated and form layers by sharing MnO_6 edges. Both Mn^{4+} and Mn^{3+} in the octahedral layers have been identified with the former type more abundant. Short-range disorder and distortions are expected from the presence of Mn^{3+} as well as vacant metal sites in the octahedral layers. The negative charge resulting from these features is compensated by incorporation of H^+, Mn^{2+}, Mn^{3+} and other metal cations in the interlayer spaces, adding further complexity. Turbostratic (stacking) disorder is also considered significant in poorly crystalline layer Mn oxides.

Zhu *et al.* (2012) presented a detailed study and PDF-derived crystal structures of synthetic analogues to Mn oxides including acid birnessite (AcidBir), δ-MnO_2, polymeric MnO_2 (PolyMnO_2) and a bacteriogenic Mn oxide (BioMnO_x) produced by a strain of Mn-oxidizing bacteria. Qualitative comparison in the range $1.5 < r < 4$ Å of the PDFs for these poorly crystalline samples and the crystalline Mn oxide references indicated similarities in local structure (*i.e.* Mn and O coordination environments and polyhedral connectivity). Further comparison at $r > 4$ Å suggested that the poorly crystalline samples were neither tunnel structures, such as todorokite, nor the type of layer structure characteristic of triclinic birnessite. Zhu *et al.* noted that the positions and relative intensities of features in the PDFs up to $r \approx 25$ Å in the poorly crystalline samples were similar to the hexagonal birnessite reference (see fig. 2C, #6-10 of Zhu *et al.*, 2012).

The authors (Zhu *et al.*, 2012) selected a structure model (space group *C*12/*m*1) based on hexagonal birnessite for the real-space analysis. Lattice parameters (*a*, *b*, *c*, β), fractional coordinates (*x* and *z* of O atoms), and the resulting calculated interatomic distances (Mn–O bond lengths and higher coordination shells) derived from real-space refinement of the model to the experimental PDFs for the poorly crystalline Mn oxide samples and the hexagonal birnessite reference were in good agreement. Based on PDF analysis results, Zhu *et al.* concluded a layer structure model belonging to the *C*12/*m*1 space group appropriate for the poorly crystalline Mn oxide samples.

Although the hexagonal birnessite model reproduced largely the interatomic distances in the sample PDFs, the authors (Zhu *et al.*, 2012) suggested that small-*r* misfits of the model and anomalies in certain fitted parameters provided evidence for deviations from periodicity (disorder). Intensity mismatches in the range r <5 Å in the PDFs for certain samples was interpreted as indicating local distortions around vacant or Mn^{3+} sites (*i.e.* Jahn-Teller distortion) and/or surface-related distortions near the sheet edges. Large anisotropic displacement parameters for Mn and O in the layers were interpreted as suggesting structural disorder along the stacking directions. The refined unit-cell parameters also suggested that the average structure of these Mn oxides deviated from hexagonal MnO_6-layer symmetry. Although highly suggestive, the unit-cell approach to fitting the PDF could not be used to test these types of disorder directly. Studies of Mn oxides (*e.g.* Petkov *et al.*, 2009; Zhu *et al.*, 2012) help to emphasize the challenge of using the unit-cell approach to capture fully atomic structure in minerals that are poorly crystalline. A key reason is that this approach requires the local structure and average long-range structure to be the same. Past work has shown that deviations in local and long-range structure occur in a number of different materials (see Egami and Billinge, 2003 and references therein).

5.3. Combined real-space and reciprocal-space analysis

The Rietveld method of analysing diffraction patterns in reciprocal space has long been the standard for determining long-range structure in crystalline materials. And as described through examples in the last section, real-space analysis by the unit-cell approach can be useful in understanding average structure in nanocrystalline and poorly crystalline materials. Several studies have attempted analysis of both the real- and reciprocal-space forms of the X-ray or neutron scattering data in order to develop a fuller understanding of materials that have different degrees of structural order. Although not reviewed here, the use of total scattering methods to analyse the structures of crystalline materials containing disorder is also noteworthy and we refer the reader to the literature for details regarding methodology (Proffen and Neder, 1997) and some selected examples relevant to mineralogy (Tucker *et al.*, 2001, Dove, 2002; Paglia *et al.*, 2006). In the remaining sections, we present the application of combined real-space and reciprocal-space analysis to analogues of natural nanoscale materials and to a sandstone that is a mixture of crystalline and amorphous phases.

5.3.1. Schwertmannite

A recent example where the unit-cell approach to fitting the PDF in real-space combined with reciprocal-space XRD analysis proved useful in understanding the structure of schwertmannite, a poorly crystalline oxy-hydroxy-sulfate of Fe $[Fe_8O_8(OH)_{8-x}(SO_4)_x$ $(1 < x < 1.75)]$ and a classic nanomineral example that forms commonly in acid-sulfate environments such as those often associated with current and former mining activities and in coastal acid-sulfate soils and sediments. The structure of schwertmannite has remained elusive despite several past studies using diffraction and spectroscopic methods. Prior diffraction studies (*e.g.* Bigham *et al.*, 1990) relied mainly on qualitative comparison of the XRD profile of schwertmannite, which consists of ~8 broad diffraction maxima, to a number of synthetic analogues. As in the previous examples, the extent of peak broadening, due primarily to the effects of particle size and structural disorder, limited the usefulness of powder XRD Rietveld refinement for more detailed structure analysis. Prior spectroscopic studies using probes such as X-ray absorption near edge spectroscopy (XANES) and Fourier transform infrared spectroscopy (FTIR) provided only local structure and speciation information for Fe and S in schwertmannite (*e.g.* Majzlan and Myneni, 2005; Boily *et al.*, 2010).

General consensus on a number of structural and chemical aspects of schwertmannite exists despite uncertainty regarding the actual crystal structure. For example, schwertmannite is considered to have FeO_6 octahedra linked mainly in edge-sharing and corner-sharing configurations. An early powder XRD study by Bigham *et al.* (1990) suggested that its structure is essentially analogous to that of akaganéite (β-FeOOH), which has a structure composed of double chains of Fe oxide (or hydroxide) octahedra that share corners to form square 2×2 tunnels. Chloride ions reside in the tunnels, and their negative charges can be compensated by hydrogen ions. In the case of schwertmannite, sulfate ions are thought to reside in the tunnels as a bidentate inner sphere complex and/or as outer-sphere complexes through hydrogen bonding (Barham, 1997; Boily *et al.* 2010).

Fernandez-Martinez *et al.* (2010) showed that the average structure of the octahedral (Fe and O) framework in schwertmannite could be resolved by analysis of high-resolution PDF measurements. The study focused on natural schwertmannite samples as well as a laboratory prepared synthetic analogue. A sulfate-doped akaganéite structure model was refined against the PDFs in real space from 1 to 30 Å. In the initial fits, lattice parameters (a, b, c, β), fractional coordinates (x and z of Fe atoms), isotropic thermal atomic displacement parameters and PDF-dependent δ and σ_Q were refined. The fractional coordinates x and z of the O atoms were added and refined in the second iteration of each refinement. The positions of S and O from the sulfate molecules were not refined due to the low contribution of these atoms to the PDF.

Fitted model parameters for the natural and synthetic samples were in reasonable agreement with those reported by Post *et al.* (2003) based on Rietveld refinement of crystalline akaganéite. Deformation in the PDF-refined unit cell was apparent, however, and attributed to the effects of structural strains induced by the presence of inner-sphere sulfate complexes in the tunnel, and possibly also the effects of surface energetics and structural defects. Strain due to sulfate binding in the tunnels was supported by density functional theory (DFT) calculations for the PDF-derived structures. Nonetheless, the authors concluded from PDF analysis that a structure model consisting of a deformed frame of FeO_6 octahedra similar to that of akaganéite could adequately describe the octahedral framework in schwertmannite.

Finally, Fernandez-Martinez *et al.* (2010) tested whether the PDF-derived structure that was obtained from fitting from 1 to 30 Å, could reproduce both short-range and long-range order in three dimensions. Real-space analysis already showed that a defective akaganéite-like structure could reproduce the PDF up to ~30 Å, which suggested that the long-range order already taken into account in the PDF analysis should also describe the observed XRD profiles for the same samples. To check for consistency, the authors calculated and compared theoretical powder diffraction patterns of the PDF/DFT-derived octahedral framework to the experimental XRD intensities of the samples. Intensity differences for certain reflections (Fig. 8) were attributed to sulfate in the structure (intentionally excluded from PDF analysis) and were further tested through comparison of theoretical and experimental data in reciprocal space. Various sulfate configurations with and without water molecules were evaluated. The best overall agreement between the theoretical and experimental XRD data was obtained for PDF-derived model of the octahedral framework containing both inner- and outer-sphere sulfate and water molecules. Thus, Fernandez-Martinez *et al.* (2010) showed that the PDF-derived model of the octahedral framework for schwertmannite could also reproduce reciprocal-space XRD data if the effects of the sulfate and water molecules were also considered.

5.3.2. Quartz/amorphous SiO_2 mixture

Materials that are a mixture of discrete crystalline and amorphous phases are another important case where structure determination can benefit from combined analysis of the real- and reciprocal-space forms of scattering data. Extracting structural information for both the crystalline and amorphous structural phases is challenging, however, using standard crystallographic methods alone. Analysis of the Bragg maxima in diffraction space using Rietveld or LeBail fitting can be used to reveal the long-range average structure. The amorphous component, however, can be difficult to detect, as it lacks well defined diffraction maxima due to an absence of long-range order. The total scattering and PDF methods can be used to evaluate the Bragg and diffuse components simultaneously (Dove, 2002), thereby making it possible to extract structural information for both crystalline and amorphous phases in a composite system.

Page *et al.* (2004) used neutron scattering and PDF methods to investigate a sample of the Fontainbleau sandstone, a natural, consolidated sandstone (>99% SiO_2)

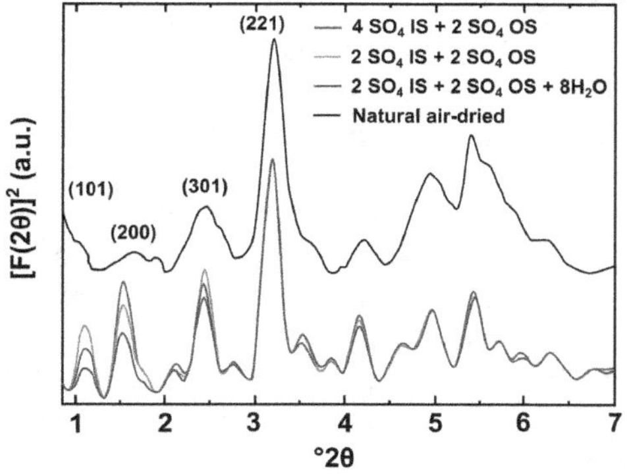

Figure 8. XRD pattern of natural air-dried schwertmannite compared to simulated patterns from the PDF-derived sulfate-akaganéite structure model. Intensity differences in the simulated patterns show the effects of loading different amounts of sulfate and water molecules in the channel. Reprinted from Fernandez-Martinez *et al.* (2010), with permission from the Mineralogical Society of America.

suspected to contain both crystalline and amorphous structural phases. Neutron diffraction data collected on a solid cylindrical core sample of the Fontainbleau showed sharp peaks characteristic of a long-range ordered structure. Rietveld refinement confirmed that all peaks in the neutron diffraction data could be indexed and fitted using the known structural model of quartz. No other crystalline phases were detected; hence, the average structure of the Fontainbleau sandstone was evidently pure crystalline quartz. Modulations in the difference curve between the calculated and experimental diffraction data, however, suggested the presence of a previously undetected amorphous SiO_2 phase.

To investigate further the possible presence of an amorphous SiO_2 component, the PDF derived from the neutron total scattering (terminated at $Q_{max} = 40$ Å^{-1}) was fitted in real space. Page *et al.* evaluated how the fit quality and residual values varied when different ranges of the PDF were included in the fitting. The scale factor, lattice parameters and atomic displacement parameters were allowed to refine from the initial values derived from Rietveld. Misfits of the model to the PDF data were observed, in particular for the region <3 Å that includes the first two peaks corresponding to the nearest Si–O and O–O atom-pair distances. This suggested that the long-range structure of crystalline quartz could not describe the PDF of the Fontainbleau over the complete distance range of the PDF. The authors attributed discrepancies in intensity at small-r values to the presence of a second SiO_2 component, which differed from crystalline quartz in terms of its extent of structural order. A two-phase quartz model was fitted to the sample PDF and the analysis showed that the addition of a second quartz phase to the model improved the overall fit, in particular for the small-r region (Fig. 9). This suggested the presence of a SiO_2 phase with only short-range order (amorphous). Refined scale factors for the two phases indicated that the Fontainbleau sandstone was ~90% crystalline quartz and ~10% amorphous SiO_2. Refined structure parameters further suggested that despite lacking long-range order, the amorphous SiO_2 phase was structurally very similar to crystalline SiO_2 (Proffen *et al.*, 2005). A recent study by Peterson *et al.* (2013) further considers the quantification of amorphous and crystalline phase content with the PDF.

5.4. Reverse Monte Carlo and whole nanoparticle simulation

As described in the previous sections, the unit-cell approach to modelling the PDF in real space is proving to be useful for evaluating structures in nanoscale and poorly crystalline materials and in mixtures of crystalline and amorphous phases. In addition, combining periodic real- and reciprocal-space analysis of the scattering data provides the means for assessing structural information over a wider range of length scales. Despite these successes, it is becoming increasingly clear that the unit-cell approach has limits in terms of generating and testing increasingly complex models of structural disorder.

Although applied historically to glasses and liquids, and more recently to crystalline materials containing disorder, the reverse Monte Carlo (RMC) method is a computational approach that is being used increasingly to investigate partially

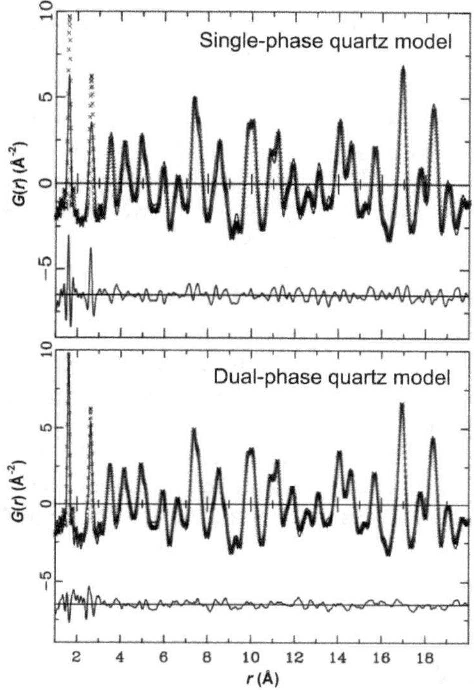

Figure 9. Single- and dual-phase modelling for Fontainebleau sandstone. Reprinted from Page *et al.* (2004), with permission from the American Geophysical Union.

disordered nanoscale materials. A Monte-Carlo type algorithm provides the basis for the RMC method in which random movements or "moves" are made in atom positions and site occupancies and agreement in the calculated reduced structure function (or PDF) with the experimental function is evaluated. If an atom movement results in an improvement in fit to the experimental function it is accepted on some statistical basis; movements that result in a poorer fit are rejected. The starting model in the case of an amorphous material consists typically of a randomized ensemble of atoms in proportions consistent with the composition of the sample. The starting model in cases of disordered crystalline materials is based usually on a periodic unit cell, but the RMC method allows for particular types of defects, such as stacking faults, short-range distortions and site vacancies, amongst others, to be incorporated and optimized. As described in the example below, a whole nanoparticle model can also be used in cases of nanoscale materials. No matter what system or model, hard constraints are introduced typically to avoid unphysical atom separations, maintain the geometry of rigid molecules, or restrict the range of variability for certain parameters. Detailed reviews of the methodology and applications were given by McGreevy (2001) and Keen *et al.* (2005).

Gilbert *et al.* (2013) recently explored structural disorder in "6-line" ferrihydrite nanoparticles and refined structure using both the real and reciprocal-space forms of X-ray scattering data. As mentioned earlier, Michel *et al.* (2007b, 2010) and Drits *et al.*

(1993) have proposed different structural models for ferrihydrite. Although both models are able to reproduce certain forms of X-ray scattering data, neither is described adequately, suggesting that certain types of structural disorder are not fully described by either model. In contrast to all previous structural studies of ferrihydrite, Gilbert *et al.* implemented a whole-nanoparticle RMC approach to evaluate types of structural variation that are not easily applied by the single unit-cell approach, including alternative configurations of iron occupancies and structural disorder. RMC models were 3.6-nm-diameter spherical particles with ~800 iron sites fully coordinated by oxygen atoms (Fig. 10). The RMC approach included relocation (swap) moves for the iron sites, displacement moves for the iron and oxygen sites, and a number of important constraints on interatomic distances. The reader should refer to Gilbert *et al.* (2013) for additional details regarding their RMC approach and specific fitting results.

Gilbert *et al.* evaluated three different starting structural configurations based on the Drits model, the Michel model and a hybrid of the two. Interestingly, the authors found that a structure model based on a hybrid of the Michel model and the defect-free portion of the Drits model was required to simulate successfully both the real- and reciprocal-space forms of the scattering data. In addition, models with tetrahedrally coordinated

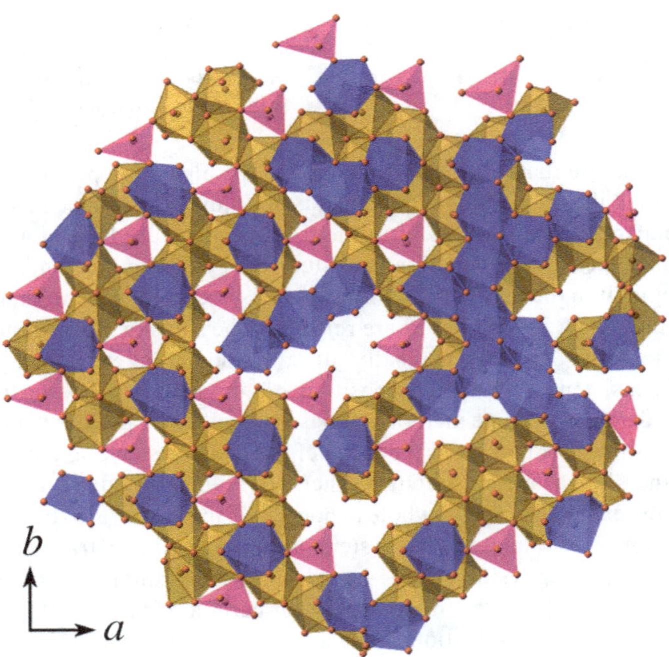

Figure 10. Example view of the 3.6-nm-diameter hybrid model nanoparticle after whole-nanoparticle RMC refinement. Cross section in the *a*–*b* plane shows disordered arrangement of Fe octahedra and tetrahedra. Reprinted from Gilbert *et al.* (2013), with permission from the Mineralogical Society of America.

iron sites consistently obtained better matches to the experimental total scattering data. Two independent studies have confirmed the presence of tetrahedrally coordinated Fe sites in ferrihydrite (Maillot *et al.*, 2011; Guyodo *et al.*, 2012). The RMC results also indicated that both long-range (nanometre-scale) defect disorder and short-range structural disorder are essential in ferrihydrite (Fig. 10). Hence, the study by Gilbert *et al.* (2013) highlights that an important advantage of whole-nanoparticle RMC is being able to detect simultaneously structural disorder on length scales extending from sub-Å to nanometre. Structural disorder, particularly on nanometre-length scales, is normally not accessible using a unit-cell approach to modelling the PDF in real-space.

6. Summary and outlook

Recent applications of X-ray and neutron total scattering methods have enhanced our understanding of structure in a variety of nanoscale solids with environmental relevance. The PDF derived from total scattering contains essential structural information, some of which can be extracted directly from the PDF in a model-independent way and interpreted. In terms of modelling, real-space Rietveld analysis of the PDF is a robust first approach to describe structural periodicity in materials that are nanocrystalline, and starting models based on crystalline counterparts are often useful in cases of mineral nanoparticles. Nanominerals, such as ferrihydrite or schwertmannite, present additional challenges to structure determination due to the lack of a crystalline counterpart. In addition, poorly crystalline materials typically have complex structural and defect disorder that occurs at a range of length scales. Several recent studies demonstrate that disorder on a range of length scales can be described when both the real- and reciprocal-space forms of the scattering are used as part of the analysis. It is also clear from recent work that combined analysis of XRD and PDF data is particularly useful in materials that are mixtures of crystalline and amorphous structural phases.

Structural studies of nanoscale and poorly crystalline solids will continue to rely on the total scattering and PDF methods for the foreseeable future. Advancements in analysis methods are required, however, in order to describe more accurately the complexity and heterogeneity of structural disorder. This is particularly true in the case of natural nanoparticles, which are often impure and even less crystalline and more disordered than their synthetic analogues. RMC and related approaches such as empirical potential structure refinement are showing promising developments, including in the simulation of whole nanoparticles. It is also clear that combined analysis of both the real- and reciprocal-space forms of the scattering data provide an important constraint during modelling. In this sense, it is important to bear in mind the importance of constraints from other studies using local structure techniques including X-ray absorption spectroscopy, nuclear magnetic resonance spectroscopy, and vibrational spectroscopies (FTIR/Raman), computational methods including molecular dynamics and density functional theory, other scattering-based methods (*e.g.* small angle X-ray or neutron scattering) and direct imaging by high-resolution transmission electron microscopy. New developments in instrumentation and in coupling these various measurements will expand the capabilities, allowing further advances in studies of natural and synthetic nanoscale materials.

Acknowledgements

This work was supported by NSF Grant EAR-1451996 and The Virginia Tech National Center for Earth and Environmental Nanotechnology Infrastructure (NSF Cooperative Agreement 1542100). F.M.M. also acknowledges gratefully funding support and the help of many collaborators in the selected studies discussed above. Use of the Advanced Photon Source (APS), an Office of Science User Facility operated for the U.S. Department of Energy (DOE) Office of Science by Argonne National Laboratory, was supported by the U.S. DOE under Contract No. DE-AC02-06CH11357. New data for synthetic ZnS was collected at X-ray Science Division beamline 1-ID at the APS, Argonne National Laboratory.

References

Aimoz, L., Taviot-Gueho, C., Churakov, S.V., Chukalina, M., Dahn, R., Curti, E., Bordet, P. and Vespa, M. (2012) Anion and cation order in iodide-bearing Mg/Zn-Al layered double hydroxides. *Journal of Physical Chemistry C*, **116**, 5460–5475.

Banfield, J.F. and Zhang, H.Z. (2001) Nanoparticles in the environment. Pp. 1–58 in: *Nanoparticles and the Environment* (J.F. Banfield and A. Navrotsky, editors). Reviews in Mineralogy, **44**. Mineralogical Society of America and the Geochemical Society, Chantilly, Virginia, USA.

Barham, R.J. (1997) Schwertmannite: A unique mineral, contains a replaceable ligand, transforms to jarosites, hematites, and/or basic iron sulfate. *Journal of Materials Research*, **12**, 2751–2758.

Barrón, V., Torrent, J. and De Grave, E. (2003) Hydromaghemite, an intermediate in the hydrothermal transformation of 2-line ferrihydrite into hematite. *American Mineralogist*, **88**, 1679–1688.

Barrón, V., Torrent, J. and Michel, F.M. (2012) Critical evaluation of the revised akdalaite model for ferrihydrite – Discussion. *American Mineralogist*, **97**, 253–254.

Bell, J.L., Sarin, P., Provis, J.L., Haggerty, R.P., Driemeyer, P.E., Chupas, P.J., van Deventer, S.J. and Kriven, W.M. (2008) Atomic structure of a cesium aluminosilicate geopolymer: A pair distribution function study. *Chemistry of Materials*, **20**, 4768–4776.

Bigham, J.M., Schwertmann, U., Carlson, L. and Murad, E. (1990) A poorly crystallized oxyhydroxysulfate of iron formed by bacterial oxidation of Fe(II) in acid mine waters. *Geochimica et Cosmochimica Acta*, **54**, 2743–2758.

Boily, J.-F., Gassman, P.L., Peretyzhko, T., Szanyi, J. and Zachara, J.M. (2010) FTIR spectral components of schwertmannite. *Environmental Science & Technology*, **44**, 1185–1190.

Chupas, P.J., Qiu, X., Hanson, J.C., Lee, P.L., Grey, C.P. and Billinge, S.J.L. (2003) Rapid-acquisition pair distribution function (RA-PDF) analysis. *Journal of Applied Crystallography*, **36**, 1342–1347.

Cismasu, A.C., Michel, F.M., Tcaciuc, P.A., Tyliszczak, T. and Brown, G.E. Jr. (2011) Composition and structural aspects of naturally occurring ferrihydrite. *Comptes Rendus Geoscience*, **343**, 210–218.

Cismasu, A.C., Michel, F.M., Stebbins, J.F., Levard, C.M. and Brown, G.E. Jr. (2012) Properties of impurity-bearing ferrihydrite. I. Effects of Al content and precipitation rate on the structure of 2-line ferrihydrite, *Geochimica et Cosmochimica Acta*, **92**, 275–291.

Dove, M.T. (2002) An introduction to the use of neutron scattering methods in mineral sciences. *European Journal of Mineralogy*, **14**, 203–224.

Drits, V.A., Sakharov, B.A., Salyn, A.L. and Manceau, A. (1993) Structural model for ferrihydrite. *Clay Minerals*, **28**, 185–207.

Dyer, L.G., Chapman, K.W., English, P., Saunders, M. and Richmond, W.R. (2012) Insights into the crystal and aggregate structure of Fe^{3+} oxide/silica co-precipitates. *American Mineralogist*, **97**, 63–69.

Egami, T. and Billinge, S.J.L. (2003) *Underneath the Bragg Peaks: Structural Analysis of Complex Materials*. Pergamon Press, Oxford, UK.

Ehm, L., Michel, F.M., Antao, S.M., Martin, C.D., Lee, P.L., Shastri, S.D., Chupas, P.J. and Parise, J.B. (2008) Structural changes in nanocrystalline mackinawite (FeS) at high pressure. *Journal of Applied Crystallography*, **42**, 1–7.

Fernandez-Martinez, A., Timon, V., Roman-Ross, G., Cuello, G.J., Daniels, J.E. and Ayora, C. (2010) The structure of schwertmannite, a nanocrystalline iron oxyhydroxysulfate. *American Mineralogist*, **95**, 1312–1322.

Gilbert, B. (2008) Finite size effects on the real-space pair distribution function of nanoparticles. *Journal of Applied Crystallography*, **41**, 554–562.

Gilbert, B., Huang, F., Zhang, H., Waychunas, G.A. and Banfield, J.F. (2004) Nanoparticles: Strained and Stiff. *Science*, **305**, 651–654.

Gilbert, B., Erbs, J.J., Penn, R.L., Petkov, V., Spagnoli, D. and Waychunas, G.A. (2013) A disordered nanoparticle model for 6-line ferrihydrite. *American Mineralogist*, **98**, 1465–1476.

Gualtieri, A.F., Ferrari, S., Leoni, M., Grathoff, G., Hugo, R., Shatnawi, M., Paglia, G. and Billinge, S. (2008) Structural characterization of the clay mineral illite-1M. *Journal of Applied Crystallography*, **41**, 402–415.

Guyodo, Y., Sainctavit, Arrio, M.-A., Carvallo, C., Penn, R.L., Erbs, J.J., Forsberg, B.S., Morin, G., Maillot, F., Lagroix, F., Bonville, P., Wilhelm, F. and Rogalev, A. (2012) X-ray magnetic circular dichroïsm provides strong evidence for tetrahedral iron in ferrihydrite. *Geochemistry, Geophysics, Geosystems*, **13**, 1–9.

Hall, B.D., Zanchet, D. and Ugarte, D. (2000) Estimating nanoparticle size from diffraction measurements, *Journal of Applied Crystallography*, *33*, 1335–1341.

Harrington, R., Hausner, D.B., Xu, Wenqian, Bhandari, N., Michel, F.M., Brown, Jr., G.E., Strongin, D.R. and Parise, J.B. (2011) Neutron pair distribution function study of two-line ferrihydrite. *Environmental Science & Technology*, **45**, 9883–9890.

Hiemstra, T. (2013) Surface and mineral structure of ferrihydrite. *Geochimica et Cosmochimica Acta*, **105**, 316–325.

Hochella, M.F., Jr., Lower, S.K., Maurice, P.A., Penn, R.L., Sahai, N., Sparks, D.L. and Twining, B.S. (2008) Nanominerals, mineral nanoparticles, and Earth systems. *Science*, **319**, 1631–1635.

Hocking, R.K., Gates, W.P. and Cashion, J.D. (2012) Comment on "Direct observation of tetrahedrally coordinated Fe(III) in ferrihydrite". *Environmental Science & Technology*, **46**, 11471–11472.

Jambor, J.L. and Dutrizac, J.E. (1998) Occurrence and constitution of natural and synthetic ferrihydrite, a widespread iron oxyhydroxide. *Chemical Reviews*, **98**, 2549–2585.

Keen, D.A., Tucker, M.G. and Dove, M.T. (2005) Reverse Monte Carlo modelling of crystalline disorder. *Journal of Physics: Condensed Matter*, **17**, S15–S22.

Klug, H.P. and Alexander, L.E. (1954) *X-ray Diffraction Procedures*. Wiley, New York.

Lennie, A.R., Redfern, S.A.T., Schofield, P.F. and Vaughan, D.J. (1995) Synthesis and Rietveld crystal structure refinement of mackinawite, tetragonal FeS. *Mineralogical Magazine*, **59**, 677–683.

Maillot, F.M., Morin, G., Wang, Y., Bonnin, D., Ildefonse, P., Chaneac, C. and Calas, G. (2011) New insight into the structure of nanocrystalline ferrihydrite: EXAFS evidence for tetrahedrally coordinated iron(III). *Geochimica et Cosmochimica Acta*, **75**, 2708–2720.

Majzlan, J. and Myneni, S.C.B. (2005) Speciation of iron and sulfate in acid waters: Aqueous clusters to mineral precipitates. *Environmental Science & Technology*, **29**, 188–194.

Manceau, A. (2009) Evaluation of the structural model for ferrihydrite from real-space modeling of high-energy X-ray diffraction data. *Clay Minerals*, **44**, 19–34.

Manceau, A. (2011) Critical evaluation of the revised akdalaite model for ferrihydrite. *American Mineralogist*, **96**, 521–533.

Manceau, A. (2012) Comment on "Direct observation of tetrahedrally coordinated Fe(III) in ferrihydrite. *Environmental Science and Technology*", **46**, 6882–6884.

McGreevy, R.L. (2001). Reverse Monte Carlo modeling. *Journal of Physics: Condensed Matter*, **13**, R877–R913.

Michel, F.M. (2014) Total scattering studies of natural and synthetic ferrihydrite. Pp. 89−135 in: *Advanced Applications of Synchrotron Radiation in Clay Science* (G.A. Waychunas, editor). CMS Workshop Lecture Series, **19**, The Clay Minerals Society, Chantilly, Virginia, USA.

Michel, F.M., Antao, S.M., Chupas, P.J., Lee, P.L., Parise, J.B. and Schoonen, M.A.A. (2005) Short- and medium-range atomic order and crystallite size of the initial FeS precipitate from pair distribution function analysis. *Chemistry of Materials*, **17**, 6246−6255.

Michel, F.M., Ehm, L., Liu, G., Han, W.Q., Antao, S.M., Chupas, P.J., Lee, P.L., Knorr, K., Eulert, H., Kim, J., Grey, C.P., Celestian, A.J., Gillow, J.B., Schoonen, M.A.A., Strongin, D.R. and Parise, J.B. (2007a) Similarities in 2- and 6-line ferrihydrite based on pair distribution function analysis of X-ray total scattering. *Chemistry of Materials*, **19**, 1489−1496.

Michel, F.M., Ehm, L., Antao, S.M., Lee, P.L., Chupas, P.J., Liu, G., Strongin, D.R., Schoonen, M.A.A., Phillips, B.L. and Parise, J.B. (2007b) The structure of ferrihydrite, a nanocrystalline material. *Science*, **316**, 1726−1729.

Michel, F.M., Barrón, V., Torrent, J., Morales, M.P., Serna, C.J., Boily, J.-F., Liu, Q., Ambrosini, A., Cismasu, A.C. and Brown, G.E., Jr. (2010) Ordered ferrimagnetic form of ferrihydrite reveals links among structure, composition, and magnetism. *Proceedings of the National Academy of Sciences of The United States of America*, **107**, 2787−2792.

Mikutta, C. (2011) X-ray absorption spectroscopy study on the effect of hydrobenzoic acids on the formation and structure of ferrihydrite. *Geochmica et Cosmochimica Acta*, **75**, 5122−5139.

Mikutta, R., Kleber, M. and Jahn, R. (2005) Poorly crystalline minerals protect organic carbon in clay subfractions from acid subsoil horizons. *Geoderma*, **128**, 106−115.

Page, K.L., Proffen, T. and McLain, S.E. (2004) Local atomic structure of Fountainebleau sandstone: Evidence for an amorphous phase? *Geophysical Research Letters*, **31**, L24606.

Paglia, G., Bozin, E.S. and Billinge, S.J.L. (2006) Fine-scale nanostructure in gamma-Al_2O_3. *Chemistry of Materials*, **18**, 3242−3248.

Peak, D. and Regier, T. (2012) Direct observation of tetrahedrally coordinated Fe(III) in ferrihydrite, *Environmental Science & Technology*, **46**, 3163−3168.

Peterson, J., TenCate, J., Proffen, T., Darling, T., Heinz, N. and Page, K. (2013) Quantifying amorphous and crystalline phase content with the atomic pair distribution function. *Journal of Applied Crystallography*, **46**, 332−336.

Petkov, V., Ren, Y., Saratovsky, I., Gurr, S.J., Hayward, M.A., Poeppelmeier, K.R. and Gaillard, J.-F. (2009) Atomic-scale structure of biogenic materials by total X-ray diffracton: A study of bacterial and fungal MnO_x. *ACS Nano*, **3**, 441−445.

Post, J.E., Heaney, P.J., Von Dreele, R.B. and Hanson, J.C. (2003) Neutron and temperature-resolved synchrotron X-ray powder diffraction study of akaganéite. *American Mineralogist*, **88**, 782−788.

Proffen, T. and Billinge, S.J.L. (1999) *PDFFIT*, a program for full profile structural refinement of the atomic pair distribution function. *Journal of Applied Crystallography*, **32**, 572−575.

Proffen, T. and Neder, R.B. (1997) *DISCUS*: a program for diffuse scattering an d defect-structure simulation. *Journal of Applied Crystallography*, **30**, 171−175.

Proffen, T., Page K.L., McLain, S.E., Clausen, B., Darling, T.W., TenCate, J.A., Lee S.-Y. and Ustundag, E. (2005) Atomic pair distribution function analysis of materials containing crystalline and amorphous phases. *Zeitschrift für Kristallographie – Crystalline Materials*, **220**, 1002−1008.

Rancourt, D.G. and Meunier, J.-F. (2008) Constraints on structural models of ferrihydrite as a nanocrystalline material. *American Mineralogist*, **93**, 1412−1417.

Scheinost, A.C., Kirsch, R., Banerjee, D., Fernandez-Martinez, A., Zaenker, H., Funke, H. and Charlet, L. (2008) X-ray absorption and photoelectron spectroscopy investigation of selenite reduction by Fe^{II}-bearing minerals. *Journal of Contaminant Hydrology*, **102**, 228−245.

Skinner, B.J. (1961) Unit-cell edges of natural and synthetic sphalerites. *American Mineralogist*, **46**, 1399−1411.

Toner, B.M., Berquó, T.S., Michel, F.M., Sorenson, J.V. and Edwards, K.J. (2012) Mineralogy of iron microbial mats from Loihi Seamount. *Frontiers in Microbiology*, **3**, 1−18.

Tucker, M.G., Dove, M.T. and Keen, D.A. (2001) Application of the reverse Monte Carlo method to crystalline materials. *Journal of Applied Crystallography*, **34**, 630–638.

Waychunas, G.A., Fuller, C.C., Rea, B.A. and Davis, J.A. (1996) Wide angle X-ray scattering (WAXS) study of "two-line" ferrihydrite structure: Effect of arsenate sorption and counterion variation and comparison with EXAFS results. *Geochmica et Cosmochimica Acta*, **60**, 1765–1781.

Wright, A.C. (1998) Diffraction studies of glass structure: The first 70 years. *Glass Physics and Chemistry*, **24**, 148–179.

Xu, W., Hausner, D.B., Harrington, R., Lee, P.L., Strongin, D.R. and Parise, J.B. (2011) Structural water in ferrihydrite and constraints this provides on structure models. *American Mineralogist*, **96**, 513–520.

Yamaguchi, G., Okumiya, M. and Ono, S. (1969) The crystal structure of tohdite. *Bulletin of the Chemical Society of Japan*, **42**, 2247–2249.

Zhang, H., Gilbert, B., Huang, F. and Banfield, J.F. (2003) Water-driven structure transformation in nanoparticles at room temperature. *Letters to Nature*, **424**, 1025–1029.

Zhang, X.V., Ellery, S.P., Friend, C.M., Holland, H.D., Michel, F.M., Schoonen, M.A.A. and Martin, S.T. (2006) Photodriven reduction and oxidation reactions of colloidal semiconductor particles: Implications for prebiotic synthesis. *Journal of Photochemistry and Photobiology A: Chemistry*, **185**, 301–311.

Zhu, M., Farrow, C.L., Post, J.E., Livi, K.J.T., Billinge, S.J.L., Ginder-Vogel, M. and Sparks, D.L. (2012) Structural study of biotic and abiotic poorly-crystalline manganese oxides using pair distribution function analyses. *Geochimica et Cosmochimica Acta*, **81**, 39–55.

Zsigmondy, R. (1920) *Kolloidchemie. Ein Lehrbuch*, 3rd edition. Spamer, Leipzig, Germany.

EMU Notes in Mineralogy, Vol. 19 (2017), Chapter 5, 213–254

Aperiodic mineral structures

L. BINDI[1] and G. CHAPUIS[2]

[1] *Dipartimento di Scienze della Terra, Università di Firenze, Via La Pira 4, I-50121 Firenze, Italy*
[2] *Laboratoire de Cristallographie, École Polytechnique Fédérale de Lausanne, CH-1015 Lausanne, Switzerland*

The three-dimensional periodic nature of crystalline structures was so strongly anchored in the minds of scientists that the numerous indications that seemed to question this model struggled to acquire the status of validity. The discovery of aperiodic crystals, a generic term including modulated, composite and quasicrystal structures, started in the 1970s with the discovery of incommensurately modulated structures and the presence of satellite reflections surrounding the main reflections in the diffraction patterns. The need to use additional integers to index such diffractograms was soon adopted and theoretical considerations showed that any crystal structure requiring more than three integers to index its diffraction pattern could be described as a periodic object in a higher dimensional space, *i.e.* superspace, with dimension equal to the number of required integers. The structure observed in physical space is thus a three-dimensional intersection of the structure described as periodic in superspace.

Once the symmetry properties of aperiodic crystals were established, the superspace theory was soon adopted in order to describe numerous examples of incommensurate crystal structures from natural and synthetic organic and inorganic compounds even to proteins. Aperiodic crystals thus exhibit perfect atomic structures with long-range order, but without any three-dimensional translational symmetry.

The discovery of modulated structures was soon followed by the discovery of composite structures consisting of structural entities with partly independent translations and finally by the discovery of quasicrystals.

In recent years, the use of CCD and imaging plate systems improved considerably the sensitivity of data collection for aperiodic structures and in particular modulated structures and, therefore, there was a need for further improvement of the methods. Today, several computer programs are able to solve and refine incommensurately modulated structures using the superspace approach.

In nature, it is uncommon to find minerals which have strong and sharp incommensurate satellites that could be used for a higher dimensional refinement. Here we describe several cases of aperiodic minerals (natrite, calaverite, melilite, fresnoite, pearceite–polybasite, cylindrite) including the first examples of natural and stable quasicrystalline structures (icosahedrite and decagonite) which settle beyond doubt any questions which remain about the long-term stability of quasicrystals.

1. Is the law of rational indices of general validity? Background information from the history of crystallography

The discovery of X-ray diffraction in 1912 followed by the first descriptions of simple crystalline structures by the Braggs in 1914 were at the origin of a new paradigm in the field of physics and chemistry in particular concerning the structural properties of

DOI: 10.1180/EMU-notes.19.1

matter. The award of the Nobel prize in physics to Max von Laue in 1914 for the discovery of X-ray diffraction by crystals followed a year later by another such prize to W. Henry and W. Lawrence Bragg for the analysis of crystal structures by means of X-rays, is direct proof of the tremendous significance and impact of diffraction on the scientific community. The elegant description of crystalline structures on the basis of a three-dimensional periodical array of atoms was so convincing at that time that no scientist dared to question this seemingly universal model.

Note that the first analyses of crystal structures, based on diffraction experiments, benefitted enormously from the pioneering works of mineralogists like Haüy, Bravais and many others. The law of rational indices and the notions of direct and reciprocal lattice vectors were already well known as well as the fact that crystalline faces were parallel to lattice planes and normal to reciprocal lattice vectors (*e.g.* Rigault, 1999). Therefore, the three (usually small) indices used to characterize any single crystalline face could easily be associated with the same three indices characterizing each single spot on the newly discovered X-ray diffraction (XRD) patterns. Obviously, these three indices were considered to be direct proof of the three-dimensional periodic character of crystalline matter.

And yet, by the beginning of the 20th century, mineralogists had already presented some striking observations pertaining to the general validity of the law of rational indices. We shall follow the evolution of these observations with the example of the mineral calaverite ($Au_{1-x}Ag_xTe_2$, with $0 \leqslant x \leqslant 0.33$) over a period of nearly one century. First Penfield and Ford (1902) pointed to the difficulties involved in indexing faces with small rational indices. Later Smith (1903) used a collection of 49 different samples and attempted to find a plausible model for indexing the crystal faces. The best he could propose was to admit the existence of three interpenetrating structures, two of them triclinic and one monoclinic! Faced with this awkward interpretation, the author proposed for the first time the idea of incommensurability in crystals.

Some 30 years later, Goldschmidt, Palache and Peacock (1931, labelled hereafter as GPP) investigated a collection of 105 different samples originating from different continents with high-precision goniometry and here again no simple indexing model could be proposed. The authors concluded that "the law of rational indices is not of general validity" and that crystalline faces might have irrational indices. Of course, from our 21st century point of view they were right but the three-dimensional periodic nature of crystals was so deep-rooted in the minds of scientists that the conclusions of their study had to wait another half century before being fully recognized! It is worth looking at the gnomonic projection published by GPP in order to see how close they were to a breakthrough in the interpretation of their data. Figure 1 is an augmented representation of the original figure. It can be interpreted as a representation of the structure in reciprocal space. The dark grid linking the open points indicated by letters is from the original figure. To each node of the grid we can thus assign three indices (the first three integers). For the remaining grey points, we observe that all of them are located on a series of parallel lines indicated by the vector, $\mathbf{q}$, represented on the figure, each of them crossing the grid points mentioned above. Moreover, each one is located on integer multiples of the same vector. Thus, it seems natural to reinterpret each point

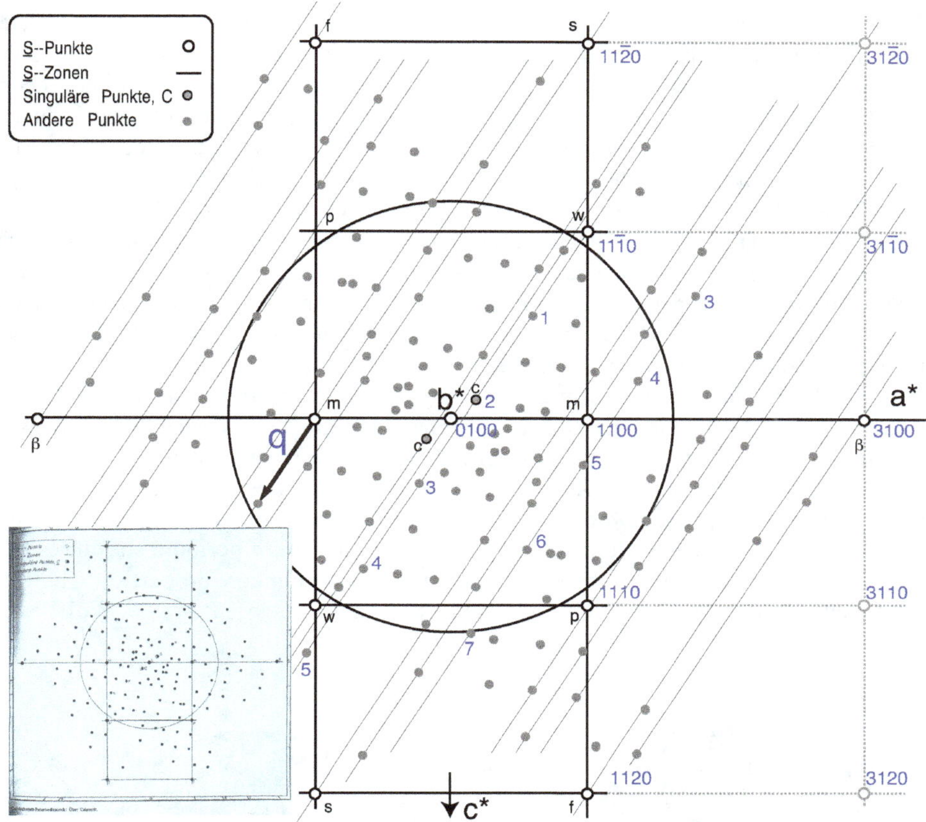

Fig. 1. Gnomonic projection of the crystal forms of calaverite according to Goldschmidt *et al.* (1931). The inset illustrates the original figure whereas the main figure is reinterpreted by indicating the three reciprocal basis vectors **a***, **b***and **c*** and introducing an additional vector **q** allowing the indexing of all the forms in terms of four indices.

of the gnomonic projection in terms of three indices provided by the lattice point containing the parallel line and a fourth index giving the multiple of the vector, **q**. Thus each point is characterized uniquely by four indices. On Fig. 1 for example, points with indices $11\bar{1}0$, $11\bar{1}1 \ldots 11\bar{1}5$ are indicated. (Note that the four indices used here are not to be confused with the indexing with four integers used for trigonal and hexagonal crystals!). The addition of a fourth reciprocal lattice vector, **q**, prevents us from using any irrational number to fully index the gnomonic projection. We shall see below how we can deal with the additional vector, **q**.

The understanding of the morphology of calaverite had to wait nearly half a century when Janner and Dam (1989) published a new and more convincing indexing of the calaverite forms (Fig. 2) based on the GPP data. All the faces could be indexed in terms of a twin and the four indices presented above.

The need for and the significance of using additional indices was only understood later when diffraction patterns of mainly single crystals started to appear. It is worth citing the work of Korekawa (1967) who published his '*Habilitationsschrift*' in German on the theory of satellite reflections. At that time, a few examples of single-crystal diffraction patterns could only meaningfully be indexed by using four indices, similar to the indexation of the calaverite forms based on GPP data. Soon the expression 'satellites' was adopted for those reflections with the fourth index different from 0. In other words, all those equidistant reflections located on a straight line crossing a main reflection located at the intersection of a three-dimensional reciprocal lattice were called 'satellite' reflections. In his work, Korekawa (1967) did show how a periodical perturbation or modulation in addition to the independent three-dimensional lattice periodicity could lead to the formation of satellite reflections, hence the name 'modulated structures' to characterize them. He studied various longitudinal and transversal periodic distortions as well as density variations and showed their effect on the diffraction patterns. In particular, he could explain the formation of satellite reflections in labradorite, a variety of the mineral anorthite, which will be mentioned below.

Unfortunately, the work of Korekawa (1967) did not have the expected impact in structural sciences probably because of its lack of symmetry treatment for modulated structures. Symmetry is so fundamental for the description of periodic structures that any theory of modulated structure has to include its symmetry properties. The first treatment of the symmetry of modulated structures was by P.M. de Wolff who

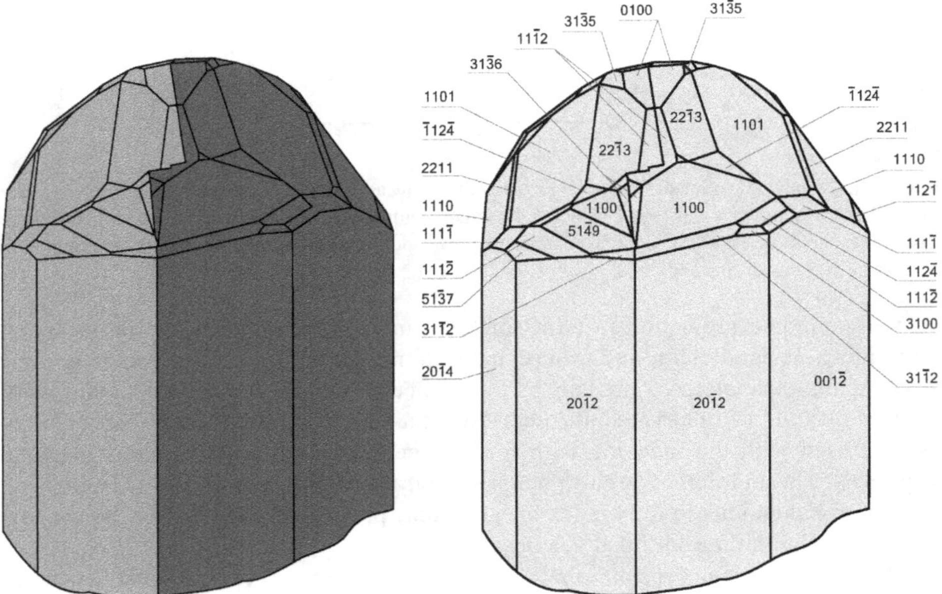

Fig. 2. A reinterpretation of the calaverite forms originally published by Goldschmidt *et al.* (1931). All the faces of the twin indicated on the left are indexed with four indices (right).

published in 1974 the first elements of a symmetry theory with the example of the modulated structure of γ-Na_2CO_3. This work is at the basis of the theory, which was later called 'superspace', a higher dimensional space, which is particularly convenient to describe modulated structures. A modulated structure is periodic in superspace and its three-dimensional section is the real-space representation of its structure. The concept of superspace will be described in more detail below.

Once structural scientists realised that the paradigm of purely three-dimensional structures had to be abandoned, the path for new discoveries on the complexity of materials was opened. In 1981, Makovicky and Hyde published an important study of non-commensurate (misfit) layer structures describing numerous examples of mineral structures consisting of two components each showing no commensurability in the periodicities of the structural components parallel to the layers but strict periodicity normal to the layers. Structures with similar characteristics are currently classified under the name 'composite' structures. It was later realised that composite structures could also be very conveniently described in terms of the superspace concept.

The publication with the largest impact on structural sciences relates to the discovery of "quasicrystals" by Shechtman, Blech, Gratias and Cahn (1984). The authors discovered the existence of stable metallic phases with long-range orientational order but without any translational symmetry. Here again, the diffraction patterns could only meaningfully be indexed by using up to six (usually small) integers. This discovery laid the basis for a new research paradigm and established definitely the superspace concept as the tool of choice to describe 'aperiodic' structures, a generic term combining modulated, composite and quasicrystalline structures.

2. An elementary introduction to the concept of superspace and redefinition of a crystal

In order to gain some insight into the concept of superspace, let us consider the indexation of the series of points located on the gnomonic projection represented in Fig. 1. We have seen that the gnomonic projection is a faithful representation of the reciprocal space similar to a reciprocal layer of a diffraction pattern. Therefore, any point can be represented by a reciprocal lattice vector, **H** (equation 1) in terms of the basis vectors, $\mathbf{a}_i^*$ (i = 1,2,3) and the modulation vector, **q**, with integers h_i and m:

$$\mathbf{H} = h_1\mathbf{a}_1^* + h_2\mathbf{a}_2^* + h_3\mathbf{a}_3^* + m\mathbf{q} \qquad (1)$$

Our next task is to represent our reciprocal space in higher dimension, *i.e.* in superspace, in order to accommodate the additional reciprocal lattice vector, **q**. This can be done according to the recipe illustrated in Fig. 3. The basic principle of this extension is to consider that the satellite reflections observed in diffraction patterns result from a projection of a higher-dimensional space onto the physical space. Let us deal first with the vector $\mathbf{a}_{S4}^*$ defined in superspace. It has a component **q** in the physical space R^* (also called 'external' *or* 'parallel space') and the unit vector, $\mathbf{e}_4^*$ in the space normal to it, which is sometimes called 'internal space' or 'perpendicular space'. The modulation vector $\mathbf{q}$ is usually expressed in terms of the reciprocal lattice vector:

$$\mathbf{q} = \alpha\mathbf{a}_1^* + \beta\mathbf{a}_2^* + \gamma\mathbf{a}_3^* \tag{2}$$

Thus $\mathbf{a}_{S4}^*$ can be expressed in the two-component form given in equation 4. The expression for the other components $\mathbf{a}_{Si}^*$ (where $i = 1, 2, 3$) is somewhat simpler, only the external component is present whereas the internal component is nil. This is illustrated in Fig. 3 for the vectors $\mathbf{a}_{S1}^*$ and $\mathbf{a}_1^*$ which are parallel and have the same length.

Once the reciprocal lattice vectors are defined in superspace, the corresponding dual vectors in superspace can be derived. Similar to the classical case in 3D, we can use the same relation between direct and reciprocal lattice vectors. The scalar product of direct and reciprocal lattice vectors is given by the Kronecker symbol δ_{ij}, *i.e.* $\delta = 1$ if i and j are identical and 0 otherwise. We have to generalize the scalar product in order to include both external and internal components of the vectors (the indices for the external components are omitted):

$$\mathbf{u}_S{\cdot}\mathbf{v}_S = \mathbf{u}{\cdot}\mathbf{v} + u_I v_I \text{ for } (\mathbf{u}, u_I) \text{ and } (\mathbf{v}, v_I) \tag{3}$$

The definition of the lattice vectors in superspace is thus given by equation 4. Each vector consists of two components given in parentheses, the external and the internal components, respectively.

$$\left.\begin{array}{l} \mathbf{a}_{S1}^* = (\mathbf{a}_1^*, 0) \\ \mathbf{a}_{S2}^* = (\mathbf{a}_2^*, 0) \\ \mathbf{a}_{S3}^* = (\mathbf{a}_3^*, 0) \\ \mathbf{a}_{S4}^* = (\mathbf{q}, 1) \end{array}\right\} \rightarrow \mathbf{a}_{Si}{\cdot}\mathbf{a}_{Sj}^* = \delta_{ij} \rightarrow \left\{\begin{array}{l} \mathbf{a}_{S1} = (\mathbf{a}_1, -\mathbf{q}{\cdot}\mathbf{a}_1) \\ \mathbf{a}_{S2} = (\mathbf{a}_2, -\mathbf{q}{\cdot}\mathbf{a}_2) \\ \mathbf{a}_{S3} = (\mathbf{a}_3, -\mathbf{q}{\cdot}\mathbf{a}_3) \\ \mathbf{a}_{S4} = (0, 1) \end{array}\right. \tag{4}$$

For simplification, the above definition is limited to four dimensions. It is straightforward to include additional dimensions with modulation vectors $\mathbf{q}_1$, $\mathbf{q}_2$, *etc.* and to derive the corresponding superspace components.

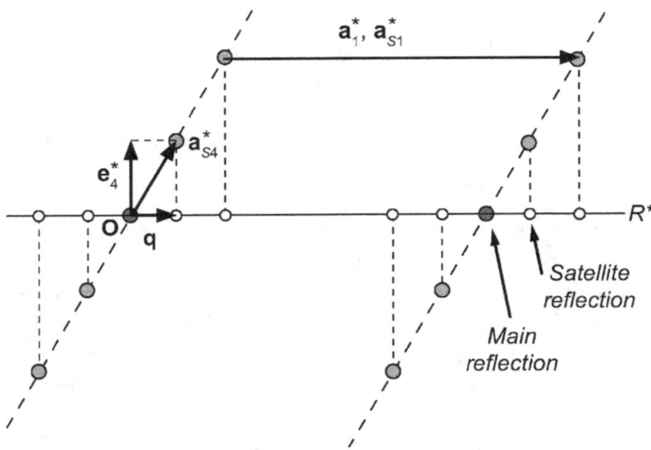

Fig. 3. Representation of main and satellite reflections in superspace. See text for explanations.

The immediate problem we are facing when dealing with the description of a structure in higher dimensional space concerns its representation. We know perfectly well how to reconstruct a three-dimensional object out of a series of two-dimensional sections. Moreover we can minimize the number of sections provided an optimal selection of planes. For example, a single 2D section is sufficient to represent any close-packed 3D structure. We shall do the same for the representation of a structure in superspace. We shall describe and analyse it in terms of a series of two-dimensional sections. Figure 4 illustrates the representation of a modulated structure with a transverse displacement wave. On the left a structure is represented by a two-dimensional array of points subject to a transverse modulation along $\mathbf{a}_2$ with a wavelength, λ, which is not generally commensurate with the periodicity of $\mathbf{a}_2$. The structure is periodic along $\mathbf{a}_1$, however. The ($\mathbf{a}_{S2}$, $\mathbf{a}_{S4}$) section is represented on the right. According to the definitions given in equation 4, the basis vector in superspace $\mathbf{a}_{S2}$ consists of the two components $\mathbf{a}_2$ and $-\mathbf{q}\cdot\mathbf{a}_2$. The component $\mathbf{a}_{S4}$ is normal to $\mathbf{a}_2$ according to the same definitions. In our representation, a string, a sinusoidal wave in the present example, which is periodically repeating, represents each atom. The 2D unit cell is illustrated in grey in the figure. The structure of the real crystal can be seen on the right figure from the intersection of the periodic array of sinusoidal waves along $\mathbf{a}_2$. The series of intersecting points are not periodic and the departure of x_2 from the average $\bar{x}_2$ varies along $\mathbf{a}_2$. We have thus verified that the real crystal is identical to the transverse modulation represented on the left. The ($\mathbf{a}_{S1}$, $\mathbf{a}_{S4}$) section is not represented but its construction is straightforward: a straight line located at $\bar{x}_\mathrm{i}$ would replace the sinusoidal wave and its intersection would be periodic.

Since the discovery of modulated crystals, many different types of periodic modulation functions have been introduced to improve their descriptions. Modulated structures rarely exhibit a purely sinusoidal wave. In most cases, the periodic perturbation can be better approximated by a series of harmonic terms if the resolution of the data permits it. Other perturbation functions have been proposed, *e.g.* step functions or saw tooth functions as illustrated in Fig. 5 (left). In the case of defects for example, the perturbation function might only exist in a limited interval of the variable x_4 defined in Fig. 4. In addition to

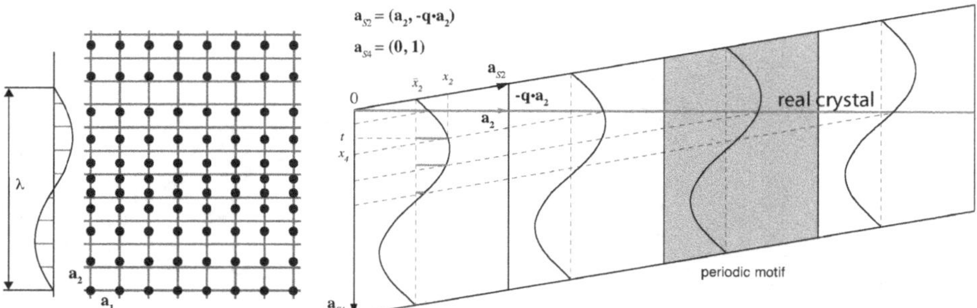

Fig. 4. Representation of a simple displacive modulation in superspace. (Left) A periodic array of points along $\mathbf{a}_1$ is subject to a transversal modulation along $\mathbf{a}_2$ with wavelength, λ, independent of $\mathbf{a}_1$ and $\mathbf{a}_2$. (Right) Representation of the displacive modulation in a section of superspace.

'displacement' modulations, cases of 'density' modulations might also exist. Such an example is illustrated in Fig. 5 (right). This example illustrates the case of layer structures where alternating sequences of compact layers in the A, B or C positions might occur. These are just a few examples and the interested reader might consult the textbooks by Janssen *et al.* (2007) and van Smaalen (2007) for others.

At the end of this section, it is appropriate to redefine the concept of crystals. A material is a crystal if it has essentially a sharp diffraction pattern. By recombining equations 1 and 2 and rewriting **H** in the form:

$$\mathbf{H} = h_1\mathbf{a}_1^* + h_2\mathbf{a}_2^* \; h_3\mathbf{a}_3^* + \cdots + h_n\mathbf{a}_n^* \tag{5}$$

where n is the 'rank', *i.e.* the minimal number of integer indices required to fully characterize any diffraction pattern in a space of d dimensions. If $n = d$, relation 5 describes a 'periodic' crystal. Otherwise, if $n > d$, the same relation describes an 'aperiodic' crystal. Note that the new definition of a crystal is done in reciprocal space and not in direct space as it used to be. Until now, any attempt to define a crystal in direct space failed. In other words, we need to explore the diffraction pattern of a crystal before we can assign it to one or the other category.

A crystal is 'incommensurately' modulated if at least one of the coefficients in equation 2 is irrational. Otherwise, the crystal is 'commensurately' modulated, which, in classical crystallography, is also called a superstructure.

3. Symmetry in superspace

The development of superspace could not have been successful without a complete development of a symmetry theory. Credit to P.M. de Wolff for having considered for the first time, in 1974, the description of a structure exhibiting satellite reflections in a space of higher dimension with the example of γ-Na_2CO_3. The theory of superspace symmetry is now well established and included in the *International Tables for X-ray Crystallography*, vol. C (Janssen *et al.*, 1992). The theory is an extension of the classical space group symmetry which is used to describe conventional structures in three dimensions. A space group symmetry operation is a transformation of space

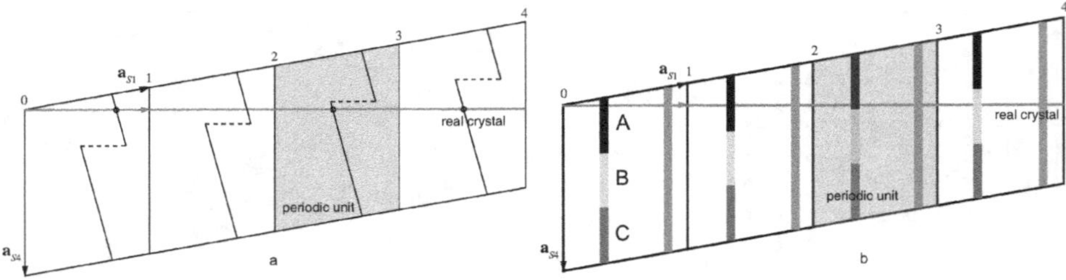

Fig. 5. Possible displacive and density modulation functions encountered in practical cases: (left) a sawtooth function; (right) a density modulation function representing different types of layers in the A, B or C positions.

leaving the density function unchanged:

$$\rho(\mathbf{r}) = \rho(R\mathbf{r} + \mathbf{v}) \tag{6}$$

This transformation consists of a rotational part in form of the matrix R acting on a point $\mathbf{r}$ followed by a translation $\mathbf{v}$. In superspace, the corresponding expression is very similar. Here the subscript S refers to the superspace:

$$\rho_S(\mathbf{r}) = \rho(R_S\mathbf{r}_S + \mathbf{v}_S) \tag{7}$$

The matrix R_S defined in equation 7 is subject to some additional conditions, *i.e.* it should be 3+1 reducible in (3+1)D superspace:

$$R_S = \begin{pmatrix} R & 0 \\ R_M & R_I \end{pmatrix} \quad \text{where } R_M = \mathbf{q}R - R_I\mathbf{q} \tag{8}$$

In general, the modulation vector $\mathbf{q}$ (equation 2) can be decomposed into a 'rational' and 'irrational' part, $\mathbf{q} = \mathbf{q}_r + \mathbf{q}_i$. For example, the rational part of $\mathbf{q} = (\alpha 0 \tfrac{1}{2})$ is $\tfrac{1}{2}$. It can be shown that for all symmetry operation in superspace, the following condition holds

$$\mathbf{q}_i R - R_I\mathbf{q}_i = 0 \tag{9}$$

With the two-component formulation, the 'external' and 'internal' space dimensions are separated and we obtain the following relation by omitting the indices for the external variables:

$$\rho_S(\mathbf{r}, \mathbf{r}_I) = \rho_S(R\mathbf{r} + \mathbf{v},\ R_I\mathbf{r}_I + \mathbf{v}_I) \tag{10}$$

In this expression, it is assumed that the selection of the pair of operators R and R_I uniquely defined the R_M term. We can gain some further simplifications by limiting ourselves to superspace transformations in (3+1)D. In this case, the internal dimension is 1 and $R_I = \varepsilon = \pm\ 1$. By definition

$$\mathbf{r}_I = t = x_4 - \mathbf{q}\cdot\mathbf{r} \tag{11}$$

(as illustrated in Fig. 4) and it can be shown that $\mathbf{v}_I = \tau - \mathbf{q}_i\cdot\mathbf{v}$. By substituting relation 11, the superspace transformation in (3+1)D becomes

$$\rho_S(\mathbf{r}, t) = \rho_S(R\mathbf{r} + \mathbf{v},\ \varepsilon t + \tau - \mathbf{q}_I\cdot\mathbf{v}) \tag{12}$$

From this expression, we can conclude that any superspace transformation consists of two parts, the external part which is similar to one of the 230 space groups in 3D in addition to an internal part. In (3+1)D, the value of ε can be deduced easily from the transformation of modulation vector $\mathbf{q}$ and thus only τ needs to be specified. With this in mind, we can generalize the 3D space group symbols by adding the internal components in order to create the corresponding superspace symbol.

As an example, let us consider a (3+1)D superspace group taken from the reference of the *International Tables for X-ray Crystallography*, vol. C (Janssen *et al.*, 1992). The symbol of the superspace group (SSG) 48.2 is $Pnnn(00\gamma)s00$. We immediately recognize the first (external) part of the 3D orthorhombic space group $Pnnn$. The second part between parentheses gives the components of the modulation vector $\mathbf{q}$ (equation 2).

Obviously, this vector is parallel to the **c** axis. The last part characterizes the internal space components of the three symmetry operations listed after the letter *P*. The symbol *s* indicates that the internal component τ of the first glide plane *n* (normal to **a**) is ½ resp. 0 for the second glide plane normal to **b**. The third *n* glide refers to the irrational part of the modulation vector. The value of τ depends on the selection of the origin of the internal space, which can be freely selected. Thus, the last digit of the symbol (0) indicates that the origin was selected on the inversion center at (¼¼¼).

Similar to 3D, where the glide components generate systematic absences in the diffraction patterns, this is also the case in superspace for both the external and internal translation components. The *International Tables for X-ray Crystallography*, vol. C (Janssen *et al.*, 1992), lists the following systematic absences for this superspace group:

$0klm$: $k+l+m = 2n$ $h0lm$: $h+l = 2n$ $hk00$: $h+k = 2n$

In most cases, the systematic absences of modulated crystals reduce the number of possible superspace groups to a unique case or generally to a very small number of alternatives.

The number of four-dimensional space groups is 4894 whereas the number of (3+1)D superspace groups is 775. Apparently (3+1)D are four-dimensional space groups with additional restrictions associated with the different properties of the external and internal space dimensions. For example, in 4D, one might consider permutations between the four space dimensions similar to the permutations of the three dimensions in the cubic space groups. This is not possible in superspace where the internal and external dimensions cannot be permuted, therefore the systematic use of the (3+1)D notation.

For simplicity, we have limited our considerations to (3+1)D superspace examples. Needless to say higher-dimensional superspace groups up to (3+3)D have not only been tabulated in the literature (e.g. Stokes *et al.*, 2011) but they have also been applied to examples of crystalline structures. In the next section, we shall deal with the measurement and solution of modulated structures in superspace.

4. Measurements and resolution of modulated structures from diffraction data

Similar to treatment of symmetry in superspace, which is an extension of the symmetry in 3D, the derivation of the electron density relation $\rho(\mathbf{r})$ in superspace is also based on an extension of the density in 3D. Independently of the rank (equation 5) of **H**, the electron density satisfies the following relation:

$$\rho(\mathbf{r}) = 1/v\Sigma_{\mathbf{H}}F_{\mathbf{H}}\ \exp(-2\pi\mathbf{i}\mathbf{H}\cdot\mathbf{r}) \qquad (13)$$

The only difference with the 3D expression is that **H** and **r** have *n* components, where *n* is the rank of **H**.

The structure factor of a structure described in (3+1)D superspace is given by the following expression:

$$F_{\mathbf{H}} = \Sigma_{\mu}f^{\mu}_{\mathbf{H}}\exp(-2\pi\mathbf{i}\mathbf{H}\cdot\mathbf{r}^{\mu}) \times \int_0^1 \mathrm{d}t\ p^{\mu}(t)\exp[\exp(-2\pi\mathbf{i}\mathbf{H}\cdot\mathbf{r}^{\mu})] \qquad (14)$$

which essentially consists of two parts. The first part is very similar to the classical treatment of the structure factor with a summation over the μ atoms in the unit cell defined in superspace. The second part deals with the internal space dimension. The integration is performed over the internal variable, *t*, defined in equation 11; p^{μ} is a population parameter, which might also express a density modulation function and the vector $\mathbf{u}^{\mu}$ gives the displacement of the μ atom from the average position along the internal coordinate *t* as illustrated in Fig. 4. In practical cases, this integration is replaced by a numerical approximation and the shape of the displacement curve is obtained during the structure solution and refinement process.

The current generation of automatic diffractometers is, in most cases, able to collect a complete set of intensities with up to six indices. This has been greatly facilitated by the diffusion of area detectors like CCD or pixel detectors. The existence of satellite reflections can be detected easily *a posteriori* from the reconstruction of diffraction layers thus allowing the identification of the modulation vector(s) and consequently the indexing of the intensities once the full data collection is performed.

Once the intensities are collected and corrected for the usual factors, the next step in the structural analysis is to solve the structure. We shall describe two methods which allow us to obtain good model structures. The first method can be considered as a trial and error method and consists of two steps. The first is to solve the so-called average structure, which is obtained by ignoring the satellite reflections. The symmetry of this structure belongs to one of the classical 230 space groups and therefore the classical methods (Patterson, direct, etc.) can be applied to derive the first structural approximation. The second part assumes that the superspace group symmetry is known from the systematic absences and uses the least-squares refinement to complete the structure by finding the modulation functions of each single atom by trial and error. In order to find the optimal solution, various modulation functions can be applied until the most plausible model satisfying the structural and chemical characteristics is found. The final step of the structure solution is obtained following the full convergence of the optimization process.

One should also mention here that the Patterson and direct methods have also been applied in higher dimensions in the early days of developments. Currently, they are used infrequently and will not be further described in this context.

The second and perhaps more elegant method is based on the charge flipping (CF) algorithm proposed by Oszlányi and Sütó (2004, 2005). The great advantage of this method is that the structure can be solved directly in superspace without the intermediary step in 3D as in the previous method. The following scheme (equation 15) illustrates the CF algorithm which is an iterative process consisting of four steps. The algorithm starts from the square root of the intensity measurements on an arbitrary scale representing the moduli of the structure factors. Initially the phases of the structure factors are missing and consequently replaced by arbitrary or zero phases. The first step uses the inverse Fourier transform of the structure factors to obtain an arbitrary density function in superspace. Of course, this density should be either positive or zero everywhere. The second step (the inversion step) replaces every density point, which is smaller than a small positive value, δ, by its inverse value in order to obtain the density $g_{\mathbf{r}}$. In the third step, this density is Fourier transformed in order to obtain a new set of structure factors, $G_{\mathbf{H}}$. The last step of the algorithm is to substitute the modulus of $G_{\mathbf{H}}$ by the experimental moduli. This process is repeated iteratively until the structure is solved.

$$\begin{array}{ccc} \delta_{\mathbf{r}} & \xrightarrow[\rho<\delta]{\text{Inversion}} & g_{\mathbf{r}} \\ FFT^{-1}\uparrow & & \downarrow FFT \\ F_{\mathbf{H}} & \xleftarrow[\text{phases}]{\text{retain}} & G_{\mathbf{H}} \end{array} \tag{15}$$

In most cases, the convergence is obtained after some hundred iterations or a small multiple of it. The only free parameter is the magnitude, δ. This parameter is linked to the standard deviation of the measurements and can easily be obtained *a priori*.

Another advantage of this method is that the superspace group symmetry of the structure is not required. Therefore, it can be deduced *a posteriori* and checked by the final least-squares refinement of the structure.

Before closing this section, we shall describe the results of a structural refinement in superspace with the example of an oxygen-deficient synthetic compound with composition $LiZnNb_4O_{11.5}$ (Morozov *et al.*, 2010). This structure is a superstructure of α-PbO_2. However, it is more convenient to treat it as a commensurately modulated structure with lattice parameters $a = 4.72$ Å, $b = 5.73$, $c = 5.03$ Å, $\gamma = 90.05°$, modulation vector $\mathbf{q} = 0.3\mathbf{a}^* + 1.1\mathbf{b}^*$ and superspace group $P112_1/n(\alpha\beta0)00$. The treatment in superspace has many advantages, one of which is to allow for a unified description and treatment of all members of the family of compounds Li_2O-ZnO-Nb_2O_5, including commensurate ($LiNb_3O_8$, $ZnNb_2O_6$, $ZnTa_2O_6$) and incommensurate (*e.g.*, $ZnTa_2O_6$) members with the same superspace symmetry. Another important advantage concerns the solution and refinement of this complex structure, which is greatly facilitated in superspace and reduces the number of refined parameters.

The model of the resulting structure determination is illustrated in Fig. 6. The unit cell in the solid line corresponds to the α-PbO_2 cell whereas the dotted line represents the 10-fold supercell. Four different types of metal-oxygen octahedra can be distinguished each containing in its centre either a single atom (dark blue: ZnO_6 or red: NbO_6) or a combination of them (yellow: various combinations of Li and Nb or light blue: ⅓Zn and ⅔Nb). The different types of octahedra form layers normal to the modulation vector $\mathbf{q}$ and the structure can be considered as a density modulation resulting from the distribution of cations, which tend to be ordered.

The structure was solved from powder diffraction data collected on a synchrotron beamline with the *Jana 2006* software (Petříček *et al.*, 2006). Figure 7 illustrates various sections of the electron density of the positional modulations of Li, Zn and Nb along the internal coordinate x_4. The colour code of the positional modulation is identical to that used for metal octahedra represented in Fig. 6. The size of the domains of existence corresponding to each type of atoms was constrained to the chemical composition known beforehand during the refinement. The largest final residual electron density values were 0.17–0.23 e/Å^3 thus illustrating the high quality of the refinement process in superspace even for powder diffraction data.

5. A reinterpretation of the law of rational indices and occurrences of aperiodic structures

Coming back to the work of GPP (1931), can we accept the conclusion that the law of rational indices is not of general validity? It would be disappointing that a law, which has been well accepted and confirmed many times over a period of 150 years, is only of limited value. The very elegant solution proposed by Janner and Dam (1989), which definitely solved the puzzle

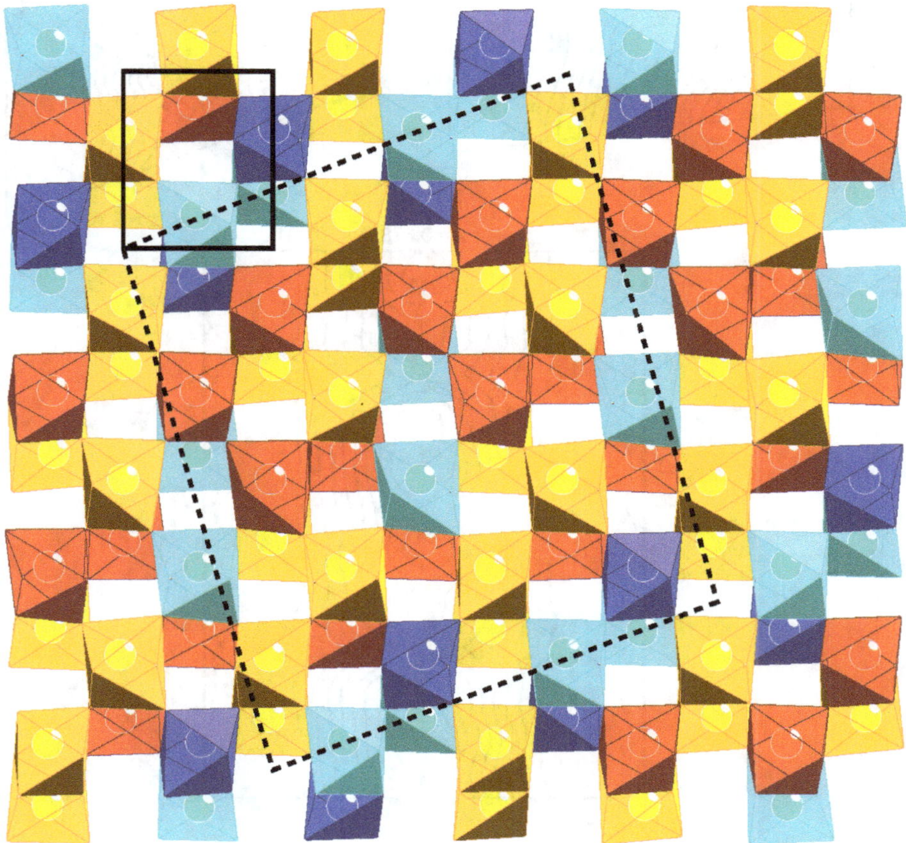

Fig. 6. Polyhedral view of $LiZnNb_4O_{11.5}$ (Morozov *et al.*, 2010). Dark blue: ZnO_6; red: NbO_6; yellow: various concentrations of LiO_6 and NbO_6; light blue: ⅓ZnO_6 and ⅔NbO_6 polyhdra. The small cell (solid line) corresponds to the α-PbO_2 cell whereas the large cell (dashed line) illustrates the supercell.

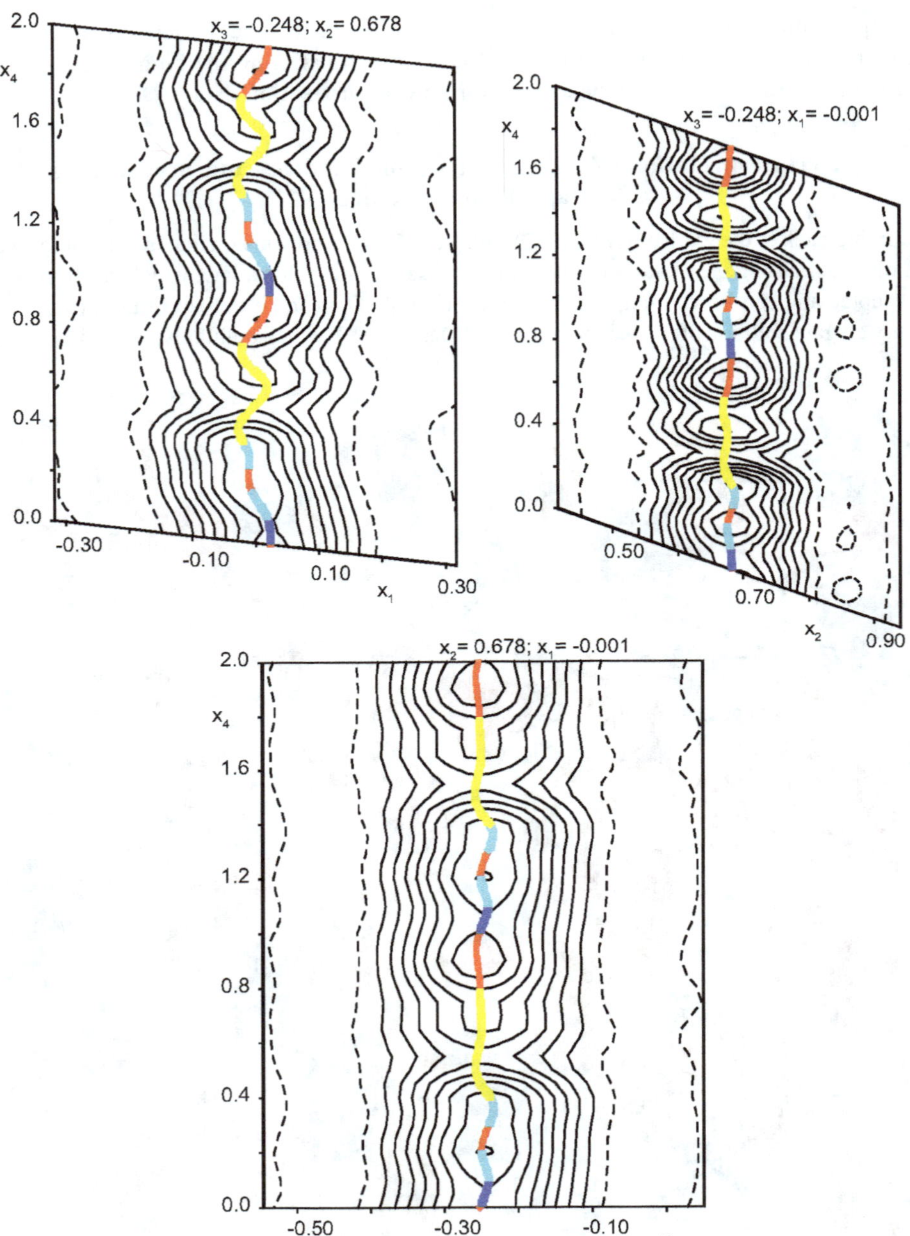

Fig. 7. Superspace representation of various electron density sections indicating the positional modulations of Li, Zn and Nb in $LiZnNb_4O_{11.5}$. The central coloured lines correspond to the calculated atomic positions with the same colour code as in Fig. 6.

of the calaverite mineral, shows that it is possible to accept the law of rational indices provided that we extend the number of indices characterizing each face (Fig. 2) of a crystal to its rank as defined in equation 5. This approach avoids not only the use of irrational indices but also the need to postulate the existence of a few interpenetrating structures belonging to different crystal classes in order to fully describe a crystalline sample. The example of calaverite is not an exception and many examples of crystal forms of modulated structures and quasicrystals have been found which require more than three indices to characterize their faces. For example, one of the phases of the metal–organic structure of $[(CH_4)_3N]_2ZnCl_4$ (Dam and Janner, 1986) which is incommensurately modulated exhibits faces which can only be indexed with four indices. We shall describe below the occurrence of the first naturally occurring quasicrystal, the mineral icosahedrite (Bindi *et al.*, 2011), the faces of which can only be indexed with six indices. The interested reader can find more examples in the textbook of Janssen *et al.* (2007).

Note that the latest developments in diffraction methods with synchrotron and highly sensitive area detectors do not favour the discovery of new aperiodic crystalline forms. Firstly, the size of the samples, which tend to reach micron scales, and the use of automatic procedures to measure samples tend to ignore the morphological aspects of crystalline forms. On the other hand, the generalization of area detectors has been an important factor for the discovery of new aperiodic structures based on single-crystal diffraction patterns. The observation and identification of satellite reflections is greatly facilitated by its use instead of point detectors and it is probably due to their diffusion that the field of aperiodic structures has recently gained some importance. The identification of aperiodic structures by powder diffraction is more difficult. The detection of weak satellite reflections is often overseen and consequently the resulting structure refinement yields models which are often and wrongly interpreted as 'disordered'. As an example, a recent study on a series of scheelite-like structures (Arakcheeva and Chapuis, 2008) has revealed that some powder diagrams published in the ICDD database were indeed incommensurately modulated structures and the modulation vectors could be derived directly from the inclusion of a few neglected peaks and a reinterpretation of the powder diagrams. These examples are by no means unique and a more careful analysis of diffraction patterns of structures with complex stoichiometry would certainly reveal many more cases.

Aperiodic structures exist in all types of crystalline materials, from organic to inorganic compounds, in minerals, in metals and alloys and even in macromolecules. They also appear in compounds subject to temperature and pressure changes. In the next sections, we shall present some specific studies of aperiodic structures occurring in nature.

6. Naturally occurring aperiodic structures

In nature, minerals with strong and sharp satellites related to an aperiodic structure which can be used for a structural refinement with superspace formalism are rarely found. Nevertheless, there are several minerals with incommensurate satellites visible with electron diffraction. Indeed, incommensurately modulated structures can be found by

TEM techniques in several rock-forming minerals including: clinopyroxenes (kosmochlor–diopside join; Sakamoto *et al.*, 2003), quartz (*e.g.* Heaney and Veblen, 1991), feldspars (*e.g.* Yamamoto *et al.*, 1984; Steurer and Jagodzinski, 1988; Kalning *et al.*, 1997; Sanchez-Munoz *et al.*, 1998), melilite (Bindi *et al.*, 2001a and references therein), fresnoite (Bindi *et al.*, 2006d), cancrinite–sodalite join (*e.g.* Hassan and Buseck, 1989, 1992; Xu and Veblen, 1995; Hassan, 2000; Hassan *et al.*, 2004; Bolotina, 2006; Bolotina *et al.*, 2006), nepheline (Withers *et al.*, 1998; Angel *et al.*, 2008; Friese *et al.*, 2011), lazurite (Rastsvetaeva *et al.*, 2002; Bolotina *et al.*, 2003a,b, 2004), trydimite (*e.g.* Pryde and Dove, 1998), and mullite (Angel *et al.*, 1991). Incommensurate modulations can also be found among sulfides, tellurides, and sulfosalts including bornite (Buseck and Cowley, 1983), pyrrhotite (Li and Franzen, 1996), calaverite (*e.g.* Caracas and Gonze, 2001), sylvanite (Krutzen and Inglesfield, 1990), rickardite (Schutte and de Boer, 1993), muthmannite (Bindi, 2008), proustite (Subramanian *et al.*, 2000), cylindrite–franckeite series (Evain *et al.*, 2006c; Makovicky *et al.*, 2008, 2011) and sartorite (Pring *et al.*, 1993). However, only a few incommensurate structures of minerals have been refined by means of the multidimensional approach namely: labradorite (Yamamoto *et al.*, 1984), andesine (Steurer and Jagodzinski, 1988), melilite (Bindi *et al.*, 2001a), lazurite (Bolotina *et al.*, 2004 and references therein), fresnoite (Bindi *et al.*, 2006d), levyclaudite (Evain *et al.*, 2006c), calaverite (Bindi *et al.*, 2009a), natrite (Arakcheeva *et al.*, 2010), franckeite (Makovicky *et al.*, 2011), nepheline (Friese *et al.*, 2011) and wagnerite-group minerals (Lazic *et al.*, 2014).

In the next part of this chapter, attention is focused on the description of outstanding examples of well characterized aperiodic mineral crystal structures. Among the incommensurately modulated structures we describe in detail the (3+1)D-modulated structure of calaverite and natrite and the (3+2)D-modulated structure of melilite and fresnoite; for the composite or intergrowth crystal structures, we analyse the minerals belonging to the pearceite–polybasite group and those of the cylindrite group, and, finally, for quasicrystals, we report briefly the story of the discovery of icosahedrite, the first naturally occurring quasicrystal.

7. Naturally occurring incommensurately modulated structures

7.1. (3+1)D incommensurately modulated structure: the cases of calaverite and natrite

Calaverite, $AuTe_2$, and natrite, Na_2CO_3, are probably the two most important minerals in the history of aperiodic structures. Calaverite, first identified by Ghent (1861), has already been mentioned in the introduction because of the difficulties mineralogists encountered in indexing crystalline specimens from different locations. It was later found that the diffraction pattern of calaverite exhibited satellite reflections, which pointed to the presence of an incommensurately modulated structure (Chapuis, 2003, and references therein).

On the basis of a transmission electron microscopy study, Van Tendeloo *et al.* (1983) discovered that the satellite reflections in calaverite were due to an incommensurately displacive modulation superimposed on the average *C*2/*m* structure. Using the information gained from the TEM study, Dam *et al.* (1985) were able to clarify the morphology of calaverite by indexing the crystal faces with four indices, the extra index applying to $\mathbf{q} = -0.4095\mathbf{a}^* + 0.4492\mathbf{c}^*$. Finally, Schutte and de Boer (1988) solved the incommensurately modulated structure of calaverite in the superspace group $C2/m(\alpha 0\gamma)0s$. These authors showed that the modulation consists mainly of displacements of Te atoms and the observed modulations were interpreted in terms of valence fluctuations between Au^+ and Au^{3+}. Later, Balzuweit *et al.* (1993) studied the variation of the modulation wave vector in synthetic calaverite as a function of temperature and silver content (up to 3.0 at.% of Ag) using X-ray diffraction and morphological techniques. More recently, Bindi *et al.* (2009a) studied various natural calaverite samples from different locations with chemical compositions $Au_{1-x}Ag_xTe_2$, with $0 \leqslant x \leqslant 0.33$, extending the study of Schutte and de Boer (1988), and shedding new light on the origin of the structural modulations observed in this family of compounds. By means of higher dimensional structure refinements, Bindi *et al.* (2009a) showed that Ag atoms present two types of distribution on the Au sites, either randomly distributed or ordered. The original study of Schutte and de Boer (1988) was concerned with an ordered distribution of Ag, which induces very strong valence fluctuations of Au in the structure. Bindi *et al.* (2009a) demonstrated another aspect of the $Au_{1-x}Ag_xTe_2$ compounds, specifically those with random distributions of Ag on the Au position. By combining the results of both studies, these authors contended that the valence fluctuation of Au produces the structure modulation in $AuTe_2$ (Fig. 8). Moreover, the ordered distribution of silver found by Bindi *et al.* (2009a) reinforces the valence fluctuation of Au. Conversely, a random distribution of Ag suppresses the valence fluctuation of Au and, therefore, their structural modulations. Apparently, natural samples can be found with the same composition but with both types of Ag ordering, which is not surprising considering the various growth conditions of the minerals.

Another interesting aspect of the $Au_{1-x}Ag_xTe_2$ series pointed out by Bindi *et al.* (2009a) is related to the (3+1)D superspace embedding of the calaverite structure type. These authors have shown that the same superspace model, independently of their 3D structures and periodicities, can describe all the samples with the same general formula. This formalism opens new perspectives, not only for a uniform description of this family of minerals, but also for other synthetic compounds related to the calaverite structure.

We mentioned earlier that the structure determination of Na_2CO_3 (Brouns *et al.*, 1964; van Aalst *et al.*, 1976) was essential in the development of the superspace approach. The structure of synthetic γ-Na_2CO_3 was later refined with greater accuracy using single-crystal XRD data (Dušek *et al.*, 2003).

The first observation of satellite reflections pointing to an incommensurately modulated structure in natural natrite was by Zubkova *et al.* (2002) during a single-crystal investigation of a crystal from Khibiny, Russia. However, the satellites were not measured and the authors refined only the average structure of this mineral sample. More recently, the incommensurate modulated structure of natural natrite, originating

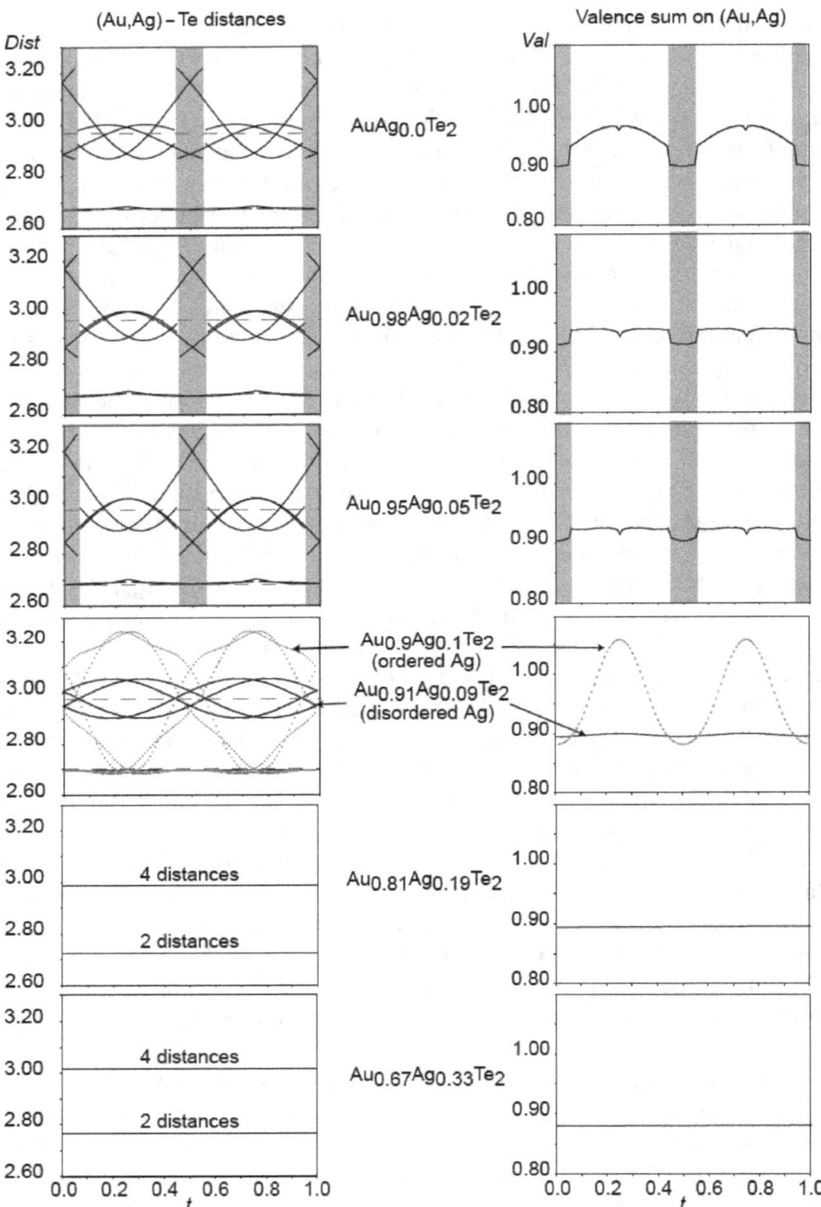

Fig. 8. Variation of the (Au,Ag)–Te distances (left column) and the bond valence sums on (Au,Ag) (right column) as functions of *t* in Ag-bearing calaverites (Bindi *et al.*, 2009a). Solid lines correspond to data from Bindi *et al.* (2009a); dashed graphs have been reconstructed using data published by Schutte and de Boer (1988). The grey fields indicate regions with CN = 2 + 2. Figure reproduced from Bindi *et al.* (2009a) with the permission of the Mineralogical Society of America.

from the Lovozero alkaline massif (Mt. Karnasurt) and the Khibiny alkaline massif (Mt. Koashva), was studied in detail by Arakcheeva *et al.* (2010) using the superspace approach. The average structure of natrite can be mainly described as graphite-like layers formed by Na3 and CO_3 ions, stacked along the **c** axis (Fig. 9). Additional Na1,2 octahedral-ions are located in the pseudo-hexagonal channels (van Aalst *et al.*, 1976; Zubkova *et al.*, 2002; Dušek *et al.*, 2003). The face-sharing Na1,2 octahedra form columns which are connected by CO_3 triangles along [010] and [100]. The modulation vector can be written as $\mathbf{q} = 0.182(1)\mathbf{a}^* + 0.322(1)\mathbf{c}^*$ and the superspace group is $C2/m(\alpha 0 \gamma)0s$ (Arakcheeva *et al.*, 2010).

The modulation in natrite is mainly associated with the Na crystal-chemical environment and with its role in participating in the second coordination sphere of the carbon atoms. As can be deduced from Fig. 10, the coordination number (CN) of Na1,2 can be better described as 4+2, where 4 distances are shorter than 2.3 Å, and 2 distances are ~2.4 Å. The Na3 site has been reported as a 7+2-fold coordinated with 7 Na–O distances in the range 2.584(1) – 2.669(1) Å and two elongated Na–O distances at 2.942(1) Å (Zubkova *et al.*, 2002). It is evident from the modulation of the Na3–O distances (Fig. 10) that eight oxygen atoms are always close to Na3 (in the range 2.35–2.9 Å) without any preference for CN = 7. Therefore, the coordination number of Na3 is equal to 8.

One final interesting remark concerns the CO_3 group and its usual planarity in the crystal structures. This group seemed to be a completely rigid unit in the average structure of natrite refined by Zubkova *et al.* (2002). Arakcheeva *et al.* (2010), however, showed that the C–O distances and the C–O2–O1–O1 torsion angle (describing the deviation from planarity of the CO_3 entity) are slightly modulated along *t*, the additional coordinate in superspace. The reasons for such a modulation were described in detail by Arakcheeva and Chapuis (2005) for synthetic Na_2CO_3. A small deviation

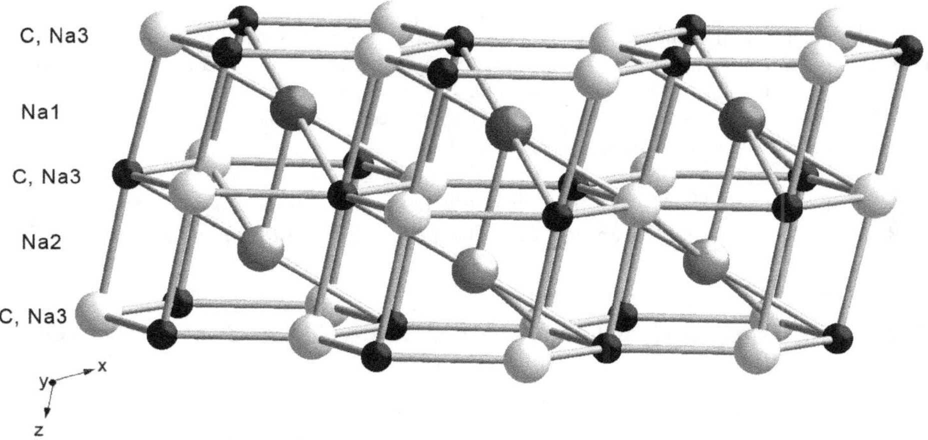

Fig. 9. Structure of the hexagonal modification of natrite, Na_2CO_3. The orientation of the structure is outlined.

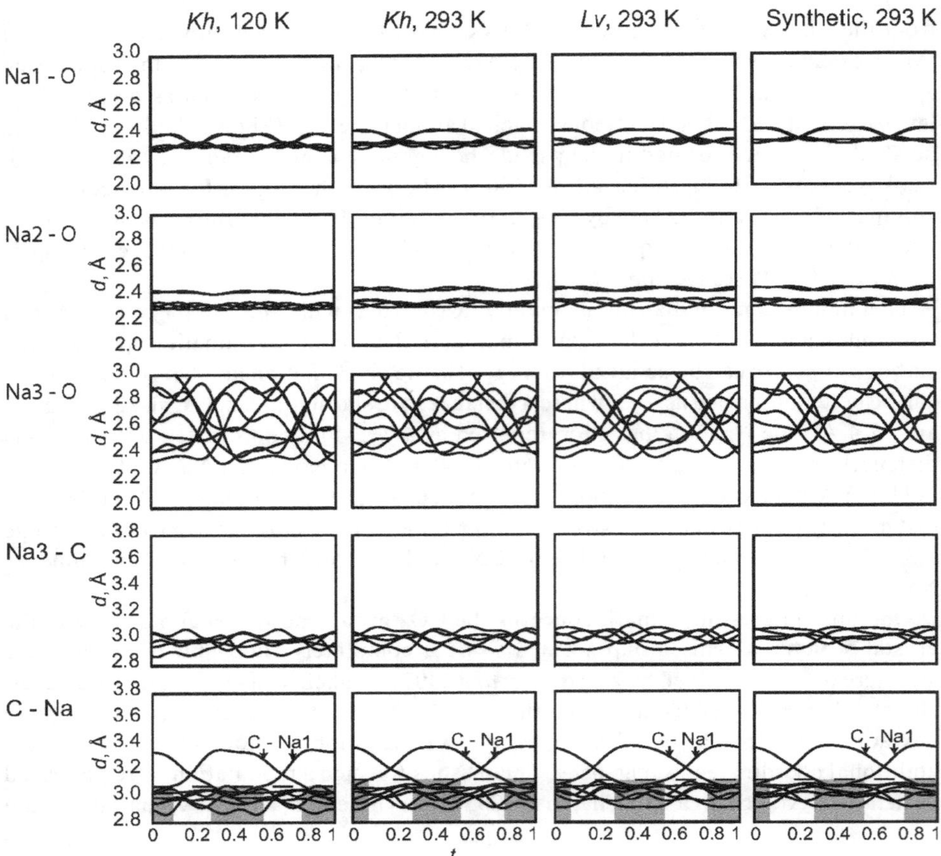

Fig. 10. Modulations of the interatomic distances along the *t* axis in the structure of natrite from Khibiny (at both RT and 120 K) and Lovozero, compared with the behaviour observed for synthetic Na_2CO_3 (Arakcheeva and Chapuis, 2005 – last column). In the C-Na *vs. t* plots, the grey areas mark the regions with 7 Na atoms in the 3.1 Å vicinity of each C (reproduced from Arakcheeva *et al.* (2010) with the permission of the Mineralogical Society of America).

(1–2°) of the C–O2–O1–O1 torsion angle from 0° should also be mentioned, thus indicating positional displacements of O atoms, resulting in the small distortions of the CO_3 group.

7.2. (3+2)D incommensurately modulated structure: the cases of melilite and fresnoite

Minerals of the melilite group exhibit a general chemical formula $X_2T1[(T2)_2O_7]$, where the cations X occupy square antiprisms and consist mainly of Ca with minor amounts of Na and other elements (*i.e.* mainly Ba and Sr). $T1$ and $T2$ tetrahedra differ from each other by their size: the larger $T1$ generally hosts Mg and other medium-size tetrahedral cations such as Fe^{2+}, Mn^{2+}, Fe^{3+}, and Al; the smaller *T2* tetrahedron is generally occupied by Si to form the $(T2)_2O_7$ groups that typically occur in sorosilicates. The structure with space group $P\bar{4}2_1m$ consists of a two-dimensional linkage of corner sharing *T1* and *T2* tetrahedra, forming irregular pentagonal rings (Fig. 11). Some synthetic melilite-type compounds with *X* occupied by calcium and *T2* by silicon, with *T1* = Mg, Zn, Co, or Fe^{2+}, exhibit an incommensurately modulated phase at room temperature (Hemingway *et al.*, 1986; Seifert *et al.*, 1987; Armbruster *et al.*, 1990; Hagiya *et al.*, 1993).

The incommensurate modulation in melilite is two-dimensional, with two modulation vectors: $\mathbf{q}_1 = \alpha(-\mathbf{a}_1^* + \mathbf{a}_2^*)$ and $\mathbf{q}_2 = \alpha(\mathbf{a}_1^* + \mathbf{a}_2^*)$, where $\mathbf{a}_1^*$ and $\mathbf{a}_2^*$ are the tetragonal reciprocal axes of the basic unit cell and $\alpha = 0.2815(3)$. As a consequence, all the reflections can be indexed by five integers, *hklmn*, corresponding to the five-dimensional basis (equation 5) $\mathbf{H} = h\mathbf{a}^* + k\mathbf{b}^* + l\mathbf{c}^* + m\mathbf{q}_1 + n\mathbf{q}_2$, where $\mathbf{a}^*$, $\mathbf{b}^*$ and $\mathbf{c}^*$ are the reciprocal axes of the basic structure. The direction of the modulation vectors is along 110 and $\bar{1}10$ and the coefficient α is approximately 2/7 of the corresponding *d* value (Fig. 12).

A similar incommensurate modulation was reported for fresnoite-type compounds (Markgraf and Bhalla, 1989; Markgraf *et al.*, 1990; Höche *et al.*, 1999). The mineral fresnoite, $Ba_2TiSi_2O_8$, is structurally related to the melilite-group minerals. The parent structure, space group *P4bm*, consists of a two-dimensional array of corner-sharing TiO_5 pyramids and Si_2O_7 groups. In the stacking of successive layers along [001], the ten-fold coordinated Ba cations are located about halfway between adjacent sheets (Fig. 13). Withers *et al.* (2002) investigated the incommensurate structure of a synthetic fresnoite by means of electron microscopy and pointed out that the primary modulation wave-vector has a ½**c*** commensurate component in addition to an incommensurate component which runs along the basis diagonal of the parent structure with modulation vectors $\mathbf{q}_1 = \alpha(-\mathbf{a}_1^* + \mathbf{a}_2^*)$ and $\mathbf{q}_2 = \alpha(\mathbf{a}_1^* + \mathbf{a}_2^*)$ with $\alpha = 0.3023(3)$, very similar to the distribution of satellites observed for melilites (Fig. 12) except for the commensurate component at ½**c***.

Prior to the studies by Bindi *et al.* (2001a,b, 2006d), the presence of additional satellite reflections due to the incommensurate structure had never been observed in natural melilites or fresnoite. By means of the superspace approach, these authors carried out (3+2)-dimensional refinements of the incommensurately modulated structure of both the minerals.

For natural melilite, Bindi *et al.* (2001a) showed that the most important structural variations as a function of the internal components *t* and *s* of the (3+2) superspace are observed for the X-polyhedra with some distances reaching values greater than 3 Å. Thus, some oxygen atoms are no longer coordinated locally by Ca. Due to this feature, six-, seven- and eight-fold coordinations of the *X* cation occur in different parts of the structure. A similar characteristic was observed by Bindi *et al.* (2006d) for natural fresnoite. As a consequence of the Ba and O positional modulation, eight-, nine-, and ten-fold Ba-coordinations occur throughout the structure. Interestingly, the change of coordination around the Ca (in melilite) and Ba (in fresnoite) atoms induces a deformation of the pentagonal rings, as seen from a projection of the two structures along [001] (Figs 11 and 13). The deformed pentagonal rings correspond to Ca

Fig. 11. The crystal structure of melilite projected down [001] (upper) and down [010] (lower). Light and dark grey tetrahedra refer to *T*1 and *T*2, respectively. White circles indicate the *X* cations.

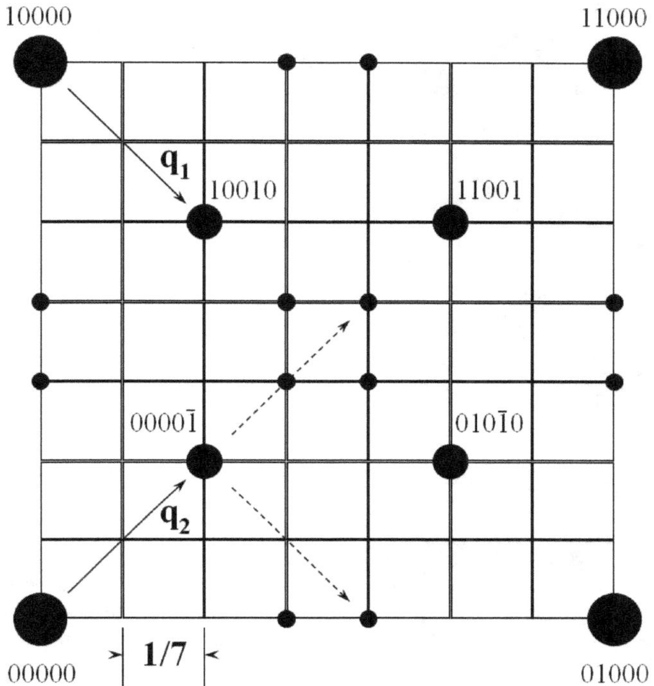

Fig. 12. Schematic view of the diffraction pattern in melilite. Main and satellite reflections (indexed on the basis of five indices) are represented with large and small black circles. The modulation **q** vectors are indicated.

(in melilite) and Ba (in fresnoite) atoms with increasing coordinations forming octagonal clusters (Fig. 14) closely resembling the configuration postulated by van Heurck *et al.* (1992), on the basis of transmission electron microscopy, for the synthetic $Ca_2ZnGe_2O_7$ melilite-type compound.

8. Naturally occurring composite modulated structures

8.1. The minerals of the pearceite–polybasite group

The members of the pearceite–polybasite group of minerals exhibit the general formula $[(Ag,Cu)_6M_2S_7][Ag_9CuS_4]$, with M = As (pearceite) or Sb (polybasite), and their structures have been characterized recently (Bindi *et al.*, 2006a,b, 2006c; Evain *et al.*, 2006a,b). On the whole, they can be described (Fig. 15) as a regular succession of two module layers stacked along the *c* axis: a first module layer *A* with composition $[(Ag,Cu)_6M_2S_7]^{2-}$, and a module layer *B* with composition $[Ag_9CuS_4]^{2+}$. In the structure, *M* atoms form isolated MS_3 pyramids typically occurring in sulfosalts, copper links two sulfur atoms in a linear coordination, and silver occupies sites with coordination ranging from quasi-linear to almost tetrahedral (Bindi *et al.*, 2006a). In

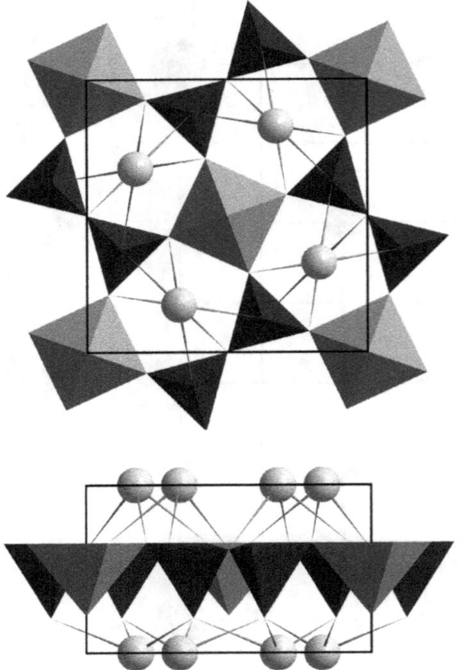

Fig. 13. The crystal structure of fresnoite projected down [001] (upper) and down [010] (lower). Interconnection of TiO_5 pyramids (light grey) and Si_2O_7 groups (dark grey); white circles represent Ba atoms.

the *B* layer the silver cations are found in various sites corresponding to the most pronounced probability density function locations of diffusion-like paths (Fig. 16, Bindi *et al.*, 2006a).

The complex crystal-chemical features observed in these minerals were initially studied by Frondel (1963) who considered that this group of minerals could be divided into two series based on their unit-cell dimensions: (1) pearceite–antimonpearceite, characterized by a relatively high Cu content and a "small" unit cell ($a \approx 7.4–7.5$, $c \approx 11.9$ Å) initially labelled '111' but more recently referred to using the more explicit polytype suffix '*Tac*' (Bindi *et al.*, 2007b), consistent with the notation of Guiner *et al.* (1984); and (2) polybasite-arsenpolybasite, characterized by a lower Cu content and doubled unit-cell parameters, initially labelled '222' but, more recently '*M2a2b2c*' (Bindi *et al.*, 2007b). Moreover, an additional unit cell of intermediate dimensions, initially labelled '221' or '*T2ac*' (Bindi *et al.*, 2007b), was discovered independently by Harris *et al.* (1965), Hall (1967), Edenharter *et al.* (1971) and Minčeva–Stefanova *et al.* (1979).

By means of LT and HT single-crystal X-ray diffraction, differential scanning calorimetry, and complex impedance spectroscopy, Bindi *et al.* (2006a) showed that all members of the pearceite–polybasite group present the same $P\bar{3}m1$ high-temperature structure and are observed at room temperature either in their high-temperature (HT) fast-ion conductivity form or in one of the low-temperature (LT) fully ordered

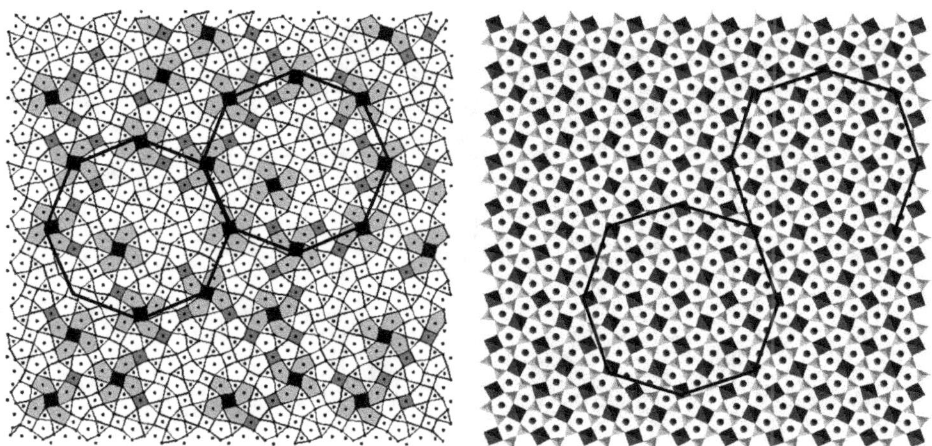

Fig. 14. Structural variations observed for the different polyhedra in melilite (left) and fresnoite (right) as a function of the fourth coordinate of the superspace (t) with the fifth coordinate (s) kept fixed at 0.00. Triangles, squares and circles indicate Si-tetrahedra, Mg-tetrahedra (left) or Ti-pyramids (right) and Ca or Ba cations, respectively. The displacements for a given value of t (or s) define a configuration that occurs somewhere in the real space. The configurations with close values of t (or s) are not necessarily neighbouring in the real space. The deformed pentagonal rings hosting the large cations (Ca in melilite and Ba in fresnoite) form octagonal arrangements indicated with solid lines (see text for explanation).

(*M*2*a*2*b*2*c* unit-cell type), partially ordered (*T*2*ac* unit-cell type) or still disordered (*Tac* unit-cell type) forms, with transition temperatures slightly above or below room temperature. Below the transition temperature, the silver ions freeze in preferred structural sites. However, the ordering is not necessarily a long-range order. In cupropearceite and cupropolybasite (Bindi *et al.*, 2007a), for example, the *Tac* unit-cell type and $P\bar{3}m1$ space group are preserved down to 100 K (Bindi *et al.*, 2006a). This peculiarity is to be related to the disorder which occurs at all temperatures within the $[(Cu,Ag)_6(As,Sb)_2S_7]^{2-}$ *A* module.

Recent TEM studies (Withers *et al.*, 2008; Bindi *et al.*, 2013) showed that these minerals exhibit a composite modulated structure, not visible either with conventional or with synchrotron X-ray diffraction. Such a composite modulated structure forms between two inherently incommensurate substructures: the first being an average primitive hexagonal, $[Ag_9]^{9+}$, 2D Ag^+ ion sub-structure (consisting of the Ag^+ ion diffusion paths within the *B* modules – Fig. 16) and the second a $[(Ag,Cu)_6(As,Sb)_2S_7]^{2-}[CuS_4]^{7-}$ framework sub-structure (formed by the $[S(Cu,Ag)_6]$ octahedra, the $(As,Sb)S_3$ trigonal pyramids, and the [S-Cu-S] dumbbells, Fig. 17).

The composite modulation is evident with the appearance of relatively strong and well defined satellite reflections in addition to the main reflections forming a hexagonal reciprocal lattice (Fig. 18). At first glance, the observed satellites seems to suggest a 2D displacive modulation of the framework structure with incommensurate modulation wave vectors of the

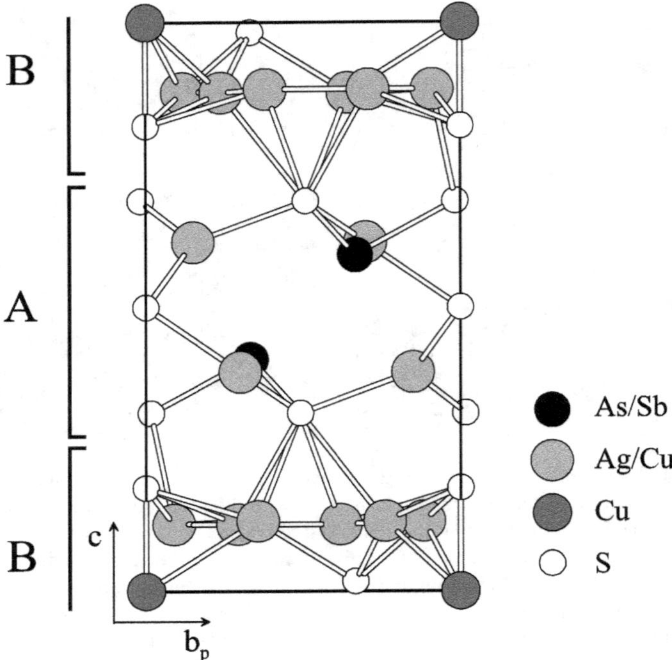

Fig. 15. Projection of the 111-structure (*Tac* polytype) of pearceite along the hexagonal *a* axis, emphasizing the succession of the $[(Ag,Cu)_6M_2S_7]^{2-}$ *A* and $[Ag_9CuS_4]^{2+}$ *B* module layers.

satellite reflections $\mathbf{q}_1 = \alpha(\mathbf{a}^*_F + \mathbf{b}^*_F)$ and $\mathbf{q}_2 = \alpha(\mathbf{a}^*_F - \mathbf{b}^*_F)$, where $\alpha \approx 0.39(1)$ and the subscript F indicates the framework structure. However, as outlined in detail by Withers *et al.* (2008) and Bindi *et al.* (2013), the apparent incommensurate modulation vectors can be interpreted more reasonably as "fundamental reciprocal lattice basis vectors of an average primitive hexagonal Ag ion substructure", with the basis vectors $\mathbf{a}^*_{Ag} = \alpha(\mathbf{a}^*_F - 2\mathbf{b}^*_F)$ and $\mathbf{b}^*_{Ag} = \alpha(2\mathbf{a}^*_F - \mathbf{b}^*_F)$ with $\alpha \approx 1.39$. This interpretation is mainly due to the observed strong intensity asymmetry of the satellite reflections (see arrows in Fig. 19) surrounding individual parent Bragg reflections, with the satellites at the high-angle side showing much higher intensity than those of the low-angle side, which are scarcely discernible. Several authors (Carter and Withers, 2004; Norén *et al.*, 2005) have provided conclusive proof of the incompatibility of this observation with a conventional incommensurate modulation and suggested, instead, a composite modulated structure.

Bindi *et al.* (2013) have studied by means of transmission electron microscopy the low- (90 K) and ultra-low- (4.2 K) temperature behaviour of the Cu-rich members of the pearceite–polybasite group, *i.e.* cupropearceite and cupropolybasite. They found a similar behaviour to that observed for Cu-poor pearceite and polybasite (*Tac* members) at room temperature (Withers *et al.*, 2008), but a new structural feature appeared at ultra-low temperature (Fig. 20). Estimates of the relevant wave vectors $\mathbf{q}_1 = \alpha(\mathbf{a}^*_F + \mathbf{b}^*_F)$ and $\mathbf{q}_2 = \alpha(\mathbf{a}^*_F - \mathbf{b}^*_F)$ of the incommensurate modulation, indeed, gave approximate α

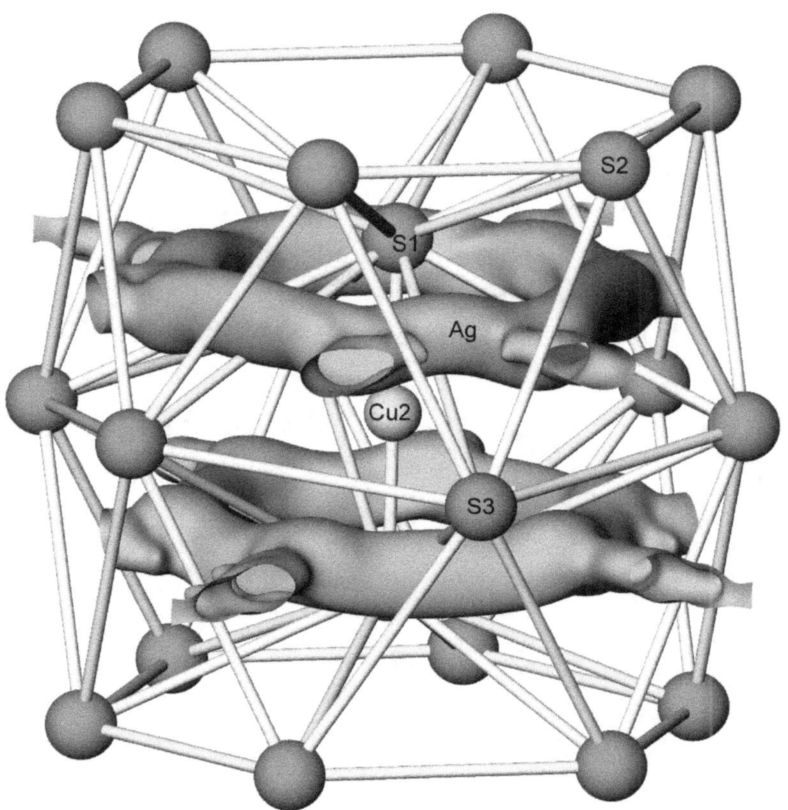

Fig. 16. Non-harmonic joint probability density isosurface for silver in the pearceite structure at RT in the $[Ag_9CuS_4]^{2+}$ layer extended with S2 atoms, exhibiting the silver diffusion in the **ab** plane (Bindi *et al.*, 2006a). Level of the map: 0.05 Å^{-3}. Sulfur and copper atoms at an arbitrary size.

values of 0.39 at room temperature, 0.40 at $T = 90$ K, and 0.50 at $T = 4.2$ K. The remarkable displacement noticed in the positions of the satellite reflections with decreasing temperature means that the modulation at 4.2 K approaches the wave vector $\alpha = 0.50$, which would correspond to a commensurate superstructure. Such behaviour is known from other compounds as, for example, certain melilite-type compounds where a transformation from the incommensurate high-temperature phase into a low-temperature commensurate lock-in phase has been observed (Riester *et al.*, 2000; Hagiya *et al.*, 2001; Schaper *et al.*, 2001; Bagautdinov *et al.*, 2002; Jia *et al.*, 2006).

8.2. The cylindrite-type minerals

Cylindrite-type minerals, which take their name from the typical cylindrical morphology of the crystals, are known as 'two-dimensional misfit' structures

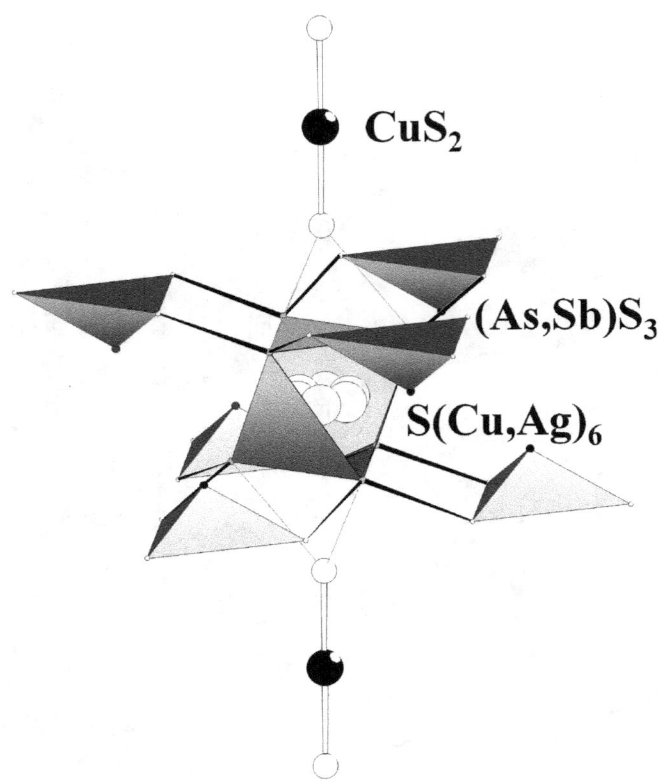

Fig. 17. A portion of the overall structure highlighting the connectivity of the fundamental building blocks [*i.e.* CuS_2, $(As,Sb)S_3$ and $S(Ag,Cu)_6$] of the framework substructure in the average $P\bar{3}m1$ crystal structure (*Tac* polytype) of pearceite and polybasite (reproduced from Withers *et al.* (2008) with the permission of Elsevier).

(Makovicky and Hyde, 1981, 1992; Evain *et al.*, 2006c; Makovicky *et al.*, 2008, 2011). According to Wiegers and Meerschaut (1992), this kind of modulated compound exhibits complex layered crystal structures of the composite type, in which there is the regular alternation of two types of layers, having one or two of their in-plane parameters in an incommensurate ratio. In these minerals, the two layers (Fig. 21) are a combination of a pseudo-quadratic layer (labelled Q) with a pseudo-hexagonal one (labelled H; Makovicky and Hyde, 1981). In general, the Q layer is a (100) slab of the PbS/NaCl archetype, two to four atoms thick (*e.g.* 2 in cylindrite, 4 in franckeite), whereas the H layer is more variable, corresponding to different kinds of layer (*i.e.* CdI_2-type, tetradymite-type or NbS_2-type). The relative orientation of the Q and H layers depends on the relative sizes of their constitutive cations (Fig. 22).

Three members of the cylindrite group have recently been characterized structurally by means of a (3+2)-dimensional superspace approach: levyclaudite, $\sim Cu_3Pb_8Sn_7(Bi,Sb)_3S_{28}$

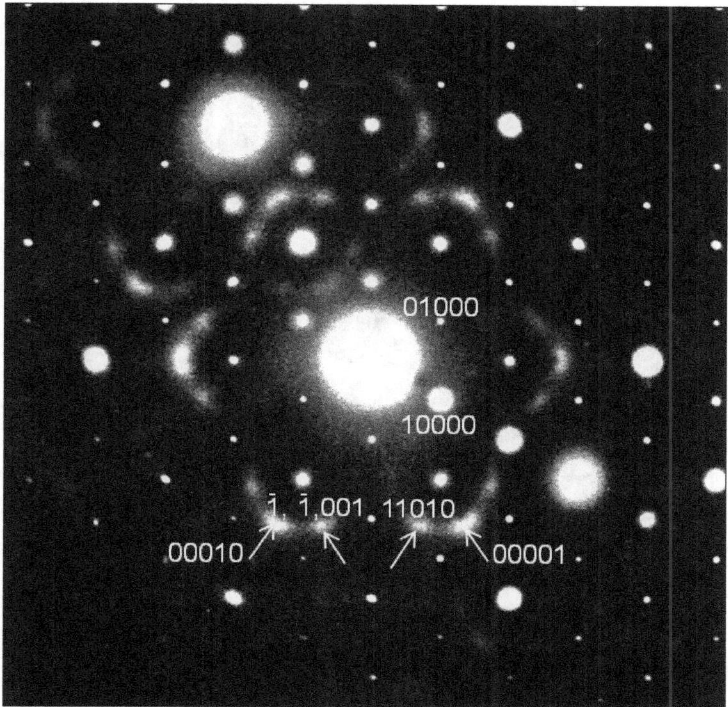

Fig. 18. The ~<001> zone axis electron diffraction pattern of a pearceite-*Tac* (Withers *et al.*, 2008) indexed in standard composite modulated fashion using the set of five reciprocal space basis vectors {$\mathbf{a}_F$*, $\mathbf{b}_F$*, $\mathbf{c}_F$*, $\mathbf{a}_{Ag}$* and $\mathbf{b}_{Ag}$*}.

(Evain *et al.*, 2006c), synthetic $Sn_{31.52}Sb_{6.23}Fe_{3.12}S_{59.12}$ (Makovicky *et al.*, 2008) and franckeite, $Pb_{21.7}Sn_{9.3}Fe_{4.0}Sb_{8.1}S_{56.9}$ (Makovicky *et al.*, 2011). All of these authors have pointed out that this family of misfit-layer compounds results from the combination of two heavily modulated triclinic Q and H subsystems with a common **q** wavevector and only one shared reciprocal axis (stacking direction). The differences and similarities among the three compounds are too complex to be outlined briefly in a general review but are related to the relationships between layer match and the modulation vector, divergence of layer stacking of the two components, as well as the pronounced disorder of the Q component.

9. Naturally occurring quasicrystals

9.1. The case of icosahedrite

Although there is an increasing number of syntheses and careful characterizations of new quasicrystals, there still has not been a general consensus concerning the status of quasicrystals as a fundamental state of matter. These materials were also considered too

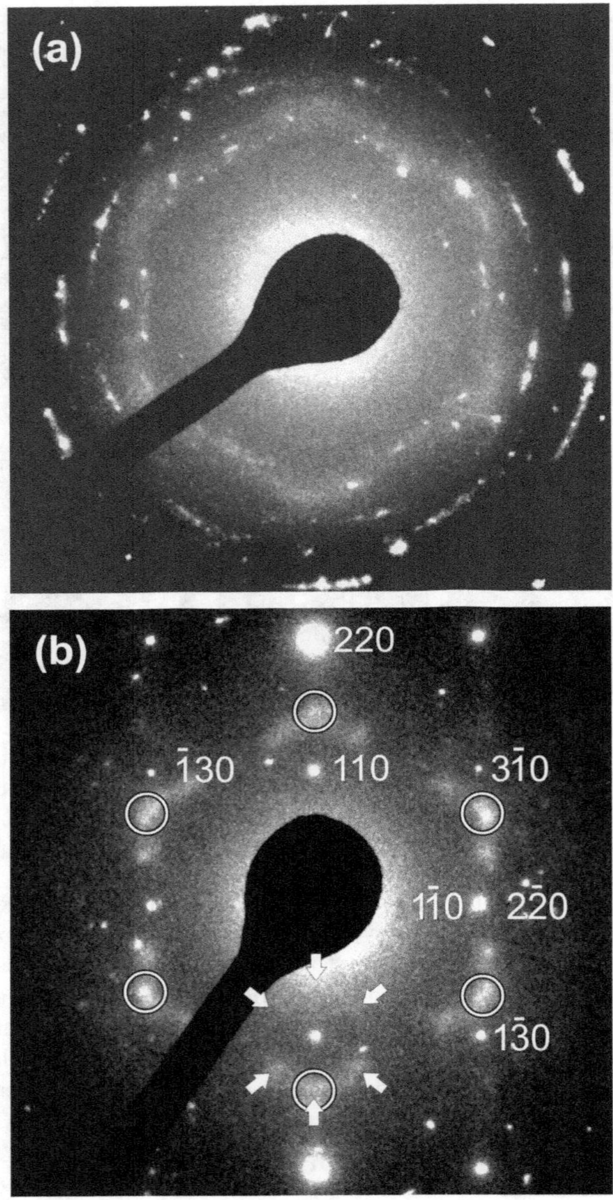

Fig. 19. Electron diffraction pattern of a thick crystal portion of cupropearceite which shows: (a) the dominant diffuse pseudo-hexagonal Ag^+ ion distribution; (b) the incommensurate satellite reflections of the Ag^+ ion distribution (arrows) with the Ag^+ sub-lattice emphasized (circle marks), along with the basic framework lattice diffraction (reproduced from Bindi *et al.* (2013) with the permission of the Mineralogical Society of America).

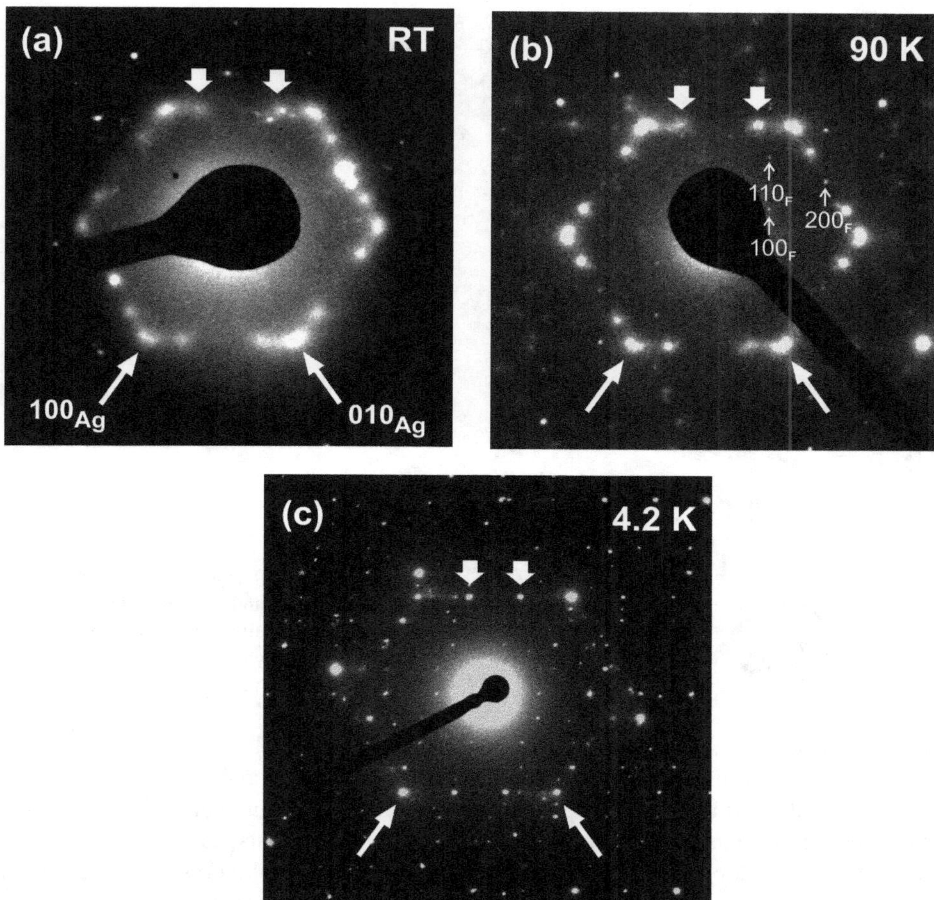

Fig. 20. Electron diffraction of the structure of cupropearceite dependent on temperature: (a) room temperature (RT); (b) $T = 90$ K; (c) $T = 4.2$ K. The composite structure consisting of the basic framework sub-structure (labelled with the subscript F) and the Ag^+ ion sub-structure (labelled with the subscript Ag) undergoes an incommensurate-to-commensurate phase transformation between RT and 4.2 K (see satellites labelled by bold arrows) (reproduced from Bindi *et al.* (2013) with the permission of the Mineralogical Society of America).

elaborate to be stable states of matter given the fact that all ground states are crystalline. This led to an alternative view to explain quasicrystals: as formed by random tilings (arrangements of Penrose tiles without matching rules) with the icosahedral symmetry appearing for entropic reasons, although not energetically preferred (Henley, 1991). According to this picture, quasicrystals are inherently delicate, metastable oddities that may only be synthesized for ideal compositions and under highly controlled laboratory conditions. On the contrary, the original theory (construction of tilings with perfect

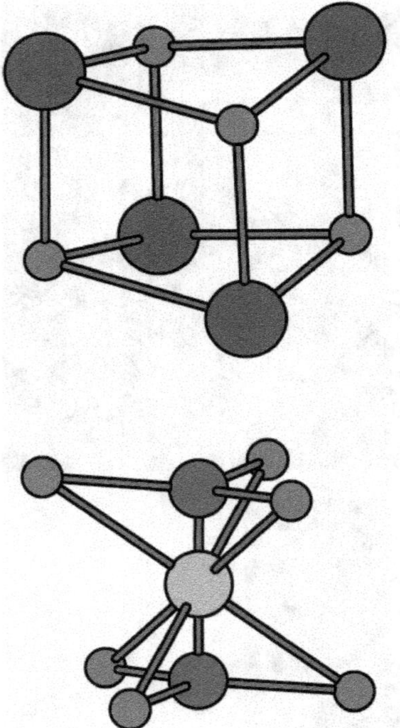

Fig. 21. Example of Q layer (with a distorted PbS-/NaCl-type structure) and H layer (with a CdI_2-type structure) in levyclaudite (modified after Evain *et al.*, 2006c). Spheres, with decreasing size, indicate Pb/Sb, Sn, Cu and S, respectively.

matching rules) considered quasicrystals as energetically stable and robust as crystals and forming even in conditions which were not ideal (Levine and Steinhardt, 1984).

In this context, it is clear that searching quasicrystals in nature could help to answer this question. If quasicrystals are energetically stable and robust as crystals, then it is conceivable that they formed under natural conditions, just like crystalline minerals. So, the question is: is it possible that quasicrystals formed through natural geological processes long before they were discovered in the laboratory?

After a decade-long search (Lu *et al.*, 2001), in 2009, a new kind of mineral was discovered in a rock labelled "khatyrkite" catalogued as coming from the Khatyrka region of the Koryak mountains in the Chukotka autonomous okrug (district) on the northeastern part of the Kamchatka peninsula (Bindi *et al.*, 2009b, 2011) belonging to the collections of the Museo di Storia Naturale of the Università degli Studi di Firenze (catalogue number 46407/G). The mineral, a naturally formed quasicrystal, exhibits a distinct atomic arrangement with rotational symmetries incompatible for ordinary crystals. The first natural quasicrystal has been accepted officially by the Commission on New Minerals, Nomenclature and Classification of the International Mineralogical Association and named icosahedrite for the icosahedral symmetry of its atomic structure (Bindi *et al.*, 2011). The mineral is classified as icosahedral (face-centred icosahedral symmetry) with diffraction peaks labelled by six indices (corresponding to the six basis vectors that define the reciprocal lattice).

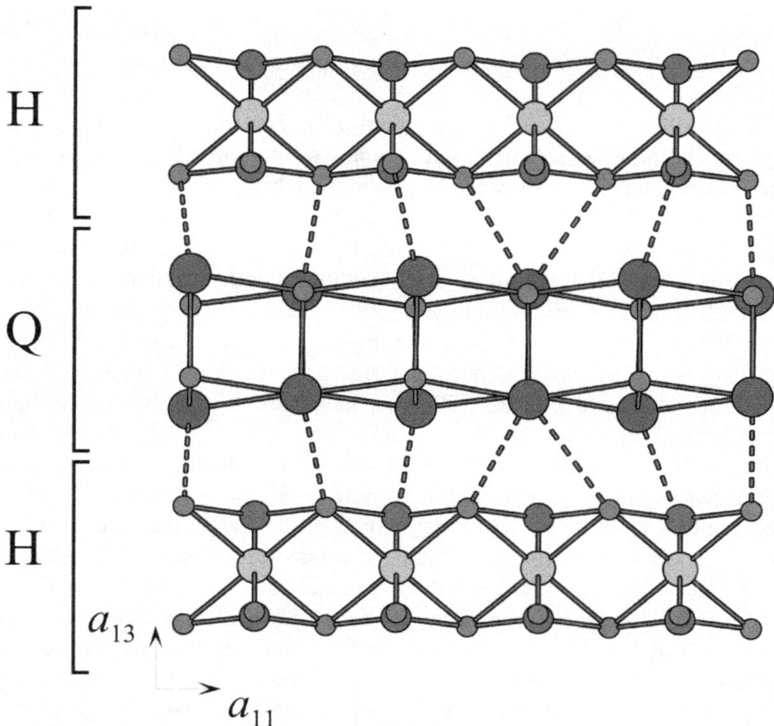

Fig. 22. Succession along the a_{13} (a_{23}) axis of the Q and H layers in levyclaudite (modified after Evain *et al.*, 2006c). The subscripts indicate the different subsystems and the different subcells (see Evain *et al.*, 2006c for explanation). Broken lines indicate the two subsystem interactions (symbols as in Fig. 1.21).

To characterize the atomic structure of icosahedrite, several grains of the mineral were examined by transmission electron microscopy. The grains were found to be mostly homogeneous, with a stoichiometry close to $Al_{63}Cu_{24}Fe_{13}$. The diffraction patterns consist of sharp peaks arranged in a quasi-periodic lattice with five-, three- and two-fold symmetry (Fig. 23), the characteristic signature of an icosahedral quasicrystal (*e.g.* Levine and Steinhardt, 1984; Bancel, 1991). In addition, the angles between the symmetry planes are consistent with icosahedral symmetry. For example, the angle between the two- and five-fold symmetry planes was 31.6(5)°, in excellent agreement with the ideal rotation angle between the two-fold and five-fold axes of an icosahedron [*i.e.* $\arctan(1/\tau) = 31.7°$, where τ is the golden ratio defined as $(1+\sqrt{5})/2$].

The cubic lattice parameter (in six-dimensional notation) obtained with the XRD experiment for icosahedrite (Bindi *et al.*, 2009b) can be written as a_{6D} = 12.64 Å (Steurer and Deloudi, 2009), with probable space group $Fm\bar{3}\bar{5}$. The six-dimensional *a* parameter for the synthetic analogue of icosahedrite first synthesized by Tsai *et al.* (1987) is reported as 4.45 Å and ~6.31 Å (Quiquandon *et al.*, 1996; Quiquandon and Gratias, 2006). These values are $a_{6D}/\sqrt{(5\tau)}$ and $a_{6D}/2$, respectively, of the value reported for icosahedrite by Bindi *et al.* (2009b); consequently, all the values are consistent, depending on the arbitrary choice of principal-axis length for the same 6D reciprocal lattice.

Interestingly, TEM and XRD experiments also demonstrated the high degree of structural perfection in icosahedrite with respect to the typical quasicrystals produced in the laboratory. Quasicrystals made by rapid quenching and/or embedded in a matrix of another phase, exhibit easily detectable phason strains (Levine *et al.*, 1985; Lubensky *et al.*, 1986). Phasons are hydrodynamic (gapless) modes that occur in incommensurate crystals and quasicrystals in addition to the usual phonon modes. If the atomic density $\rho(x)$ is decomposed into a sum of incommensurate density waves, a phason shift corresponds to uniform phase shifts between density waves the periods of which have an irrational ratio. A phason strain corresponds to a spatial gradient in the phase shift. According to Bancel (1991), at the atomic level, phason strains produce rearrangements of atoms relative to the ideal, ground state configuration that require diffusive motion to relax away, so the phason strains remain after solidification unless the sample is grown under a very carefully controlled protocol. In reciprocal space, the signature of phason strain is a systematic shift in the Bragg peak positions from the ideal by an amount that increases as the peak intensity decreases. The effect is easily observed by holding the diffraction pattern at a grazing angle and viewing down rows of peaks. The phason strain can be observed as deviations of the dimmer peaks from straight lines (Lubensky *et al.*, 1986). The fact that icosahedrite shows no visible phason strain (Fig. 23) means either the mineral sample formed without phason strain in the first place, or subsequent annealing was sufficient for phason strains to relax away.

Beside the crystallographic description of the first naturally occurring quasicrystal, a puzzling, but, at the same time, very exciting aspect for geologists is the presence of metallic aluminium in the quasicrystal grains as well as some of the other mineral phases (*i.e.* khatyrkite, $CuAl_2$, and cupalite, CuAl) found in the rock hosted in the mineralogical collections of the Florence Museum (Bindi *et al.*, 2009b, 2011). Indeed, aluminium normally oxidizes unless placed in a highly reducing environment, as is done in the laboratory or in industry. Hence, serious consideration had to be given to the possibility that the sample is slag or an accidental byproduct of some anthropogenic process. Some of the significant evidence against this possibility includes: the remote, uninhabited region where the sample was found being very far from any industry; the presence of typical rock-forming minerals (forsterite and diopside) in direct contact with metal alloys; the absence of glass or bubbles; unusual zoning of phosphorus and chromium in the forsterite grains (Bindi *et al.*, 2012; Steinhardt and Bindi, 2012); the concentration of nickel in the forsterite but not in the metal alloys (Bindi *et al.*, 2012); and, most importantly, grains of quasicrystal trapped within stishovite (Bindi *et al.*, 2012; Steinhardt and Bindi, 2012). Each of these features is individually inconsistent with an anthropogenic origin for different reasons; collectively, they make an overwhelming case that some natural process led to the formation of the rock.

As the quasicrystal and the other metallic aluminium alloys cannot be explained by any conventional geological processes, determining how they formed has been both

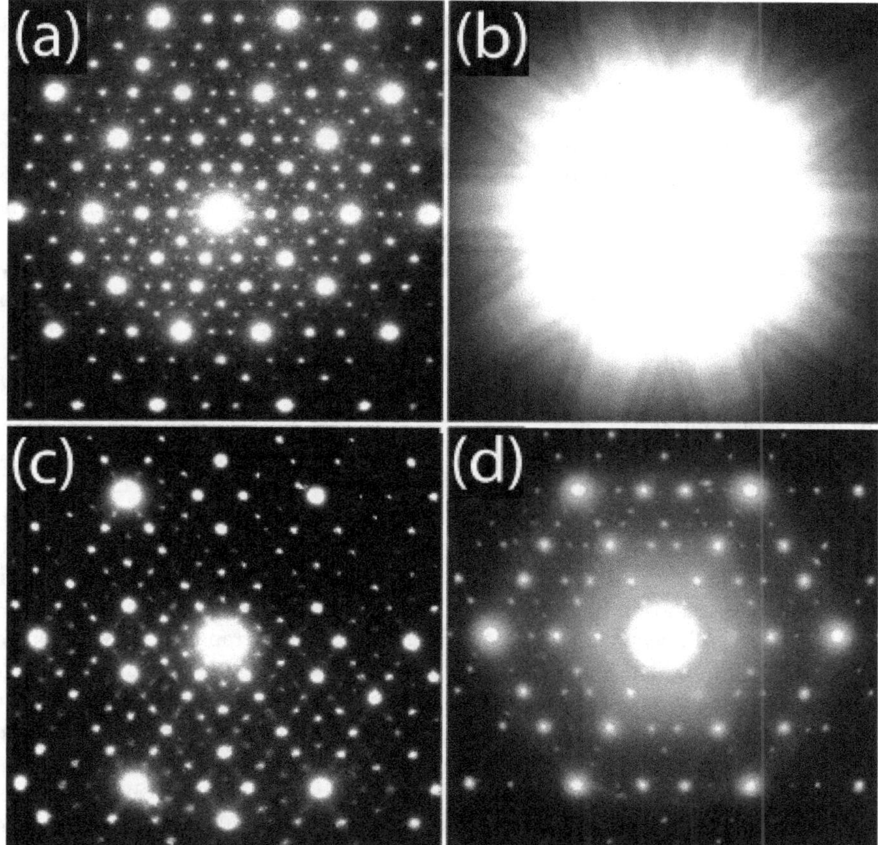

Fig. 23. The five-fold (a), two-fold (c) and three-fold (d) electron diffraction patterns for icosahedrite. The patterns correspond perfectly, up to experimental resolution, with the five-, two-, and three-fold patterns predicted for a face-centred icosahedral quasicrystal. Figure (b) shows the convergent beam (Kikuchi) diffraction pattern obtained along the five-fold axis, providing further physical evidence of the symmetry and high degree of translational order (reproduced from Bindi *et al.* (2011) with the permission of the Mineralogical Society America).

challenging and important. The discovery of the stishovite grain noted above (Bindi *et al.*, 2012; Steinhardt and Bindi, 2012) indicated that the mineral formed at pressures that occur under exotic conditions, either very deep under the surface of the Earth, near the boundary between core and mantle, or in space, through the collisions of meteors and asteroids. To distinguish between the two possibilities, a series of ion-probe experiments that measure the ratio of oxygen isotopes was carried out (Bindi *et al.*, 2012; MacPherson *et al.*, 2013). The results were unmistakable: the oxygen isotopes matched precisely the known abundances of the carbonaceous chondrites, among the oldest meteorites to have formed in our solar system.

Establishing that the sample is indeed natural took nearly 2 years of laboratory (Bindi *et al.*, 2012) and detective work, and, ultimately, a geological expedition to the Koryak Mountains of Chukotka in far eastern Russia to find additional samples. The extraordinary story, recounted elsewhere (Bindi and Steinhardt, 2012; Steinhardt and Bindi, 2012) established that the quasicrystal and the rock containing it are definitely part of a carbonaceous chondrite, with calcium aluminium inclusions, that dates back 4.5 Gya to the formation of the solar system (MacPherson *et al.*, 2013). The new meteorite find has been named Khatyrka (MacPherson *et al.*, 2013). The name derives from the Khatyrka river, which is one of the main rivers draining the Koryak Mountains. That river is also the basis for the name of the mineral, khatyrkite, which gives an added symmetry to the meteorite name. Khatyrka has been approved by the Nomenclature Committee of the Meteoritical Society and representative specimens have been deposited at the U.S. National Museum of Natural History, Smithsonian Institution, Washington D.C.

In the Khatyrka meteoritic fragments recovered during the expedition, which present a range of evidence indicating that an impact shock generated a heterogeneous distribution of pressures and temperatures in which some portions of the meteorite reached at least 5 GPa and 1200°C (Hollister *et al.*, 2014), the second natural quasicrystal has been found (Bindi *et al.*, 2015a). The quasicrystal has the composition $Al_{71}Ni_{24}Fe_5$ and is the first known natural quasicrystal with decagonal symmetry, a periodic stacking of layers containing quasiperiodic atomic arrangements with ten-fold symmetry. The mineral and its name have been approved by the Commission on New Minerals, Nomenclature and Classification of the International Mineralogical Association (Bindi *et al.*, 2015b). As already observed for icosahedrite (Bindi *et al.*, 2009b, 2011), decagonite exhibits a high degree of structural perfection, particularly the absence of significant phason strains. This is unusual because such a high degree of perfection was obtained in a quasicrystal intergrown with other phases under conditions that are clearly far from equilibrium and not under controlled laboratory conditions. We think that either the mineral samples formed without phason strain in the first place, or subsequent annealing was sufficient for phason strains to relax away.

So, what is the significance of the discovery of the first minerals with forbidden symmetry in Nature? In geology, the discovery of natural quasicrystals has opened a new chapter in the study of mineralogy, forever altering the conventional classification of mineral forms. In condensed-matter physics, the discovery has pushed back the age of the oldest quasicrystals by orders of magnitude and has had an impact on our view of how difficult it is for quasicrystals to form. Finding natural quasicrystals has also been a way of studying quasicrystal stability over annealing times and conditions not accessible in the laboratory. Identifying materials that form quasicrystals has always relied significantly on trial-and-error and serendipity, and searching through nature has proved to be an effective complement to laboratory methods. Finally, the discovery has suggested new geological or extra-terrestrial processes of formation.

10. Conclusions

The field of aperiodic crystallography is now well established. The discovery of natural minerals covering all the variants of aperiodic structures, *i.e.* incommensurate and composite structures as well as quasicrystals is definite proof of the stability of their structures over a long period of time. The theory of superspace including the symmetry properties is now commonly used to resolve and describe aperiodic structures.

The recent development of intensive synchrotron sources along with the rapid evolution of two-dimensional detectors has led to the discovery of several incommensurate crystals, and indicated that incommensurability cannot be ignored in any field of solid-state sciences. Indeed, incommensurate modulation has been found in almost all families of structures that include synthetic inorganic and organic compounds, as well as proteins. Modulated structures have also been found widely in various minerals. However, structural studies of naturally occurring incommensurately modulated structures, using the superspace formalism, are few in number. Studies limited to electron diffraction are not sufficient because accurate knowledge of the modulation functions is a prerequisite for understanding the crystal chemistry of incommensurate crystals. The structural analysis of modulated minerals is still very underdeveloped, and is probably one of the major future challenges in the field of crystallography.

Acknowledgements

The authors give special thanks to Paola Bonazzi, Silvio Menchetti (Università di Firenze, Italy), Alla Arakcheeva (Swiss Federal School of Technology, Switzerland), Vladimir Dmitrienko (Shubnikov Institute of Crystallography, Russia), Michel Evain (Université de Nantes, France), Michal Dušek, Vaclav Petříček (Institute of Physics, Czech Republic), Ray Withers (Australian National University, Australia), Andreas Schaper (Philipps University, Germany) Lincoln S. Hollister, Paul J. Steinhardt, Nan Yao (Princeton University, USA), Peter J. Lu, Will Steinhardt (Harvard University, USA), Glenn MacPherson (Smithsonian Institution, USA), Christopher Andronicos (Purdue University, USA), Vadim Distler, Marina Yudovskaya (IGEM, Russia), Alexander Kostin (BHP Billiton, USA), Valery Kryachko (Voronezh, IGEM) and Michael Eddy (MIT, USA) for fruitful discussions and for production of some of the data reported here.

The authors also appreciate helpful reviews by Vaclav Petříček and Stefano Merlino.

References

Angel, R.J., McMullan, R.K. and Prewitt, C.T. (1991) Substructure and superstructure of mullite by neutron diffraction. *American Mineralogist*, **76**, 332–342.

Angel, R.J., Gatta, G.D., Boffa-Ballaran, T. and Carpenter, M. (2008) The mechanism of coupling in the modulated structure of nepheline. *The Canadian Mineralogist*, **46**, 1465–1476.

Arakcheeva, A., Bindi, L., Pattison, P., Meisser, N., Chapuis, G. and Pekov, I. (2010) The incommensurately modulated structures of natural natrites at 120 and 293 K from synchrotron X-ray data. *American Mineralogist*, **95**, 574–581.

Arakcheeva, A. and Chapuis, G. (2005) A reinterpretation of the phase transitions in Na_2CO_3. *Acta*

Crystallographica, **B61**, 601–607.

Arakcheeva, A. and Chapuis, G. (2008) Capabilities and limitations of a (3 + d)-dimensional incommensurately modulated structure as a model for the derivation of an extended family of compounds: example of the scheelite-like structures. *Acta Crystallographica* **B64**, 12–25.

Armbruster, T., Röthlisberger, F. and Seifert, F. (1990) Layer topology, stacking variation, and site distortion in melilite-related compounds in the system CaO-ZnO-GeO_2-SiO_2. *American Mineralogist*, **75**, 847–858.

Bagautdinov, B., Hagiya, K., Noguchi, S., Ohmasa, M., Ikeda, N., Kusaka, K. and Iishi, K. (2002) Low-temperature studies on the two-dimensional modulations in åkermanite-type crystals: $Ca_2MgSi_2O_7$ and $Ca_2ZnSi_2O_7$. *Physics and Chemistry of Minerals*, **29**, 346–350.

Balzuweit, K., Hovestad, A., Meekes, H. and de Boer, J.L. (1993) The temperature and composition dependence of the modulation wave vector of incommensurate calaverite. *Journal of Crystal Growth*, **131**, 518–532.

Bancel, P.A. (1991) Order and disorder in icosahedral alloys. Pp. 17–56 in: *Quasicrystals: The State of the Art* (D. DiVincenzo and P.J. Steinhardt, editors). World Scientific, Singapore.

Bindi, L. (2008) Commensurate-incommensurate phase transition in muthmannite, $AuAgTe_2$: First evidence of a modulated structure at low-temperature. *Philosophical Magazine Letters*, **88**, 533–541.

Bindi, L. and Steinhardt, P.J. (2012) The discovery of the first natural quasicrystal. A new era for mineralogy? *Elements*, **8**, 13–14.

Bindi, L., Bonazzi, P., Dušek, M., Petříček, V. and Chapuis, G. (2001a) Five dimensional structure refinement of natural melilite $(Ca_{1.89}Sr_{0.01}Na_{0.08}K_{0.02})(Mg_{0.92}Al_{0.08})(Si_{1.98}Al_{0.02})O_7$. *Acta Crystallographica*, **B57**, 739–746.

Bindi, L., Czank, M., Röthlisberger, F. and Bonazzi, P. (2001b) Hardystonite from Franklin Furnace: a natural modulated melilite. *American Mineralogist*, **86**, 747–751.

Bindi, L., Evain, M., Pradel, A., Albert, S., Ribes, M. and Menchetti, S. (2006a) Fast ionic conduction character and ionic phase-transitions in disordered crystals: The complex case of the minerals of the pearceite–polybasite group. *Physics and Chemistry of Minerals*, **33**, 677–690.

Bindi, L., Evain, M. and Menchetti, S. (2006b) Temperature dependence of the silver distribution in the crystal structure of natural pearceite, $(Ag,Cu)_{16}(As,Sb)_2S_{11}$. *Acta Crystallographica*, **B62**, 212–219.

Bindi, L., Evain, M. and Menchetti, S. (2006c) Complex twinning, polytypism and disorder phenomena in the crystal structures of antimonpearceite and arsenpolybasite. *The Canadian Mineralogist*, **45**, 321–333.

Bindi, L., Dušek, M., Petříček, V. and Bonazzi, P. (2006d) Superspace-symmetry determination and multidimensional refinement of the incommensurately modulated structure of a natural fresnoite. *Acta Crystallographica*, **B62**, 1031–1037.

Bindi, L., Evain, M., Spry, P.G., Tait, K.T. and Menchetti, S. (2007a) Structural role of copper in the minerals of the pearceite-polybasite group: The case of the new minerals cupropearceite and cupropolybasite. *Mineralogical Magazine*, **71**, 641–650.

Bindi, L., Evain, M., Spry, P.G. and Menchetti, S. (2007b) The pearceite-polybasite group of minerals: Crystal chemistry and new nomenclature rules. *American Mineralogist*, **92**, 918–925.

Bindi, L., Arakcheeva, A. and Chapuis, G. (2009a) The role of silver on the stabilization of the incommensurately modulated structure in calaverite, $AuTe_2$. *American Mineralogist*, **94**, 728–736.

Bindi, L., Steinhardt, P.J., Yao, N. and Lu, P.J. (2009b) Natural quasicrystals. *Science*, **324**, 1306–1309.

Bindi, L., Steinhardt, P.J., Yao, N. and Lu, P.J. (2011) Icosahedrite, $Al_{63}Cu_{24}Fe_{13}$, the first natural quasicrystal. *American Mineralogist*, **96**, 928–931.

Bindi, L., Eiler, J.M., Guan, Y., Hollister, L., MacPherson, G.J., Steinhardt, P.J. and Yao, N. (2012) Evidence for the extra-terrestrial origin of a natural quasicrystal. *Proceedings of the National Academy of Sciences, USA*, **109**, 1396–1401.

Bindi, L., Schaper, A.K., Kurata, H. and Menchetti, S. (2013) The composite modulated structure of cupropearceite and cupropolybasite and its behavior toward low temperature. *American Mineralogist*, **98**, 1279–1284.

Bindi, L., Yao, N., Lin, C., Hollister, L.S., Andronicos, C.L., Distler, V.V., Eddy, M.P., Kostin, A., Kryachko, V., MacPherson, G.J., Steinhardt, W.M., Yudovskaya, M. and Steinhardt, P.J. (2015a) Natural quasicrystal with decagonal symmetry. *Scientific Reports*, **5**, 9111.

Bindi, L., Yao, N., Lin, C., Hollister, L.S., Andronicos, C.L., Distler, V.V., Eddy, M.P., Kostin, A., Kryachko, V., MacPherson, G.J., Steinhardt, W.M., Yudovskaya, M. and Steinhardt, P.J. (2015b) Decagonite,

$Al_{71}Ni_{24}Fe_5$, a quasicrystal with decagonal symmetry from the Khatyrka CV3 carbonaceous chondrite. *American Mineralogist*, **100**, 2340–2343.

Bolotina, N.B. (2006) Isotropic lazurite: A cubic single crystal with an incommensurate three-dimensional modulation of the structure. *Crystallography Reports*, **51**, 968–976.

Bolotina, N.B., Rastsvetaeva, R.K., Sapozhnikov, A.N., Kashaev, A.A., Schönleber, A. and Chapuis, G. (2003a) Incommensurately modulated structure of isotropic lazurite as a product of twinning of two-dimensionally modulated domains. *Crystallography Reports*, **48**, 721–727.

Bolotina, N.B., Rastsvetaeva, R.K., Sapozhnikov, A.N., Kashaev, A.A., Schönleber, A. and Chapuis, G. (2003b) Three-dimensionally modulated incommensurate crystal structure of lazurite from the Baikal region. *Crystallography Reports*, **48**, 8–11.

Bolotina, N.B., Rastsveteva, R.K., Chapuis, G., Schönleber, A., Sapozhnikov, A.N. and Kashaev, A.A. (2004) On the symmetry of optically isotropic modulated lazurites from the Baikal region. *Ferroelectrics*, **305**, 95–98.

Bolotina, N.B., Rastsvetaeva, R.K. and Sapozhnikov, A.N. (2006) Average structure of incommensurately modulated monoclinic lazurite. *Crystallography Reports*, **51**, 589–595.

Brouns, E., Visser, J.W. and de Wolff, P.M. (1964) An anomaly in the crystal structure of Na_2CO_3. *Acta Crystallographica*, **A17**, 614.

Buseck, P.R. and Cowley, J.M. (1983) Modulated and intergrowth structures in minerals and electron microscope methods for their study. *American Mineralogist*, **68**, 18–40.

Caracas, R. and Gonze, X. (2001) *Ab initio* determination of the valence electron distribution in the average structure of the incommensurately modulated calaverite $AuTe_2$. *Acta Crystallographica*, **B57**, 633–637.

Carter, M.L. and Withers, R.L. (2004) An electron and X-ray diffraction study of the compositely modulated barium nickel hollandite $Ba_x(Ni_xTi_{8-x})O_{16}$, $1.16 < x < 1.32$, solid solution. *Zeitschrift für Kristallographie*, **219**, 763–767.

Chapuis, G. (2003) Crystallographic excursion in superspace. *Crystal Engineering*, **6**, 187–195.

Dam, B. and Janner, A. (1986) A Superspace approach to the structure and morphology of tetramethylammonium tetrachlorozincate, $2C_4H_{12}N^+$. $ZnCl_4^{2-}$. *Acta Crystallographica*, **B42**, 69–77.

Dam, B., Janner, A. and Donnay, J.D.H. (1985) Incommensurate morphology of calaverite ($AuTe_2$) crystals. *Physical Review Letters*, **55**, 2301–2304.

Dušek, M., Chapuis, G., Meyer, M. and Petříček, V. (2003) Sodium carbonate revisited. *Acta Crystallographica*, **B59**, 337–352.

Edenharter, A., Koto, K. and Nowacki, W. (1971) Uber Pearceit, Polybasit und Binnit. *Neues Jahrbuch für Mineralogie Monatshefte*, **1971**, 337–341.

Evain, M., Bindi, L. and Menchetti, S. (2006a) Structure and phase transition in the Se-rich variety of antimonpearceite, $(Ag_{14.67}Cu_{1.20}Bi_{0.01}Pb_{0.01}Zn_{0.01}Fe_{0.03})_{15.93}(Sb_{1.86}As_{0.19})_{2.05}$ $(S_{8.47}Se_{2.55})_{11.02}$. *Acta Crystallographica*, **B62**, 768–774.

Evain, M., Bindi, L. and Menchetti, S. (2006b) Structural complexity in minerals: twinning, polytypism and disorder in the crystal structure of polybasite, $(Ag,Cu)_{16}(Sb,As)_2S_{11}$. *Acta Crystallographica*, **B62**, 447–456.

Evain, M., Petříček, V., Moëlo, Y. and Maurel, C. (2006c) First (3 + 2)-dimensional superspace approach to the structure of levyclaudite-(Sb), a member of the cylindrite-type minerals. *Acta Crystallographica*, **B62**, 775–789.

Friese, K., Grzechnik, A., Petříček, V., Schönleber, A., van Smaalen, S. and Morgenroth, W. (2011) Modulated structure of nepheline. *Acta Crystallographica*, **B67**, 18–29.

Frondel, C. (1963) Isodimorphism of the polybasite and pearceite series. *American Mineralogist*, **48**, 565–572.

Ghent, F.A. (1861) Calaverite. *American Journal of Science*, **45**, 314–317.

Goldschmidt, V., Palache, C. and Peacock, M. (1931) Ueber Calaverit. *Neues Jahrbuch für Mineralogie Monatshefte*, **63**, 1–58.

Guinier, A., Bokij, G.B., Boll-Dornberger, K., Cowley, J.M., Durovic, S., Jagodzinski, H., Krishna, P., de Wolff, P.M., Zvyagin, B.B., Cox, D.E., Goodman, P., Hahn, Th., Kuchitsu, K. and Abrahams, S.C. (1984) Nomenclature of polytype structures. *Acta Crystallographica*, **A40**, 399–404.

Hagiya, K., Ohmasa, M. and Iishi, K. (1993) The modulated structure of synthetic Co-åkermanite, $Ca_2CoSi_2O_7$. *Acta Crystallographica*, **B49**, 172–179.

Hagiya, K., Kusaka, K., Ohmasa, M. and Iishi, K. (2001) Commensurate structure of $Ca_2CoSi_2O_7$, a new twinned orthorhombic structure. *Acta Crystallographica*, **B57**, 271–277.

Hall, H.T. (1967) The pearceite and polybasite series. *American Mineralogist*, **52**, 1311–1321.

Harris, D.C., Nuffield, E.W. and Frohberg, M.H. (1965) Studies of mineral sulphosalts: XIX-Selenian polybasite. *The Canadian Mineralogist*, **8**, 172–184.

Hassan, I. (2000) Transmission electron microscopy and differential thermal studies of lazurite polymorphs. *American Mineralogist*, **85**, 1383–1389.

Hassan, I. and Buseck, P.R. (1989) Incommensurate-modulated structure of nosean, a sodalite-group mineral. *American Mineralogist*, **74**, 394–410.

Hassan, I. and Buseck, P.R. (1992) The origin of the superstructure and modulations in cancrinite. *The Canadian Mineralogist*, **30**, 49–59.

Hassan, I., Antao, S.M. and Parise, J.B. (2004) Haüyne: phase transition and high-temperature structures obtained from synchrotron radiation and Rietveld refinements. *Mineralogical Magazine*, **68**, 499–513.

Heaney, P.J. and Veblen, D.R. (1991) Observations of the α-β phase transition in quartz: A review of imaging and diffraction studies and some new results. *American Mineralogist*, **76**, 1018–1032.

Hemingway, B.S., Evans, Jr. H.T., Nord, G.L. Jr., Haselton, H.T. Jr., Robie, R.A. and McGee, J.J. (1986) Åkermanite: phase transitions in heat capacity and thermal expansion, and revised thermodynamic data. *The Canadian Mineralogist*, **24**, 425–434.

Henley, C. (1991) Random tiling models. Pp. 429–524 in: *Quasicrystals: the State of the Art* (D. DiVincenzo and P.J. Steinhardt, editors). World Scientific, Singapore.

Hollister, L.S., Bindi, L., Yao, N., Poirier, G.R., Andronicos, C.L., MacPherson, G.J., Lin, C., Distler, V.V., Eddy, M.P., Kostin, A., Kryachko, V., Steinhardt, W.M., Yudovskaya, M., Eiler, J.M., Guan, Y., Clarke, J.J. and Steinhardt, P.J. (2014) Impact-induced shock and the formation of natural quasicrystals in the early solar system. *Nature Communications*, **5**: 3040 doi:10.1038/ncomms5040.

Höche, T., Rüssel, C. and Neumann, W. (1999) Incommensurate modulations in $Ba_2TiSi_2O_8$, $Sr_2TiSi_2O_8$ and $Ba_2TiGe_2O_8$. *Solid State Communications*, **110**, 651–656.

Janner, A. and Dam, B. (1989) The morphology of calaverite ($AuTe_2$) from data of 1931 – solution of an old problem of rational indexes. *Acta Crystallographica*, **A45**, 115–123.

Janssen, T., Janner, A., Looijenga-Vos, A. and Wolff de, P.M. (1992) Incommensurate and commensurate modulated structures. *International Tables for X-ray Crystallography*, Vol. C, p. 899–937. Kluwer Academic Publishers, Dordrecht, The Netherlands.

Janssen, T., Chapuis, G. and de Boissieu, M. (2007) *Aperiodic Crystals: From Modulated Phases to Quasicrystals*. Oxford University Press, Oxford, UK.

Jia, Z.H., Schaper, A.K., Massa, W., Treutmann, W. and Rager, H. (2006) Structure and phase transitions in $Ca_2CoSi_2O_7$-$Ca_2ZnSi_2O_7$ solid-solution crystals. *Acta Crystallographica*, **B62**, 547–555.

Kalning, M., Dorna, V., Press, W., Kek, S. and Boysen, H. (1997) Profile analysis of the superlattice reflections in labradorite. *Zeitschrift für Kristallogrographie*, **212**, 545–549.

Korekawa, M. (1967) Theorie der satellitenreflexe. *Hohen Naturwissenschaftlichen Fakultät*, p. 153. Ludwigs-Maximilians-Universität, Munich.

Krutzen, B.C.H. and Inglesfield, J.E. (1990) First-principles electronic structure calculations for incommensurately modulated sylvanite. *Journal of Physics: Condensed Matter*, **2**, 4829–4848.

Lazic, B., Armbruster, T., Chopin, C., Grew, E.S., Baronnet, A. and Palatinus, L. (2014) Superspace description of wagnerite-group minerals $(Mg,Fe,Mn)_2(PO_4)(F,OH)$. *Acta Crystallographica*, **B70**, 243–258.

Levine, D. and Steinhardt, P.J. (1984) Quasicrystals: A new class of ordered structures. *Physical Review Letters*, **53**, 2477–2480.

Levine, D., Lubensky, T.C., Ostlund, S., Ramaswamy, A. and Steinhardt, P.J. (1985) Elasticity and defects in pentagonal and icosahedral quasicrystals. *Physical Review Letters*, **54**, 1520–1523.

Li, F. and Franzen, H.F. (1996) Ordering, incommensuration, and phase transitions in pyrrhotite: Part II: A high-temperature X-ray powder diffraction and thermomagnetic study. *Journal of Solid State Chemistry*, **126**, 108–120.

Lu, P.J., Deffeyes, K., Steinhardt, P.J. and Yao, N. (2001) Identifying and indexing icosahedral quasicrystals from powder diffraction patterns. *Physical Review Letters*, **87**, 275507-1–275507-4.

Lubensky, T.C., Socolar, J.E.S., Steinhardt, P.J., Bancel, P.A. and Heiney, P.A. (1986) Distortion and peak broadening in quasicrystal diffraction patterns. *Physical Review Letters*, **57**, 1440–1443.

MacPherson, G.J., Andronicos, C., Bindi, L., Distler, V.V., Eddy, M., Eiler, J., Guan, Y., Hollister, L.S., Kostin, A., Kryachko, V., Steinhardt, W., Yudovskaya, M. and Steinhardt, P.J. (2013) Khatyrka, a new CV3 find from the Koryak Mountains, Eastern Russia. *Meteoritics and Planetary Science*, **48**, 1499–1514

Makovicky, E. and Hyde, B.G. (1981) Non-commensurate (misfit) layer structures. *Inorganic Chemistry*, **46**, 101–170.

Makovicky, E. and Hyde, B.G. (1992) Incommensurate, two-layer structures with complex crystal chemistry: Minerals and related synthetics. *Materials Science Forum*, **100–101**, 1–100.

Makovicky, E., Petříček, V., Dušek, M. and Topa, D. (2008) Crystal structure of a synthetic tin-selenium representative of the cylindrite structure type. *American Mineralogist*, **93**, 1787–1798.

Makovicky, E., Petříček, V., Dušek, M. and Topa, D. (2011) The crystal structure of franckeite, $Pb_{21.7}Sn_{9.3}Fe_{4.0}Sb_{8.1}S_{56.9}$. *American Mineralogist*, **96**, 1686–1702.

Markgraf, S.A. and Bhalla, A.S. (1989) Low temperature phase transition in $Ba_2TiGe_2O_8$. *Phase Transitions*, **18**, 55–76.

Markgraf, S.A., Randall, C.A., Bhalla, A.S. and Reeder, R.J. (1990) Incommensurate phase in $Ba_2TiGe_2O_8$. *Solid State Communications*, **75**, 821–824.

Minčeva-Stefanova, I., Bonev, I. and Punev, L. (1979) Pearceite with an intermediate unit cell – first discovery in nature. *Geokhimiya, Mineralogiya I Petrologiya*, **11**, 13–34 (in Bulgarian).

Morozov, V.A., Arakcheeva, A.V., Konovalova, V.V., Pattison, P., Chapuis, G., Lebedev, O.I., Fomichev, V.V. and Van Tendeloo, G. (2010) $LiZnNb_4O_{11.5}$: A novel oxygen deficient compound in the Nb-rich part of the Li_2O-ZnO-Nb_2O_5 system. *Journal of Solid State Chemistry*, **183**, 408–418.

Norén, L., Withers, R.L. and Brink, F.J. (2005) Where are the Sn atoms in $LaSb_2Sn_x$, $0.1 \leqslant x \leqslant \sim 0.75$? *Journal of Solid State Chemistry*, **178**, 2133–2143.

Oszlányi, G. and Sütö, A. (2004) *Ab initio* structure solution by charge flipping. I. *Acta Crystallographica*, **A60**, 134–141.

Oszlányi, G. and Sütö, A. (2005) *Ab initio* structure solution by charge flipping. II. Use of weak reflections. *Acta Crystallographica*, **A61**, 147–152.

Penfield, S.L. and Ford, W.E. (1902) Ueber den Calaverit. *Zeitschrift für Kristallographie*, **35**, 430–451.

Petříček, V., Dušek, M. and Palatinus, L. (2006) *Jana2006*. The crystallographic computing system. Institute of Physics, Praha, Czech Republic.

Pring, A., Williams, T. and Withers, R. (1993) Structural modulation in sartorite: An electron microscope study. *American Mineralogist*, **78**, 619–626.

Pryde, A.K.A. and Dove, M.T. (1998) On the sequence of phase transitions in tridymite. *Physics and Chemistry of Minerals*, **26**, 171–179.

Quiquandon, M. and Gratias, D. (2006) Unique six-dimensional structural model for Al-Pd-Mn and Al-Cu-Fe icosahedral phases. *Physical Review*, **B74**, article #214205.

Quiquandon, M., Quivy, A., Devaud, J., Faudot, F., Lefebvre, S., Bessière, M. and Calvayrac, Y. (1996) Quasicrystal and approximant structures in the Al-Cu-Fe system. *Journal of Physics: Condensed Matter*, **8**, 2487–2512.

Rastsvetaeva, R.K., Bolotina, N.B., Sapozhnikov, A.N., Kashaev, A.A., Schönleber, A. and Chapuis, G. (2002) Average structure of cubic lazurite with a three-dimensional incommensurate modulation. *Crystallography Reports*, **47**, 404–407.

Riester, M., Böhm, H. and Petříček, V. (2000) The commensurately modulated structure of the lock-in phase of synthetic Co-åkermanite, $Ca_2CoSi_2O_7$. *Zeitschrift für Kristallographie*, **215**, 102–109.

Rigault, G. (1999) Il reticolo polare di Bravais e il reticolo reciproco. *Atti Scienze Fisiche, Accademia delle Scienze di Torino*, **133**, 1–8 (in Italian).

Sakamoto, S., Shimobayashi., N. and Kitamura, M. (2003) Incommensurate phase in the kosmochlor-diopside join: A new polymorph of clinopyroxene. *American Mineralogist*, **88**, 1605–1607.

Sanchez-Munoz, L., Nistor, L., Van Tendeloo, G. and Sanz, J. (1998) Modulated structures in $KAlSi_3O_8$: a study by high resolution electron microscopy and ^{29}Si MAS-NMR spectroscopy. *Journal of Electron Microscopy*, **47**, 17–28.

Schaper, A.K., Schosnig, M., Kutoglu, A., Treutmann, W. and Rager, H. (2001) Transition from the incommensurately modulated structure to the lock-in phase in Co-åkermanite. *Acta Crystallographica*, **B57**, 443–448.

Schutte, W.J. and de Boer, J.L. (1988) Valence fluctuations in the incommensu¡rately modulated structure of calaverite $AuTe_2$. *Acta Crystallographica*, **B44**, 486–494.

Schutte, W.J. and de Boer, J.L. (1993) Determination of the incommensurately modulated structure of $Cu_{3-x}Te_2$. *Acta Crystallographica*, **B49**, 398–403.

Seifert, F., Czank, M., Simons, B. and Schmahl, W. (1987) A commensurate-incommensurate phase transition in iron-bearing åkermanite. *Physics and Chemistry of Minerals*, **14**, 26–35.

Shechtman, D., Blech, I., Gratias, D. and Cahn, J.W. (1984) Metallic phase with long-range orientational order and no translational symmetry. *Physical Review Letters*, **53**, 1951–1953.

Smith, H. (1903) Ueber das bemerkenswerthe Problem der Entwicklung der Krystallformen des Calaverit. *Zeitschrift für Kristallographie*, **37**, 209–234.

Steinhardt, P.J. and Bindi, L. (2012) In search of natural quasicrystals. *Reports on Progress in Physics*, **75**, 092601–092611.

Steurer, W. and Deloudi, S. (2009) *Crystallography of Quasicrystals. Concepts, Methods and Structures.* Springer, Berlin.

Steurer, W. and Jagodzinski, H. (1988) The incommensurately modulated structure of an andesine (An_{38}). *Acta Crystallographica*, **B44**, 344–351.

Stokes, H.T., Campbell, B.J. and van Smaalen, S. (2011) Generation of (3 + d)-dimensional superspace groups for describing the symmetry of modulated crystalline structures. *Acta Crystallographica*, **A67**, 45–55.

Subramanian, R.K., Muntean, L., Norcross, J.A. and Ailion, D.C. (2000) ^{109}Ag NMR investigation of atomic motion in the incommensurate and paraelectric phases of proustite (Ag_3AsS_3). *Physical Review*, **B61**, 996–1002.

Tsai, A.P., Inoue, A. and Masumoto, T. (1987) A stable quasicrystal in Al-Cu-Fe system. *Japanese Journal of Applied Physics*, **26**, L1505.

van Aalst, W., den Hollander, J., Peterse, W.J.A.M. and de Wolff, P.M. (1976) The modulated structure of γ-Na_2CO_3 in a harmonic approximation. *Acta Crystallographica*, **B32**, 47–58.

van Heurck, C., van Tendeloo, G. and Amelinckx, S. (1992) The modulated structure in the melilite $Ca_2ZnGe_2O_7$. *Physics and Chemistry of Minerals*, **18**, 441–452.

van Smaalen, S. (2007) *Incommensurate Crystallography*. Oxford University Press, Oxford, UK.

van Tendeloo, G., Gregoriades, P. and Amelinckx, S. (1983) Electron microscopy studies of modulated structures in $(Au,Ag)Te_2$: Part I. Calaverite $AuTe_2$. *Journal of Solid State Chemistry*, **50**, 321–334.

Wiegers, G.A. and Meerschaut, A. (1992) Misfit layer compounds $(MS)_nTS_2$ (M = Sn,Pb,Bi, rare earth metals; T = Nb,Ta,Ti,V,Cr; $1.08 < n < 1.23$): Structures and physical properties. *Materials Science Forum*, **100–101**, 101–172.

Withers, R.L., Thompson, J.G., Melnitchenko, A. and Palethorpe, S.R. (1998) Cristobalite-related phases in the $NaAlO_2$-$NaAlSiO_4$ system. II. A commensurately modulated cubic structure. *Acta Crystallographica*, **B54**, 547–557.

Withers, R.L., Tabira, Y., Liu, Y. and Höche, T. (2002) A TEM and RUM study of the inherent displacive flexibility of the fresnoite framework structure type. *Physics and Chemistry of Minerals*, **29**, 624–632.

Withers, R.L., Norén, L., Welberry, T.R., Bindi, L., Evain, M. and Menchetti, S. (2008) A composite modulated structure mechanism for Ag^+ fast ion conduction in pearceite and polybasite mineral solid electrolytes. *Solid State Ionics*, **179**, 2080–2089.

Wolff, P.M. de (1974) The pseudo-symmetry of modulated crystal structures. *Acta Crystallographica*, **A30**, 777–785.

Xu, H. and Veblen, D.R. (1995) Transmission electron microscopy study of optically anisotropic and isotropic haüyne. *American Mineralogist*, **80**, 87–93.

Yamamoto, A., Nakazawa, H., Kitamura, M. and Morimoto, N. (1984) The modulated structure of intermediate plagioclase feldspar $Ca_xNa_{1-x}Al_{1+x}Si_{3-x}O_8$. *Acta Crystallographica*, **B40**, 228–237.

Zubkova, N.V., Pushcharovsky, D.Yu., Ivaldi, G., Ferraris, G., Pekov, I.V. and Chukanov, N.V. (2002) Crystal structure of natrite, γ-Na_2CO_3. *Neues Jahrbuch für Mineralogie Monatshefte*, **2**, 85–96.

EMU Notes in Mineralogy, Vol. 19 (2017), Index, 255–258

Subject and Author index

Numbers refer to the page where a definition or explanation of, and/or a *figure* or a **table** for, a given subject is found. Author names are given with the number of the first page of the paper in which they are involved.

DOI: 10.1180/EMU-notes.19.index

www.ingramcontent.com/pod-product-compliance
Lightning Source LLC
LaVergne TN
LVHW061221100826
845148LV00004B/813

* 9 7 8 0 9 0 3 0 5 6 5 9 5 *